Y0-EHV-119

Biotransformations:
Microbial degradation of health-risk compounds

Vol. 14 (1978) edited by M.J. Bull (1st reprint 1983)
Vol. 15 (1979) edited by M.J. Bull
Vol. 16 (1982) edited by M.J. Bull
Vol. 17 (1983) edited by M.E. Bushell
Vol. 18 (1983) Microbial Polysaccharides, edited by M.E. Bushell
Vol. 19 (1984) Modern Applications of Traditional Biotechnologies, edited by M.E Bushell
Vol. 20 (1984) Innovations in Biotechnologie, edited by E.H. Houwink and R.R. van der Meer
Vol. 21 (1989) Statistical Aspects of the Microbiological Analysis of Foods, by B. Jarvis
Vol, 22 (1986) Moulds and Filamentous Fungi in Technical Microbiology, by O. Fassatiová
Vol. 23 (1986) Micro-organisms in the Production of Food, edited by M.R. Adams
Vol. 24 (1986) Biotechnology of Animo Acid Production; edited by K. Aida, I. Chibata, K. Nakayama, K. Takinama and H. Yamada
Vol. 25 (1988) Computers in Fermentation Technology, edited by M.E. Bushell
Vol. 26 (1989) Rapid Methods in Food Microbiology, edited by M.R. Adams and C.F.A. Hope
Vol. 27 (1989) Bioactive Metabolites from Microorganisms, edited by M.E. Bushell and U. Gräfe
Vol. 28 (1993) Micromycetes in Foodstuffs and Feedstuffs; edited by Z. Jesenská
Vol. 29 (1994) *Aspergillus:* 50 years on; edited by S.D. Martinelli and J.R. Kinghorn
Vol. 30 (1994) Bioactive Secondary Metabolites of Microorganisms, edited by V. Betina
Vol. 31 (1995) Techniques in Applied Microbiology, edited by B. Sikyta

Biotransformations:
Microbial degradation of health-risk compounds

EDITED BY

VED PAL SINGH

Department of Botany, University of Delhi, Delhi, India

progress in industrial microbiology

ELSEVIER

Amsterdam – Lausanne – New York – Oxford – Shannon– Tokyo 1995

ELSEVIER SCIENCE B.V.
Sara Burgerhartstraat 25
P.O. Box 211, 1000 AE Amsterdam, The Netherlands

Library of Congress Cataloging-in-Publication Data

Biotransformation : microbial degradation of health-risk compounds / edited by Ved Pal Singh.
p. cm. -- (Progress in industrial microbiology ; v. 32)
Includes bibliographical references and index.
ISBN 0-444-81977-0
1. Xenobiotics--Biodegradation. 2. Microbial metabolism. 3. Biotransformation (Metabolism.) 4. Xenobiotics--Metabolic detoxication. I. Singh, Vedpal, 1951- . II. Series.
QR97.X46B57 1995
628.5'2--dc20 95-11660
CIP

ISBN 0-444-81977-0 (Vol. 32)
ISBN 0-444-41668-8 (Series)

© 1995 Elsevier Science B.V. All rights reserved

No part of this publication may be reproduced, stored in a retrieval system or transmitted, in any form or by any means, electronic, mechanical, photocopying, recording or otherwise, without the prior written permission of the publisher, Elsevier Science B.V., Copyright & Permissions Department, P.O. Box 521, 1000 AM Amsterdam, The Netherlands.

Special regulations for readers in the U.S.A. - This publication has been registered with the Copyright Clearance Center Inc. (CCC), 222 Rosewood Drive, Danvers, MA 01923. Information can be obtained from the CCC about conditions under which photocopies of parts of this publication may be made in the U.S.A. All other copyright questions, including photocopying outside of the USA, should be referred to the copyright owner, Elsevier Science B.V., unless otherwise specified.

No responsibility is assumed by the publisher for any injury and/or damage to persons or property as a matter of products liability, negligence or otherwise, or from any use or operation of any methods, products, instructions or ideas contained in the material herein.

This book is printed on acid-free paper.

Printed in The Netherlands

FOREWORD

It is well known that the advancements made in scientific, agricultural, and industrial fields have been responsible for releasing into the environment, waste products, which serve as xenobiotics that pose a potential threat to both human and animal health. Keeping in view the need for tackling this problem, many thought-provoking ideas from eminent scientists have been compiled in this book, titled "Biotransformations: Microbial Degradation of Health-Risk Compounds", edited by Dr Ved Pal Singh, a scholar of great distinction. Based on his extensive research experience in the field of Applied Microbiology and Biotechnology, Dr Singh has successfully put together, in 14 contributed chapters by recognized experts, a comprehensive and consolidated account of how microorganisms can play a significant role in degrading and detoxifying toxic, carcinogenic, mutagenic, and teratogenic compounds, such as nitrogenous xenobiotics, dimethyl nitrosamine, toxins, haloaromatics, coal-tar, rubbers, tannins, herbicides, pesticides, plastics, polyesters, dyes, and detergents.

I strongly feel that the book will have a wide readership, as it will attract academicians, industrialists, professionals, and scientists from various disciplines of diverse interests as well as the students of biology and medicine. Dr Singh's book is the first of its kind, and it will open new vistas of research in the field of Applied Microbiology and Biotechnology in general, and Biotransformations in particular. The book is of very high standard, and I congratulate both Dr Ved Pal Singh and the Elsevier Science Publishers for having brought out such an excellent piece of work through their joint venture.

Professor A.S. Paintal, FRS, FRCP (London), FNA
Former Director General, Indian Council of Medical Research

DST Centre for Visceral Mechanisms
Vallabhbhai Patel Chest Institute
University of Delhi
Delhi-110007, India

Delhi, February 6, 1995

About the editor

Dr Ved Pal Singh is a Senior Lecturer at the Department of Botany, University of Delhi. He has 13 years of teaching and research experience in the field of Applied Microbiology and Biotechnology, and has about 50 publications to his credit. He is one of the editors for the International Review Series 'Frontiers in Applied Microbiology' and 'Concepts in Applied Microbiology and Biotechnology'.

Dr Singh received Young Scientist Awards from the Indian National Science Academy (INSA) and from the United Nations Educational Scientific and Cultural Organization (UNESCO). He was awarded with the INSA-COSTED Travel Fellowship to visit Hungary (1985). He was a British Council Visitor to the U.K. and Germany (1987). He worked as a Commonwealth Academic Staff fellow at the Royal College, Glasgow (1990, 1991).

He chaired sessions and delivered lectures at a number of symposia/seminars/workshops in India and abroad. He has been honoured by the International Society of Conservators and Explorers of Natural Resources (ISCENR), conferring on him the Founder Fellowship with the title FNRS.

This book is dedicated to my teachers

Abbas Musavi
John Smith
and
Late Umakant Sinha

PREFACE

In addition to the biological sources of undesirable organochemicals, agricultural and industrial wastes introduce a great variety of xenobiotic compounds in the biosphere and pollute it. Therefore, there is an urgent need to look for the possibilities to tackle this situation of increased agro-industrial wastes generated by fast increasing global population. Microorganisms have tremendous potential to degrade an array of compounds. Owing to their biotechnological potential in degrading and eliminating the hazardous organochemicals, microorganisms occupy a key position in health and environmental protection programmes. "Biotransformations: Microbial Degradation of Health-Risk Compounds" helps us to understand how microbes, following their degradative processes, contribute to the benefit of mankind. It provides a clear understanding of the biotechnological implications of microbial degradation of health-risk compounds, so as to assist in environmental protection and improve human and animal health.

In this book, fourteen chapters contributed by leading scientists from different parts of the world cover a wide variety of xenobiotics such as toxic, carcinogenic, teratogenic, and mutagenic compounds. Moreover, they deal with all aspects of microbial degradation, ranging from screening methods for the degradative microorganisms, processes of degradation, strain improvement for enhanced biodegradation, and elimination of undesirable compounds to improving health and environmental protection strategies.

The book intends to provide an opportunity for scientists in the areas of microbiology, biochemistry, engineering, food science, biotechnology, and environmental science to obtain a clear understanding of microbial biotransformations of xenobiotics, and provides an interface between industry and the academic world.

I hope that it will provide new dimensions to identify major problems and prospects in Applied Microbiology and Biotechnology, with special reference to Biotransformations and that it will generate new thought-provoking ideas for scientists of future generations.

I am grateful to the scientists, who accepted my invitation and contributed their valuable review articles for this book. I am greatly indebted to my colleagues Professor N.S. Rangaswamy, Professor K.R. Shivanna, Dr Sudhir Sawhney, Dr A.K. Bhatnagar, Dr S.S. Bhojwani, Dr S.N. Raina, and Dr P.D. Sharma for their valuable suggestions during the course of preparation of the manuscript.

I am thankful to Professor A.S. Paintal, Professor Bilquis Musavi, Professor R.P. Roy, and Dr S.K. Chawla for encouragement, and to Dr B.D. Vashishtha, Dr Sarla, and Dr Tripat Kapoor for helping me in various ways. My sincere appreciation is extended to Mr M.S. Sejwal,

Mr R.K. Gupta, Mr Krishan Lal, Mr S.K. Dass, Mr L.K. Verma, Mr B.K. Sharma, Mr Ram Pal Giri, and Mr Jai Prakash for technical help.

The help rendered by Mr Satish Kumar Sundan and Mrs Mohini Sundan in preparing the manuscript is gratefully acknowledged.

I am extremely grateful to the Elsevier Science Publishers B.V., The Netherlands, and especially to Dr Ingrid van de Stadt, the Publishing Editor and her Secretary, Ms Ursula Isaacs for having taken keen interest in my proposal about this book and keeping me well informed about its status from time to time.

To complete this book expeditiously, I have received much inducement from my parents Mr Gajadhar Singh and Mrs Anandi Devi, my parents-in-law Mr Biri Singh Bhoj, Mrs Satyawati Devi and Mrs Sushila Devi, and my brothers Mr Prem Pal Singh and Mr Ram Pal Singh.

My wife Kusum, daughter Sandhya, and sons Sudhir and Hemant deserve my warmest appreciations for bearing with me throughout the period of my work on this book.

VED PAL SINGH

Delhi, 25 January, 1995

LIST OF ABBREVIATIONS

AFB_1	Aflatoxin B_1
AFB_2	Aflatoxin B_2
AFG_1	Aflatoxin G_1
AFG_2	Aflatoxin G_2
AFM_1	Aflatoxin M_1
AFQ_1	Aflatoxin Q_1
AFR_0	Aflatoxin R_0
BOAA	β-N-oxalyl L-α,β-diamino propionic acid
CAAT	2-Chloro-1,3,5-triazine-4,6-diamine
CBAs	Chlorinated benzoic acids
CBS	Cyclohexyl benzothiazyl sulphenamide
CF	Chloroform
CFCs	Chlorofluoro carbons
CIPC	Isopropyl-N-3-chlorophenyl-carbamate
CT	Carbon tetrachloride
2,4-D	2,4-Dichlorophenoxyacetic acid
1,2-DCA	1,2-Dichloroethane
DCAA	Dichloroacetate
1,1-DCE	1,1-Dichloroethylene
1,2-DCE	1,2-Dichloroethylene
DCM	Dichloromethane
DCP	Dicumyl peroxide
DDS	Drug delivery system
DDT	1,1,1-Trichloro-2,2'-bis(4-chlorobiphenyl) ethane
DFP	Diisopropylfluorophosphate
DHBA	Dihydroxybenzoate
2,3-DHBPO	2,3-Dihydroxybiphenyl dioxygenase
DMNA	Dimethyl nitrosamine
1,3-DNB	1,3-Dinitrobenzene
2,4-DNP	2,4-Dinitrophenol
2,6-DNP	2,6-Dinitrophenol
2,4-DNT	2,4-Dinitrotoluene
DPNA	N-nitrosodipropylamine
DTT	Dithiothreitol
FAD	Flavin adenine dinucleotide
FIFRA	Federal Insecticide, Fungicide and Rodenticide Act
FMN	Flavin mononucleotide
GCMS	Gas chromatographic-mass spectrometry
GPC	Gel permeation chromatography
GS/GST	Glutathione-S-transferase
4HB	4-Hydroxybutyrate
ICI	Imperial Chemical Industries

LAE	Linear alcohol ethoxylate
LAS	Linear alkylbenzene sulphonate
MCAA	Monochloroacetate
MCPA	4-Chloro-2-methylphenoxyacetic acid
m-DCB	*m*-Dichlorobenzene
MFO	Mixed-function oxidase
MMO	Methane monooxygenase
MNC	4-Methyl-5-nitrocatechol
MNP	*m*-Nitrophenol
NAD	Nicotinamide adenine dinucleotide
NR	Natural rubber
o-DCB	*o*-Dichlorobenzene
ONP	*o*-Nitrophenol
PAGE	Polyacrylamide gel electrophoresis
PAHs	Polycyclic aromatic hydrocarbons
p-CB	*p*-Chlorobiphenyl
PCBs	Polychlorinated biphenyls
PCE	Tetrachloroethylene
PCP	Pentachlorophenol
p-DCB	*p*-Dichlorobenzene
PHA	Poly(hydroxyalkanoate)
P(3HB)	Poly(3-hydroxybutyrate)
PLFA	Phospholipid fatty acid
pMMO	Particulate type of methane monooxygenase
PMSF	Phenylmethylsulphonyl fluoride
PNP	*p*-Nitrophenol
PVC	Polyvinyl chloride
SDS	Sodium dodecyl sulphate
sMMO	Soluble type of methane monooxygenase
SMO	Styrene monooxygenase
SOI	Styrene oxide isomerase
2,4,5-T	2,4,5-Trichlorophenoxyacetic acid
2,4,6-TNP	2,4,6-Trinitrophenol
TCA	1,1,1-Trichloroethane
TCAA	Trichloroacetate
TCE	Trichloroethylene
TMTD	Tetramethyl thiuram disulphide
TNT	2,4,6-Trinitrotoluene
TSCA	Toxic Substances Control Act
USEPA	United States Environmental Protection Agency
VC	Vinyl chloride

LIST OF CONTRIBUTORS

Chapter numbers are shown in parentheses following the address of each contributor

Todd A. Anderson, Pesticide Toxicology Laboratory, Department of Entomology, Iowa State University, Ames, IA, U.S.A. (10)

V. Andreoni, Dipartimento di Scienze e Tecnologie Alimentari e Microbiologiche, Università degli Studi di Milano, 20133 Milano -Via G. Celoria 2, Italy (1)

G. Baggi, Dipartimento di Scienze e Tecnologie Alimentari e Microbiologiche, Università degli Studi di Milano, 20133 Milano -Via G. Celoria 2, Italy (1)

S. Bernasconi, Dipartimento di Chimica Organica e Industriale, Università degli Studi di Milano, 20133 Milano - Via G. Celoria 2, Italy (1)

Manzoor A. Bhat, Department of Biochemistry and UGC Centre of Advanced Study, Indian Institute of Science, Bangalore 560012, India (6)

John A. Bumpus, Centre for Bioengineering and Pollution Control and Department of Chemistry and Biochemistry, University of Notre Dame, Indiana 46556, U.S.A. (7)

Craig S. Criddle, National Science Foundation Center for Microbial Ecology, Michigan State University, East Lansing MT 48824, U.S.A. (4)

Yoshiharu Doi, Head, Polymer Chemistry Laboratory, The Institute of Physical and Chemical Research (RIKEN), Hirosawa, Wako-Shi, Saitama 351-01, Japan (9)

Jörg Fiedler, Universität Bielefeld, Facultät für Biologie, Gentechnologie/Mikrobiologie, Postfach 100131, Universitätsstraβe, D-33594 Bielefeld 1, Germany (5)

Karl-Heinz Gartemann, Universität Bielefeld, Facultät für Biologie, Gentechnologie/Mikrobiologie, Postfach 100131, Universitätsstraβe, D-33594 Bielefeld 1, Germany (5)

Erwin Grund, GBF - Gesellschaft für Biotechnologische Forschung mBH, Mascheroder Weg 1, D-38124 Braunschweig, Germany (5)

S. Hartmans, Division of Industrial Microbiology, Department of Food Science, Wageningen Agricultural University, P.O. Box 8129, 6700 EV Wageningen, The Netherlands (11,12)

Mukesh K. Jain, Department of Civil and Environmental Engineering, Michigan State University, East Lansing MI 48824, U.S.A. (4)

S.L. Mehta, Head, Division of Biochemistry, Indian Agricultural Research Institute, New Delhi - 110012, India (13)

Katsuyuki Mukai, Polymer Chemistry Laboratory, The Institute of Physical and Chemical Research (RIKEN), Hirosawa, Wako-Shi, Saitama 351-01, Japan (9)

I.M. Santha, Division of Biochemistry, Indian Agricultural Research Institute, New Delhi - 110012, India (13)

R.K. Saxena, Head, Department of Microbiology, University of Delhi, South Campus, New Delhi-110021, India (14)

Annegret Schmitz, Universität Bielefeld, Fakultät für Biologie, Gentechnologie/Mikrobiologie, Postfach 100131, Universtätsstraβe, D-33594 Bielefeld 1, Germany (5)

P. Sharmila, Department of Microbiology, University of Delhi, South Campus, New Delhi-110021, India (14)

Ved Pal Singh, Department of Botany, University of Delhi, Delhi - 110007, India (3,14)

Akio Tsuchii, National Institute of Bioscience and Human-Technology, Agency of Industrial Science and Technology, Tsukuba City, Ibaragi 305, Japan (8)

C.S. Vaidyanathan, Department of Biochemistry and UGC Centre of Advanced Study, Indian Institute of Science, Bangalore 560012, India (6)

Barbara T. Walton, Environmental Sciences Division, Oak Ridge National Laboratory, Oak Ridge, TN, U.S.A. (10)

David C. White, Center for Environmental Biotechnology, The University of Tennessee, Knoxville, TN, U.S.A. (10)

Tadashi Yoshinari, Wadsworth Center for Laboratories and Research, New York State Department of Health and School of Public Health, State University of New York at Albany, Empire State Plaza, P.O. Box 509, Albany, NY 12201-0509, U.S.A. (2)

CONTENTS

Biotransformations: Microbial Degradation of Health Risk Compounds
Ved Pal Singh, editor
© 1995 Elsevier Science B.V. All rights reserved.

Microbial degradation of nitrogenous xenobiotics of environmental concern

V. Andreoni[a], G. Baggi[a] and S. Bernasconi[b]

[a]Dipartimento di Scienze e Tecnologie Alimentari e Microbiologiche, Università degli Studi di Milano, 20133 Milano - Via G. Celoria 2, Italy

[b]Dipartimento di Chimica Organica e Industriale, Università degli Studi di Milano, 20133 Milano - Via G. Celoria 2, Italy

INTRODUCTION

Nitrogen forms a variety of functional groups in combination with carbon, hydrogen, and oxygen. These functional groups have been particularly useful for adapting and activating aromatic compounds to be used as chemical intermediates in synthetic processes [1]. In addition, many final products, such as pesticides, explosives, drugs, dyes, antioxidants and antiozonants, contain nitrogen functionalities [2-7]. Consequently, these compounds are contaminants of rivers, ground water, soils treated with pesticides, and atmosphere [8-11].

Exposure to amines and related compounds has large impact on human health: workers exposed to benzidine and naphthylamine have developed cancer of the bladder [12,13]; the Food and Drug Administration (F.D.A.) has found that aromatic amines [14] and nitroaromatics [15] can enter the food chain. In addition, many degradation products from nitroaromatics are easily polymerized, in presence of oxygen, to persistent macromolecules [16]. The wide distribution of these compounds in the environment, coupled with their toxicity, has given rise to concern about their environmental fate.

The complete degradation of nitrogenous compounds is mainly the result of microbial attack and represents one of the primary mechanisms by which these pollutants are eliminated from the environment.

The term *degradation* is often used only to indicate the disappearance of a compound, which, in turn, is transformed into another, with no evidence of the extent of degradation. The degree of degradation depends on the nature of the compounds: some are resistant to microbial attack, others are partially broken down into persistent intermediates, or transformed into more toxic products. On the contrary, complete biodegradation will result in mineralization to carbon dioxide or methane with release of nitrite or ammonium ion.

In aerobic environment, O_2 is both the terminal electron acceptor and a reactant in the initial reactions. In the absence of oxygen, however, organic compounds, like nitrate, sulphate and carbonate, are alternate electron acceptors in the microbial degradation of organic material, and the presence or absence of these electron acceptors plays a crucial role in biodegradability and influences microbial activity and diversity.

The intent of this review is to present a broad and updated overview of the physiological, biochemical, and genetic basis of biodegradation of nitrogenous compounds by aerobic and anaerobic microorganisms.

NITROAROMATICS

Nitroaromatic compounds are produced industrially on a large scale. Such chemicals are widely used as pesticides, or for other chemical uses [2-7]. Nitroaromatics may be produced enzymatically in microbial cultures [17], or photochemically in urban air [18,19].

Nitroaromatics are highly *toxic* to man and mammals, being easily reduced by enzymes to nitroso and hydroxylamine derivatives. These derivatives may lead to the formation of either metahemoglobin, which is unable to bind oxygen, or of nitrosoamines, which are *carcinogenic* [14,20,21]. Some nitroaromatics, such as nitropyrene [19], are *mutagenic* and several nitrophenols have an uncoupling effect on oxidative phosphorylation [22]. Most nitroaromatics are highly toxic also to bacteria and, consequently, may inhibit microbial growth. In activated sludges, their presence may destabilize the continuous process of sewage treatments.

Nitroaromatics are slowly degraded by microorganisms both under aerobic and anaerobic conditions, and the metabolic steps involved in the degradation have been poorly documented until now. Two major catabolic pathways are involved in the degradation of nitroaromatics (Figure 1) [23]. In the first pathway, the nitro group is reduced to an aniline intermediate, which is further degraded to ammonium ion and catechol [5,24-28].

A reduction of nitro substituent, under both aerobic and anaerobic conditions, seems to be a common enzymatic mechanism in the environment [5,29]. Such reduction has been demonstrated in various organisms which utilize the nitro compound as an electron acceptor. The activity of nitroreductases, many of which have a broad substrate specificity, has been demonstrated in cell-free systems, and some enzymes have been purified and characterized [5,19,29]. The resulting aromatic amines are often further transformed into persistent azo compounds or polymers by biotic or abiotic processes [1,30,31]. In the second pathway, the nitro substituent is directly removed as nitrite [24,32,33], with the formation of catechol.

The *microbial degradation* of nitroaromatics to catechol involves a series of reductions and oxidations, generally catalyzed by reductases and oxygenases.

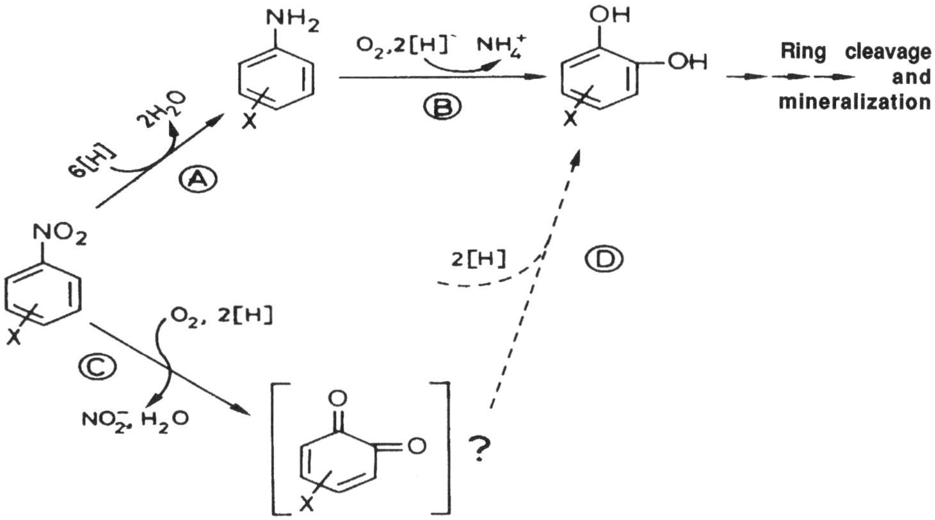

Figure 1. Microbial degradation of nitroaromatic compounds [23]. —, steps demonstrated;_ _ , steps postulated; A, nitroreductase; B, aniline oxygenase; C, nitrophenyl oxygenase; D, chinoreductase?

Nitrotoluenes

2,4,6-Trinitrotoluene (TNT) is the predominant conventional *explosive* used by military forces [34], and the disposal of wastes containing TNT leads to soil, sediment, and water contamination [35]. This is of great concern because TNT causes liver injury and marked changes in the hemopoietic system, producing anemia in humans and other mammals [36]. Moreover, TNT is toxic to certain fish at concentrations greater than 2 µg/ml [37] and to certain green algae [38]; finally, TNT is mutagenic [5].

TNT was shown to be oxidized by three *Pseudomonas*-like bacteria; the degradation was accelerated by the addition of glucose or yeast extract and proceeded through the formation of several intermediates: dinitrohydroxylaminotoluene, dinitroamino toluene, nitrodiaminotoluene, and azoxy toluenes [39]. Among these metabolites, only nitrodiaminotoluene, and dinitroaminotoluene were not degraded further.

The same reduced and azoxy compounds have been isolated by McCormick et al. [5] (Figure 2). They found that the nitro groups of TNT were reduced by both aerobic and anaerobic systems, and that the number of the nitro groups reduced depended on the reducing potential of the system and on the species utilized. Cell-free extracts of *Veillonella alcalescens* utilized 3 moles of H_2 to reduce 1 mole of nitro group. The

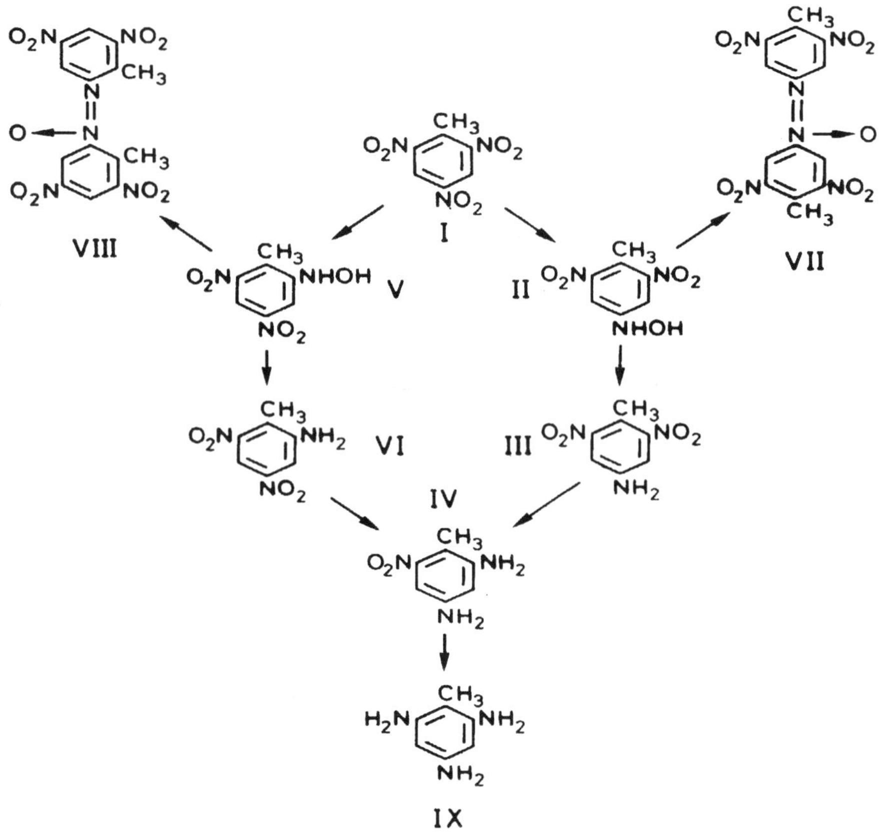

Figure 2. Proposed pathway for transformation of 2,4,6-TNT [5]. I, 2,4,6-TNT; II, 4-hydroxylamino-2,6-dinitrotoluene; III, 4-amino-2,6-dinitrotoluene; IV, 2,4-diamino-6-nitrotoluene; V, 2-hydroxylamino-4,6-dinitrotoluene; VI, 2-amino-4,6-dinitrotoluene; VII, 4,4'-azoxycompound; VIII, 2,2'-azoxycompound; IX, 2,4,6-triaminotoluene.

oxidation of the hydroxylamino derivatives to azoxy compounds may also occur non-enzymatically in anaerobic environment [40].

The same reduced products are present in the urine excreted by rabbits, rats, or human volunteers fed with TNT [41]. TNT is also degraded by different fungi [42,43]. *Rhizopus stolonifer* was able to degrade almost all TNT, when present in cultural broth at a concentration of 100 mg/l [42], and extensive biodegradation of [^{14}C]-TNT by the white rot fungus *Phanerochaete chrysosporium* was also observed [44]. Biodegradation of [^{14}C]-TNT occurs even in a mixture of soil and corncobs inoculated with *P. chrysosporium* [44]. However, substantially less [^{14}C]-TNT was converted to $^{14}CO_2$ in soil cultures, compared with liquid cultures.

2,4-Dinitrotoluene (2,4-DNT), listed as a *priority pollutant* by the United States Environmental Protection Agency (USEPA) [45], is the major impurity resulting from the manufacture of TNT, and is a starting material for the synthesis of toluenediisocyanate, used in the production of polyurethane foam. 2,4-DNT was transformed to 2-amino-4-nitrotoluene, 4-amino-2-nitrotoluene, 2-nitroso-4-nitrotoluene and 4-nitroso-2-nitrotoluene by a mixed culture derived from activated sludge only under anaerobic conditions and with an exogenous carbon source [46]. The two nitroso compounds were unstable and could be detected between 48 and 72 h of incubation.

A *Pseudomonas* sp., which is able to degrade aerobically 2,4-DNT, using the latter as the sole source of carbon and energy with stoichiometric release of nitrite, has been described. 4-Methyl-5-nitrocatechol (MNC) accumulated transiently when cells grown on acetate were transferred to medium containing 2,4-DNT. Conversion of 2,4-DNT to MNC was catalyzed by a dioxygenase [47] (Figure 3). MNC was then rapidly oxidized with the removal of the second nitro group as nitrite.

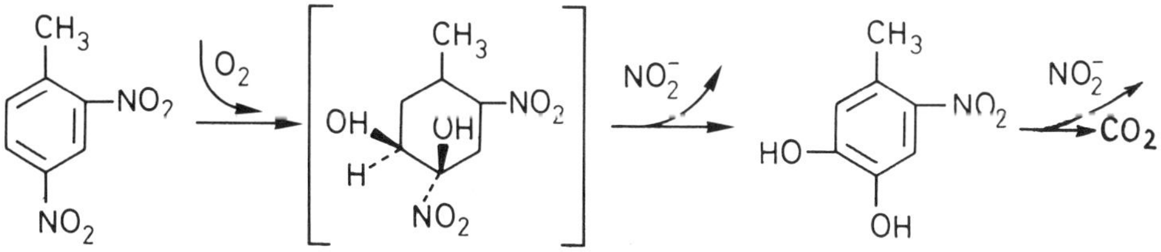

Figure 3. Initial steps in 2,4-DNT degradation pathway [47].

Finally, nitrotoluenes largely used in the manufacture of azo and sulphur dyes, and in the production of explosives [48], have been detected at high levels in waste waters from paper mills and chemical plants [45]. Delgado et al. [49] showed that the *biotransformation* of nitrotoluenes into more oxidized nitroaromatic chemicals is mediated by the upper pathway of the TOL-plasmid. In fact, they found that the TOL-upper-pathways enzymes recognize nitroaromatics as substrates, although the regulator, the XyLR protein, does not recognize nitrotoluenes as effectors. The TOL-encoded toluene monooxygenase enzyme biotransformed 3-nitrotoluene and 4-nitrotoluene into their corresponding benzyl alcohols and benzaldehydes, but not 2-nitrotoluene. The transformation of nitrobenzyl alcohol into the corresponding nitro benzaldheyde was carried out by the same toluene monooxygenase, in agreement with Harayama et al. [50], who have reported that this enzyme, in addition to its primary oxidative activity, also shows an alcohol dehydrogenase activity. Recently, it has been reported that cells of *Pseudomonas putida* F1 and *Pseudomonas* sp. strain JS150 were capable of degrading nitrotoluenes into 3-methyl-6-nitrocatechol and 2-methyl-5-nitrophenol through initial oxidation into 4-nitrotoluene-2,3-dihydrodiol by toluene dioxygenase [51].

Nitrobenzenes

Nitrobenzenes are widely used in the manufacture of aniline and pyroxylin compounds, in the refinery of lubricant oils and in the production of soap and shoe polishes [52]. 1,3-Dinitrobenzene (1,3-DNB) is the main impurity of TNT [53].

These non polar nitroaromatic compounds are considered recalcitrant to microbial attack [54] for their resistance to the reduction of electron density in the aromatic ring, which can hinder electrophilic attack by oxygenase, and for their toxicity against microorganisms [4,5].

However, several bacterial strains, capable of degrading toluene and to oxidize nitrobenzene, have been isolated. While in cells of *Pseudomonas putida* F1 and *Pseudomonas* sp. strain JS150, a dioxygenase mechanism converts nitrobenzene into the corresponding dihydrodiol, in other microorganisms, a monooxygenase is instead responsible for the initial attack on nitrobenzene [55].

Recently, a new *Rhodococcus* species, isolated under nitrogen limiting conditions from contaminated soils, and capable of utilizing 1,3-DNB, has been described. 0.5 mM of 1,3-DNB was completely and immediately metabolized by induced cells with release of 2 moles of nitrite per mole of 1,3-DNB via 4-nitrocatechol [56] (Figure 4). According to the mechanism

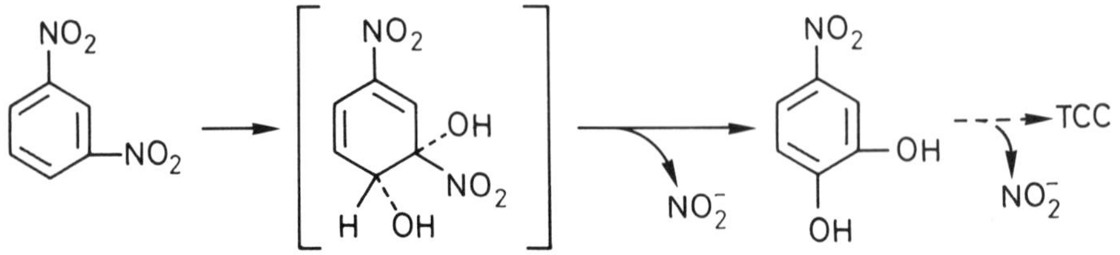

Figure 4. Proposed pathway for degradation of 1,3-DNB by *Rhodococcus* sp. QT-1 [56].

reported in Figure 4, 4-nitrocatechol could be generated by an initial 3,4-dioxygenation with subsequent elimination of 1 mole of nitrite. Interestingly, *Rhodococcus* utilized 1,3-DNB as source of nitrogen in the absence as well as in the presence of high amounts of ammonium ion.

Nitrophenols

Nitrophenols, used in the manufacture of dyes, explosive, and pesticides [2,5,6], are released into the environment during the hydrolysis of several organophosphorous pesticides, such as parathion. 2-Nitrophenol, 4-nitrophenol, and 2,4-dinitrophenol are priority pollutants according to the USEPA [57].

2,4-Dinitrophenol (2,4-DNP) is an uncoupler of electron transport [22,58,59] and its structural analogues, 4,6-dinitro-2-methylphenol, 2-sec-butyl-4,6-dinitrophenol and Dinoseb are important pesticides [60]. Dinoseb, which is also the major degradation product of the herbicide, Acrex by soil microorganisms [61], is responsible for health and environmental *hazards* [62].

p-Nitrophenol (PNP) was degraded either by resting or growing cells of a *Flavobacterium* strain with stoichiometric release of nitrite and formation of 4-nitrocatechol [63]. The degradation of PNP was accompanied by the disappearance of the characteristic yellow colour in the medium, indicating consumption of the nitrogen, when nitrophenols were tested as nitrogen source. The cells grown on PNP also oxidized *m*-nitrophenol into nitrohydroquinone, but did not use *m*-nitrophenol as carbon source for growth.

Rapid biodegradation of PNP was shown to occur in a pond in 6 days; a second treatment of the pond with PNP enhanced its biodegradation which began immediately [64].

Differently, a *Moraxella* strain degraded PNP by replacing the nitrogroup with a hydroxyl group and accumulating traces of hydroquinone in the medium. Hydroquinone was then converted into β-ketoadipic acid via γ-hydroxymuconic semialdehyde [65].

A *Pseudomonas putida* utilized *o*-nitrophenol (ONP) and *m*-nitrophenol (MNP) as source of carbon and nitrogen, but not PNP. Growing cells of these organisms degraded ONP and MNP, releasing nitrite and ammonium, respectively. The enzymes involved in the metabolism of ONP or MNP were inducible. Only the degradation pathway of ONP has been described by using a crude enzyme extract. The crude extract converted ONP to nitrite and catechol, which were further metabolized through the *ortho*-cleavage [32].

Zeyer et al. subsequently observed that *P. putida* did not utilize *para*-substituted derivatives of ONP and, in the same study, they characterized the inducible nitrophenol oxygenase responsible for the release of nitrite [66]. The enzyme was found to be soluble, NADPH-dependent and its activity was stimulated by magnesium and manganese ions, but not by FAD, and consisted of a single polypeptide chain with a molecular weight of 58,000 (determined by gel filtration) or 65,000 (determined on a sodium dodecyl sulphate polyacrylamide gel) [67].

The degradation of PNP was found to occur also under cometabolic conditions by a *Pseudomonas* sp. in presence of glucose, but not in presence of phenol that, instead, inhibited PNP mineralization [68].

The oxidative elimination of nitrite ions by dinitrophenols has been detected also during the mineralization of 2,6-dinitrophenol (2,6-DNP) by *Alcaligenes eutrophus* JMP 134 [69]; a total degradation of 2,6-DNP was performed by a *Pseudomonas* strain, which, although grew scant, decolourized the culture medium within 3 to 4 days of incubation [70].

Under nitrogen-limiting conditions, two *Rhodococcus erythropolis* strains, able of mineralizing 2,4-DNP as the sole source of carbon, have been isolated. Both strains metabolized 2,4-DNP, present at a final concentration lower than 0.5 mM with liberation of stoichiometric amounts of nitrite and of low amount of 4,4-dinitrohexanoate [71]. The identification of the last compound as the only organic metabolite of 2,4-DNP suggested the involvement of a reductive mechanism in the degradation pathway. The initial reduction of the aromatic ring gives evidence for a nucleophilic attack in consequence of the highly electrophilic character of the aromatic nucleus of 2,4-DNP, that favours an initial reductive reaction.

A different reductive mechanism was described for the degradation of 2,4-DNP by a *Fusarium oxysporum* strain [72]. Although 2,4-DNP is used extensively in fungicidal preparations, *Fusarium* reduced it to the less toxic 2-amino-4-nitrophenol and its isomer 4-amino-2-nitrophenol. The reduction of the nitro groups occur in successive stages by the intermediate formation of the nitroso and hydroxyleamino groups.

Finally 2,4,6-trinitrophenol (2,4,6-TNP), picric acid, is used as explosive under the form of ammonium 2,4,6-trinitrophenoxide. Both picric acid and picramic acid (2-amino-4,6-dinitrophenol), deriving from the bioconversion of picric acid under anaerobic conditions, are well-known mutagenic compounds [73].

The degradation of picric acid proceeded by nucleophilic attack of the aromatic ring [74]. A spontaneous mutant of *Rhodococcus erythropolis* HL 24-2, originally isolated for its ability to degrade 2,4-dinitrophenol [14], could also utilize picric acid as nitrogen source. The mutant HL-PM-1 transiently accumulated an orange-red metabolite, which was identified as a hydride-Meisenheimer complex of picric acid. This complex was further converted with release of nitrite. 2,4,6- Trinitrocyclohexanone was the dead-end metabolite of the degradation of picric acid.

Nitro and aminobenzoates

Like all nitroaromatics, these compounds enter industrial waste streams, and may accumulate in the environment. In aerobic metabolism of 2- and 4-nitrobenzoate, reduction of the nitro group via nitroso and hydroxylaminobenzoate was demonstrated [75-78]. In this case, the nitro group was used as a terminal electron acceptor. The amino intermediate was transiently accumulated during the growth on the corresponding nitrobenzoate, but it was not clear whether or not amines were intermediates in aerobic degradation of nitrobenzoates. Ke et al. [79], working with a strain of *Flavobacterium* capable of mineralizing 2-nitrobenzoate, indicated that the cells growing on 2-nitrobenzoate were not simultaneously adapted to 2-aminobenzoate. A similar consideration was made in the case of the degradation of 4-nitrobenzoate [76]; 4-aminobenzoate did not lie on the direct oxidative pathway because no appreciable oxidation of this substrate was observed with cells grown on 4-nitrobenzoate. Involvement of amines, however, is often implied. The incubation of cells of *Pseudomonas* sp. CBS3 with the isomers of nitrobenzoate, under aerobic conditions, allowed to detect the presence of the corresponding amines in cultural broths. In the course of the conversion of 4-nitrobenzoic acid, 4-nitrosobenzoic acid appeared shortly after the start of the reaction, and was no longer detectable after 24 h of incubation, when the major product present was instead 4-aminobenzoic acid. As the same conversion of 4-nitrobenzoate by *Pseudomonas* sp. CBS3 occurred in absence of molecular oxygen, the nitroaromatic compound may be electron acceptor, according to the scheme proposed in Figure 5 [80].

Recently, the degradation of 4-nitrobenzoate by a *Comamonas acidovorans* NBA-10 was shown to proceed via reduction of the nitro group, but without the formation of 4-aminobenzoate. Cell extracts, in presence of NADPH, degraded 4-nitrobenzoate into 4-hydroxylamino-

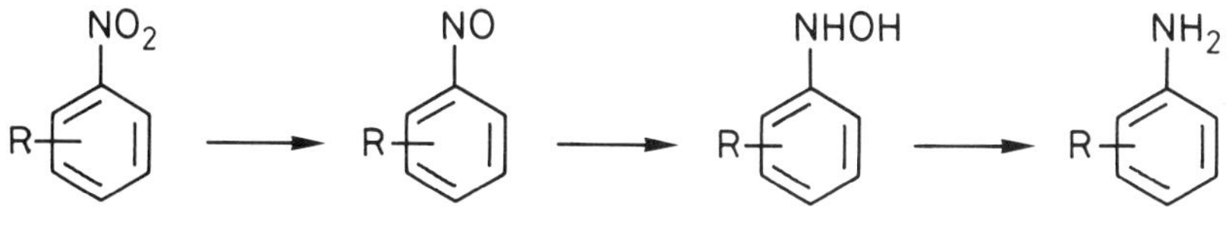

NO_2 → NO → NHOH → NH_2 (R-substituted benzene rings)

Figure 5. Proposed pathway for the reduction of nitroaromatic compounds by *Pseudomonas* sp. CBS3: R, -OH, -COOH, -Cl, $-NO_2$ [80].

benzoate and 3,4-dihydroxybenzoate [81] (Figure 6), evidencing a new pathway for its aerobic degradation which did not involve molecular oxygen [82]. The conversion of 2-nitrobenzoate by an oxygenative reaction, yielding nitrite, is reported by Andreoni et al. [83]. An *Achromobacter* strain utilized 2-nitrobenzoate as only carbon and nitrogen source with release of nitrite in the culture medium. Simultaneous oxidation of 2-nitrobenzoate and 2,3-dihydroxybenzoate by resting cells of *Achromobacter* suggests the involvement of an inducible dioxygenase in the degradation of 2-nitrobenzoate.

2-Aminobenzoic (anthranilic) acid is an important intermediary metabolite in both biosynthetic and catabolic pathways of microorganisms. In fact, it serves as precursor for tryptophane for many bacteria and molds [84,85], and is formed by bacterial reduction of 2-nitrobenzoic acid [76,77], and in the degradation of compounds containing an indole moiety, both aerobically [86, 87] and anaerobically [88]. While 2- and 4-aminobenzoates occur in nature, 3-aminobenzoate is a *xenobiotic* compound, produced mainly for the synthesis of azo dyes. A toxic effect of 3-aminobenzoate has been reported for man and many mammals, which metabolize it partly to 3-ureidobenzoic acid and 3-aminohippuric acid [89].

The aerobic catabolism of aminobenzoates proceeds via catechol [90] or gentisic acid [91], and requires molecular oxygen. Brown and Gibson reported some denitrifying organisms belonging to the genus *Pseudomonas,* that degraded 2-aminobenzoate completely into CO_2 and NH_4^+ [92]. Nitrate, the terminal electron acceptor, was first reduced to nitrite and then to nitrogen. Aerobically, 3-aminobenzoate was degraded through hydroxylation to 5-aminosalicylic acid which was further metabolized (Russ, Stoltz and Knackmuss, Abstr. Ann. Meetg. VAAM, Freiburg, 1991; Poster P 136). Anaerobically, 3-aminobenzoate was found to be decomposed both by a sulphate-reducing bacterium in pure culture, and by a methanogenic and a nitrate-reducing enrichment culture [93]. The

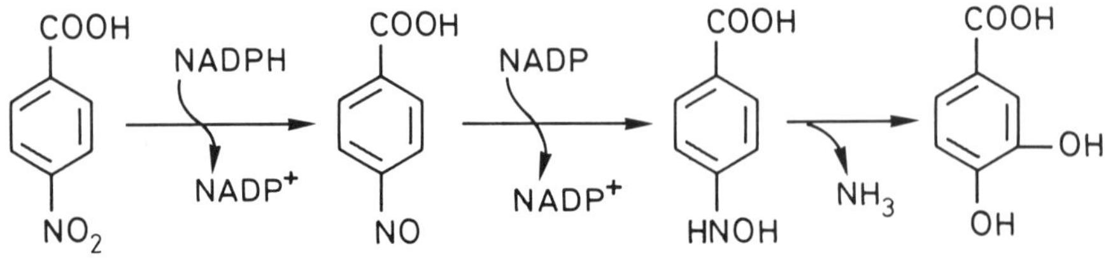

Figure 6. Proposed degradative pathway of 4-NBA by *Comamonas acidovorans* NBA-10 [81].

sulphate reducer oxidized 3-aminobenzoate completely to CO_2, with concomitant reduction of sulphate to sulphide and release of ammonium. In the absence of an external electron acceptor, 3-aminobenzoate was degraded by a methanogenic culture, consisting of three types of bacteria into CO_2, CH_4, and NH_4^+. The consortium was constituted by a short rod able to ferment 3-aminobenzoate, a H_2-degrading *Methanospirillum*-like bacterium, and an acetate-degrading *Methanothrix*-like bacterium.

NITRILES

Nitriles are cyanide-substituted carboxylic acids, which occur naturally and synthetically, and are of the general structure, R-CN. The naturally occurring nitriles are found in higher plants [94-97], bone oils, insects [98], and microorganisms [99,100]; the synthetic ones are used industrially in benzonitrile *herbicides* [101], as organic solvents and in the synthesis of polymers, plastics [102,103], synthetic fibers, resins, and dye stuffs.

There is little information on the ecological impact of these compounds, most of which are highly toxic, and some mutagenic and carcinogenic. Particularly, acrylonitrile is a largely used vinyl cyanide of which the production in U.S. was estimated at 2.5 x 10^9 pounds [104-106]. Because of its acute neurotoxicity, mutagenicity, and teratogenicity, the USEPA has targeted acrylonitrile as a priority pollutant [56].

Microbial degradation of nitrile compounds was performed by different microorganisms, capable of growing on various aliphatic and aromatic nitriles [107-109]. They are degraded through two pathways (Figure 7): one is the direct hydrolysis of nitriles to carboxylic acid and ammonia, catalyzed by nitrilases. Nitrilases, that utilize benzonitrile and related aromatic nitriles as substrates, have been purified from *Pseudomonas* sp. [110,111], *Nocardia* sp. strains NCIB 11215 [112] and NCIB 11216

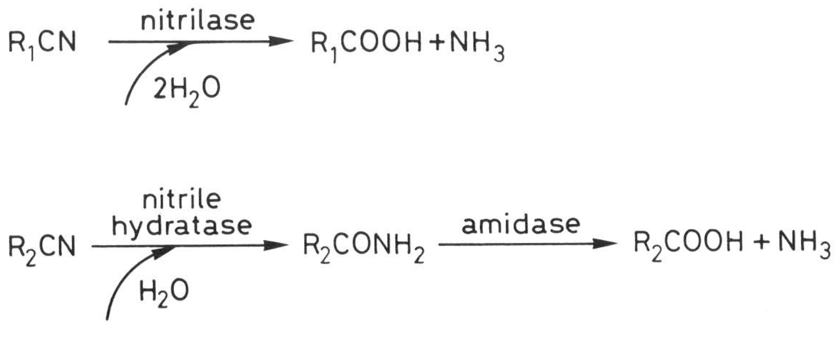

$$R_1CN \xrightarrow[2H_2O]{\text{nitrilase}} R_1COOH + NH_3$$

$$R_2CN \xrightarrow[H_2O]{\text{nitrile hydratase}} R_2CONH_2 \xrightarrow{\text{amidase}} R_2COOH + NH_3$$

Figure 7. Proposed pathway for the degradation of nitriles: R1, -phenyl; -α-β-alkenyl; R2, -alkyl.

[113], *Fusarium solani* [114], *Arthrobacter* sp. [115], *Escherichia coli,* transformed with a *Klebsiella ozaenae* plasmid DNA [116], *Rhodococcus rhodochrous* J1 [117,118], and *Alcaligenes faecalis* JM3 [119, 120]. The enzyme of *Rhodococcus rhodochrous* J1 was employed for the production of *p*-aminobenzoic acid from *p*-aminobenzonitrile [121], and nicotinic acid from 3-cyanopyridine [122]. The conversion of 3-cyanopyridine to nicotinic acid, by a nitrilase of *Nocardia rhodochrous* LL100-2, was also reported by Vaughan et al. [123]. These nitrilases were usually inactive on aliphatic nitriles. More recently, a new nitrilase, that acts preferentially on aliphatic nitriles, was purified and characterized in *Rhodococcus rhodochrous* K22 [124].

The other pathway, working preferentially on aliphatic nitriles, is a two-step degradation process, involving nitrile hydratase and amidase, via an amide as intermediate. The corresponding amides are then hydrolyzed into the respective carboxylic acids and ammonia [108,125-127].

Aliphatic nitrile hydratases, that catalyzed the hydratation of nitriles to amides, were purified and characterized in *Arthrobacter* sp. J1 [128], *Brevibacterium* R312 [129], and *Rhodococcus* sp. N774 [130]. In the first strain, the activity of an amidase, which forms acetic acid and ammonia stoichiometrically from acetamide, was also detected [131]. *Bioconversion* of dinitrile to mononitrile catalyzed by nitrile hydratase and amidase was obtained from *Corynebacterium* sp. C5. The two enzymes were constitutively formed in cells [132].

The production of amides from nitriles has been studied by several workers, and most of them focused on the accumulation of acetamide from acetonitrile [126,133-136]. The enzymatic production of acrylamide from acrylonitrile by nitrile hydratase of *P. chlororaphis* B23, *Rhodococcus* sp. N-774, and *Klebsiella pneumoniae*, respectively has been reported [137-142]. These microorganisms exhibited a high nitrile hydratase activity and a low amidase activity, allowing the accumulation of the corresponding amide. Nagasawa et al. optimized the reaction conditions for the production of nicotinamide by a nitrile hydratase, found in *Rhodococcus rhodochrous* J1. The enzyme contains cobalt, and shows high activity towards 3-cyanopyridine [143,144].

Recently, the potential of bacterial enzymes for the synthesis of aromatic, optically active amides, and carboxylic acids from racemic nitriles was evaluated. An enantiomer-selective amidase, active on several 2-aryl and 2-aryloxy propionamides, was identifided and purified from *Brevibacterium* sp. strain R312 [145]. A nitrilase, found in *Acinetobacter* sp. strain AK226 and able to hydrolyze efficiently both aromatic and aliphatic nitriles, was reported to hydrolyze racemic nitriles to optically active 2-aryl propionic acids [146]. Enzyme system of *Rhodococcus butanica* could be successfully adapted for the kinetic resolution of α-arylpropionitriles resulting in the formation of (R)-

amides and (S)-carboxylic acids [147]. Finally, from racemic mandelonitrile and its acetylated derivatives, R(-)-mandelic and R(-) acetylmandelic acids were obtained by using enantioselective nitrilases of *Alcaligenes faecalis* ATCC 8750, and *Pseudomonas*, respectively [148,149].

The interest of some workers has been focused on the biodegradative process of halogenated aromatic nitriles, herbicides, widely used in post-emergence control of seedling broadleef weeds in a number of tolerant plants [150]. The di-*ortho*-substituted Dichlobenil (2,6-dichlorobenzonitrile) is a persistent compound [151], presumably because the double *ortho*-substitution in the aromatic ring is incompatible with enzyme attack. To confirm this hypothesis, Dichlobenil was not attacked by nitrilase either of *Nocardia* [113] or of *Fusarium solani* [114], both found capable of growing on benzonitrile as sole carbon and nitrogen source. On the other hand, the same fungal enzyme was tested on different benzonitriles such as Yoxynil (3,5-diiodo-4-hydroxybenzonitrile) and Bromoxynil (3,5-dibromo-4-hydroxybenzonitrile) [114], hydrolyzed at significant rates.

The metabolism of Bromoxynil has been studied in some microorganisms, revealing the presence of amide and acid products [3,152-154] (Figure 8). *Klebsiella pneumoniae* subsp. *ozaenae* was shown to transform completely the herbicide, with the involvement of a nitrilase enzyme to

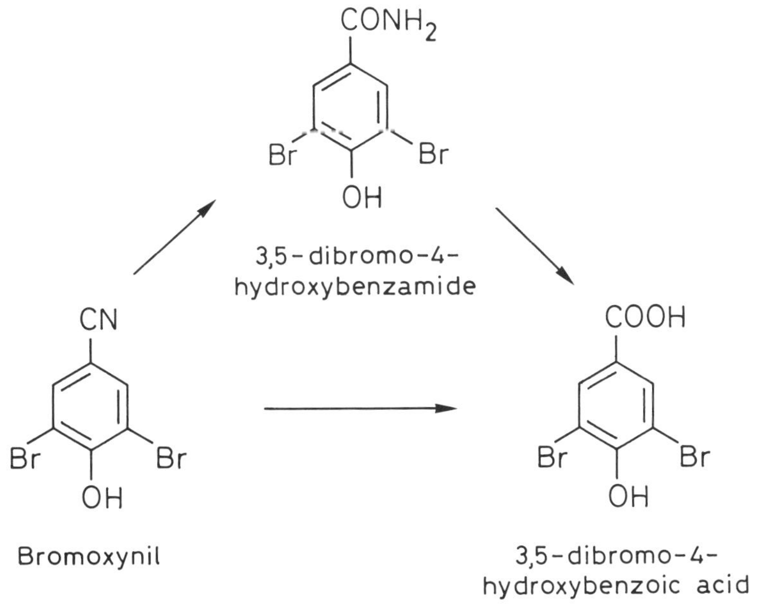

$CONH_2$ Br Br OH

3,5-dibromo-4-hydroxybenzamide

CN Br Br OH

Bromoxynil

COOH Br Br OH

3,5-dibromo-4-hydroxybenzoic acid

Figure 8. Initial reactions in the metabolism of the herbicide, Bromoxynil [3].

corresponding acid, and to utilize the liberated ammonia as sole nitrogen source. As the cyano-moiety of the molecule is important for the toxic properties of the herbicide, its removal is essential for the *detoxification* of the compound [3].

HETEROCYCLIC COMPOUNDS

Indole

Indole and its derivatives form a class of toxic recalcitrant compounds released into the environment through cigarette smoke, coal-tar, and sewage. Many of these compounds, which are present in several edible plants, are responsible for different diseases in cattle and goats [155, 156], and have shown to be toxic and mutagenic [157,158].

The studies on the microbial metabolism of indole always indicate anthranilic acid as intermediate, but the degradative pathways observed, leading to this compound, were found to be different.

By carrying out oxygen uptake studies, Sakamoto et al. proposed indoxyl, dihydroxyindole, and isatin as metabolites during the *bacterial degradation* of indole [159], whereas another group suggested that indole is metabolized via 2,3-dihydroxyindole, through the initial formation of an epoxide. Dihydroxyindole is converted into anthranilic acid by dihydroxyindole oxygenase, an inducible enzyme, which appears only when the organism has grown on indole. This conversion is apparently a single enzymatic step [160]. More recently, Ensley et al., by using cloned naphthalene dioxygenase and other dioxygenases, reported that indole oxidation is a property of bacterial dioxygenases, that form *cis*-dihydrodiols from other aromatic hydrocarbons by a dioxygenation reaction [161]. Spontaneous elimination of water from the *cis*-dihydrodiol would yield indoxyl, the precursor of indigo. Indoxyl was found as an intermediate also in the metabolism of indole by *Alcaligenes*. On the basis of its respiratory activity, isatin is considered the next intermediate, while in broth cultures with indole, anthranilic acid was found. This latter intermediate was catabolized via gentisate, which is cleaved by gentisate 1,2-dioxygenase, present in cells grown on indole, to maleylpyruvate [162]. A strain of *Aspergillus niger*, which cometabolized indole in the presence of glucose and nitrate, monohydroxylated this compound into indoxyl (Figure 9). This mechanism is considered prevalent in fungi and other higher organisms as well as in anaerobic bacteria, which hydroxylate indole to oxindole, whereas in aerobic bacteria unhydroxylated aromatic compounds are attacked by dioxygenases [161]. Indoxyl was further converted to N-formylanthranilate by a dioxygenase, but this activity was not demonstrated for the instability of the substrate. In the cytosolic fraction, N-formylanthranilate deformylase, anthranilate hydroxylase, dihydroxybenzoate (DHBA) decarboxylase, and catechol

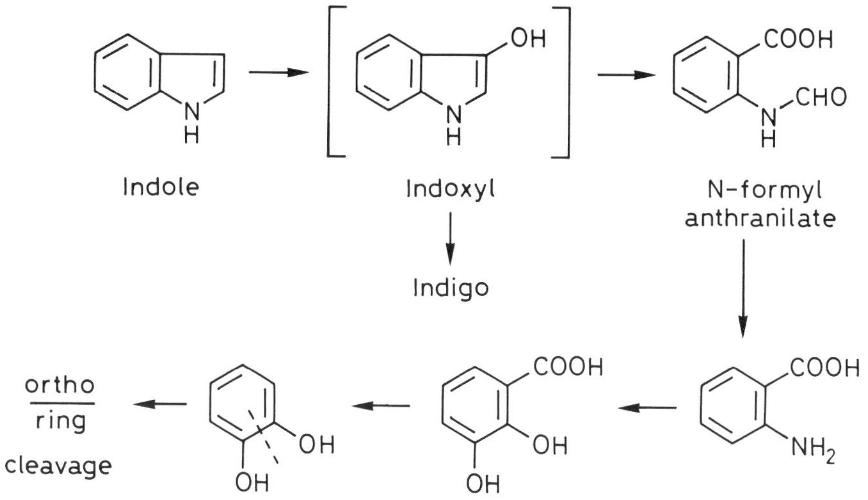

Figure 9. Proposed pathway for the degradation of indole by *A. niger*. Indoxyl is a proposed intermediate [87].

dioxygenase, induced by growth on the glucose plus indole, were detected. DHBA decarboxylase has been found only in fungal systems.

The anaerobic metabolism of indole and its derivatives has been studied more recently, and relatively little is known about the fate of these compounds in anaerobic conditions. Wang et al., using methanogenic consortia, reported biodegradation of indole into methane and carbon dioxide under strict anaerobic conditions [163], and oxindole was recognized as the initial intermediate [164,165] (Figure 10). The increase in net methane production indicates that indole was mineralized presumably through the formation of anthranilic acid. The ability of sediment and sewage sludge microcosmos to degrade indole was dependent upon several factors, including incubation temperature and the amount of sediment or sludge inoculum used [165]. The effect of substituent groups on indole hydroxylation reactions was studied by Gu and Berry with an indole-degrading methanogenic consortium, that was also able to degrade 3-methylindole and 3-indolyl acetate [166]. Oxindole, 3-methyloxindole, and indoxyl were identified as metabolites of indole

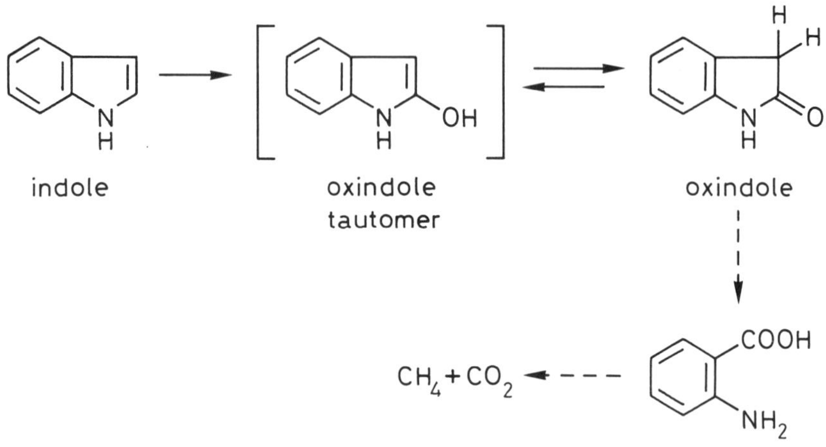

indole → [oxindole tautomer] ⇌ oxindole → $CH_4 + CO_2$

Figure 10. Suggested pathway of methanogenic indole fermentation [164].

and its two derivatives, respectively. Isatin was produced as intermediate when the consortium was amended with oxindole, providing evidence that degradation of indole proceeded through successive hydroxylations of C-2 and C-3 atoms prior to ring cleavage. The 3-substituted indoles were not further metabolized by the consortium. The metabolism of indole was also explored with a denitrifying microbial community. After oxindole and isatin formation, the addition of two hydroxyl groups has been postulated to yield isatoic acid, which is chemically unstable and spontaneously decarboxylates to form anthranilic acid. Dioxindole was also isolated from the denitrifying cultures exposed to isatin, presumably by reducing agents present in the sewage-sludge inoculum; however, whether dioxindole was formed by chemical or by microbiological process remains uncertain, and it has not been clarified if the dioxindole formed is reoxidized to isatin or if the reduction step represents a branch point in the metabolism of indole [167].

The degradation of indole, under sulphate-reducing conditions, has not been elucidated todate; only Bak and Widdle have isolated, from a marine enrichment with indole as sole electron donor and carbon source, a sulphate-reducing bacterium, ascribed as new species of *Desulfobacterium indolicum*, which oxidized completely indole to CO_2 with sulphate as electron acceptor [168].

S-Triazines

Among heterocyclic compounds, S-triazines are of great environmental concern, as different compounds, largely used, derive from their nucleus. All S-triazines, used as herbicides, are diamino-S-triazines bearing a chloro, methoxy or methylthio group, linked directly to the nucleus. Among these compounds, atrazine (2-chloro-4-ethylamino-6-isopropylamino-1,3,5-triazine) has often been detected in groundwater and soil because of its persistence [169]. Though physico-chemical decomposition plays an important role in the removal of these compounds from the environment [170], microbial degradation is the principal mechanism of *detoxification.*

The metabolism of chloro-S-triazines in soil involves reactions of dealkylation, deamination, hydroxylation, and ring cleavage [171]. Dealkylation of chloro-S-triazines does not remove their toxicity, which has been, instead, attributed to the release by chemical hydrolysis of active chlorine [172]. In a study on the effects of atrazine and its degradation products on phototrophic microorganisms, the most toxic degradation product was deethylated atrazine, which was 2 to 7 times more effective towards cyanobacteria than deisopropylated atrazine. On the contrary, diamino and hydroxyatrazine were non-toxic [173].

The first demonstration of the microbial degradation of chloro-S-triazines is the fungal *dealkylation* of simazine (2-chloro-4,6-bis-ethylamino-1,3,5-triazine) to deethylsimazine, and to another product, tentatively identified as 2-chloro-1,3,5-triazine-4,6-diamine (CAAT) [174,175]. Analogously, the same fungus, *Aspergillus fumigatus* Fres. monodealkylates atrazine to 2-chloro-4-amino 6-isopropylamino-S-triazine and 2-chloro-4-ethylamino-6-amino-S-triazine [176] (Figure 11). Cook and Hütter reported that deethylsimazine was quantitatively utilized as a nitrogen source by a strain of *Rhodococcus corallinus,* yielding ethylamino-dihydroxy-triazine which is utilized in co-culture with a *Pseudomonas* to yield cell material [177]. The same authors confirmed that the initial reaction is a quantitative hydrolytic ring dechlorination by two isofunctional, but different, hydrolases [178]. CAAT, the other product of fungal dealkylation of simazine, was degraded to 2-chloro-4-amino-1,3,5-triazine-6(5H)-one, a new product in the biodegradation of chlorinated S-triazines, which appeared to be further metabolized [179]. In this mechanism, the initial *dechlorination,* observed with deethylsimazine, does not occur [177]. Giardina and coworkers [4,180,181], working with a *Nocardia* strain capable of growing on atrazine, isolated dealkylated and deaminated products of the herbicide, among which 4-amino-2-chloro-1,3-5-triazine represented the most highly degraded compound, still having chlorine in the triazinic ring. This metabolite does not accumulate in the medium, as it undergoes rapid hydrolysis, causing the cleavage of the heterocyclic ring. These results suggest that the microbial attack is the necessary step for further chemical transformations, leading to ring cleavage. Behki and Khan isolated *Pseudomonas* species, that caused N-dealkylation by removing either the isopropyl or the ethyl moiety of atrazine [182]. The same organisms carried out the dehalogenation, when incubated with both mono-N-dealkylated derivates, suggesting that such dechlorination occurred after the elimination of one or both alkyl groups (Figure 11). The mechanism of dechlorination of chloro-S-triazines, formerly attributed to chemical hydrolysis which occurs spontaneously in soil [175,183, 184], is now demonstrated to occur also in microorganisms, and this hydrolytic mechanism is responsible for the removal of the toxicity of these compounds. A new degradation product of atrazine in the soil was identified by Tafuri et al. as 2-isopropopylamino-4-methoxy-6-methylamino-1,3,5-triazine [185]. However, the substitution of -Cl by -OCH_3 seems to be a non biological process, and was in agreement with Pape and Zabik [186], who reported the formation of methoxy analogues of chloro-1,3,5-triazine herbicides when *photolyzed* in methanol under laboratory conditions in natural sunlight. Although anaerobic conditions may provide a much more favourable environment than do aerobic conditions for hydrolytic dehalogenation which favour chloro-S-triazines degradation, the topic of anaerobic degradation of these compounds has not been widely studied till todate. Only Jessee et al. have reported the isolation of a facultative anaerobic bacterium which degrades atrazine under anaerobic conditions in a defined medium [187]. With regard to non-chlorinated triazines, such as ammelide, ammeline, melamine, cyanuric acid (CA) etc., a conclusive report for rapid and complete bacterial degradation was presented by Cook and Hütter, who isolated

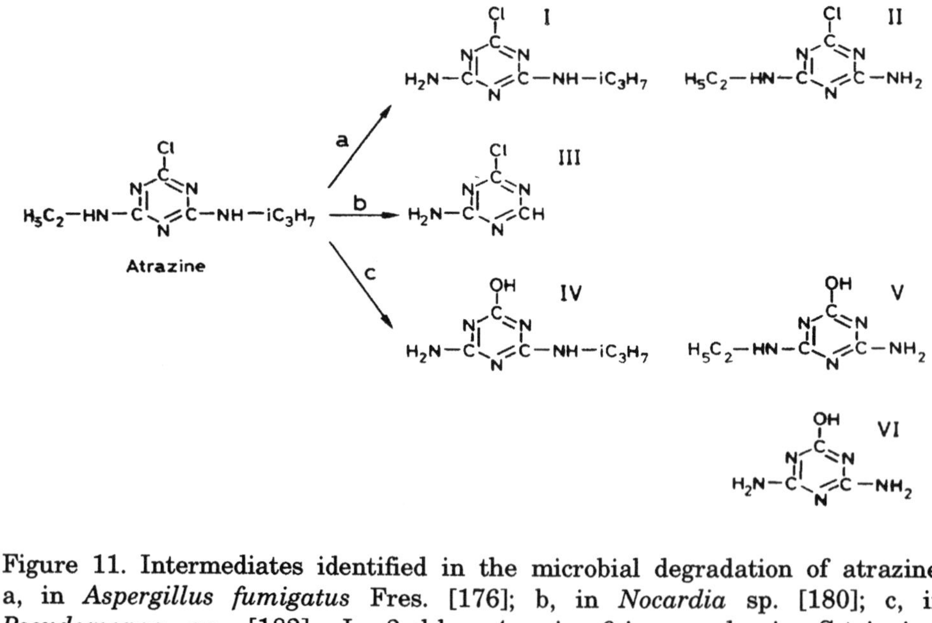

Figure 11. Intermediates identified in the microbial degradation of atrazine: a, in *Aspergillus fumigatus* Fres. [176]; b, in *Nocardia* sp. [180]; c, in *Pseudomonas* sp. [182]. I, 2-chloro-4-amino-6-isopropylamino-S-triazine; II, 2-chloro-4-ethylamino-6-amino-S-triazine; III, 4-amino-2-chloro-S-triazine; IV, 2-hydroxy-4-amino-6-isopropylamino-S-triazine; V, 2-hydroxy-4-ethylamino-6-amino-S-triazine; VI, 2-hydroxy-4,6-diamino-S-triazine.

organisms, utilizing S-triazines as nitrogen source [188]. Melamine (triamino-S-triazine) is a compound, widely used as commercial chemical in the synthesis of industrial polymers. For this compound, Yutzi et al. proposed a degradative pathway with a *Pseudomonas* sp. strain A, involving three successive deaminations leading to the formation of CA, which is further metabolized [189]. The mechanism of each of the three deaminations and ring cleavage appears to be hydrolytic, as the reactions proceed in the absence of O_2. The four hydrolytic activities have been separated, exhibiting different characteristics: the reaction products from melamine were ammeline, ammelide, and CA. At each step, NH_4^+ in equimolar amount was liberated. An analogous metabolic pathway was demonstrated by Cook et al. for N-cyclopropylmelamine, a representative of the most highly aminated groups of mono-N-alkylated S-triazines [190]. This compound was shown to be degraded quantitatively to ammonium ion by two *Pseudomonas* spp. strains A and D. The first two reactions were quantitative stoichiometric deaminations catalyzed, exclusively, by strain A to give N-cyclopropylammelide. This intermediate was degraded exclusively by strain D in a quantitative stoichiometric hydrolytic dealkylamination process to CA, which is further converted into ammonium ion by both strains A and D. In this case, the alkylamino substituent is replaced by -OH and not by -H [180].

CA is the last S-triazine ring system in the degradation pathway of S-triazine herbicides. This compound is known to be utilized as the sole and growth-limiting source of nitrogen by bacteria [174]. Growth of a facultative anaerobic bacterium under anaerobic conditions on CA is reported by Jessee et al. [187]. Its degradative pathway was conclusively demonstrated by Beilstein and Hütter in *Klebsiella pneumoniae* [191], and by Cook et al. in a *Pseudomonas* sp. strain D [192] (Figure 12).

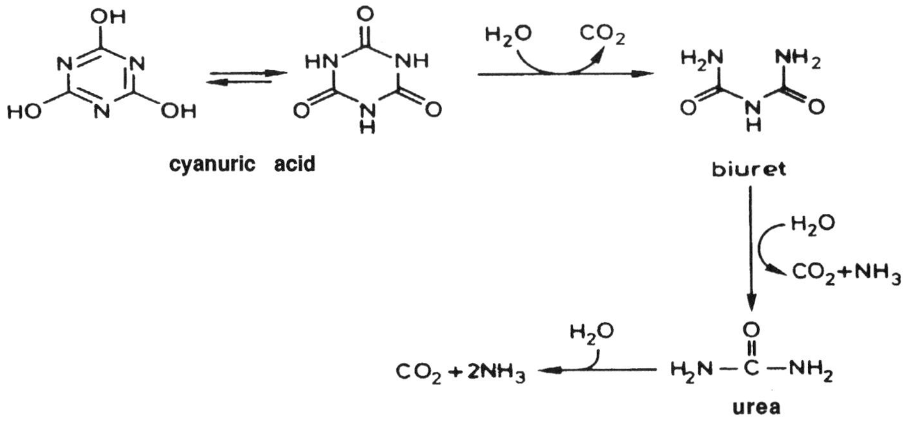

Figure 12. Metabolic pathway of cyanuric acid in *Pseudomonas* sp. [192].

Crude extracts of strain D degraded CA, biuret, and urea quantitatively to NH_4^+ and CO_2 through three successive reactions, which occurred under either aerobic or anoxic conditions, and were presumed to be hydrolytic. From the results above reported, chlorinated and not S-triazines seem to be completely degraded to ammonia and carbon dioxide through different steps carried out, in some cases, by different organisms.

CARBAMATES, PHENYLUREAS AND ANILIDES

Among pesticides and herbicides largely employed, carbamates and phenylureas play an important role, and their *biotransformations* are considered important for ecological and health reasons [193-195]. N-hydroxylated derivatives of carbamates have shown to be mutagenic [195]. Carbamates and phenylureas are derivatives of carbamic acid. While carbamates, also known as urethans, are esters, phenylureas are amides. Carbamic acid and N-substituted derivatives are unstable; they decompose spontaneously to carbon dioxide and ammonia or amine. Anilides are arylamides of carboxylic acids. For all these compounds, hydrolysis is the most important reaction involved in the biodegradation of ester or amide linkage (Figure 13) [196].

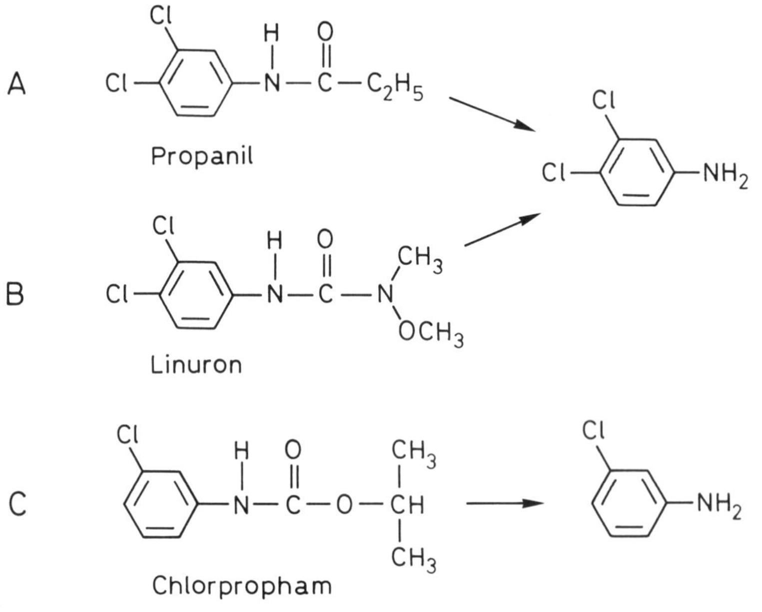

Figure 13. Hydrolysis of herbicides: A, phenylamide; B, phenylurea; C, phenylcarbamate [196].

Carbamates

This class of xenobiotics is structurally and physiologically heterogeneous: Carbaryl and Carbofuran are acetylcholin esterase inhibitors, whereas Chlorpropham (CIPC) is a herbicide, interfering with cell division [197]. The structure of carbamates is very different, as the amine can be aromatic, like in CIPC, or aliphatic, like in Carbaryl and Carbofuran. Everywhere, these microorganisms are able to hydrolyze the carbamate linkage by producing CO_2, alcohol, and the corresponding amine, aromatic or aliphatic.

Kaufman and Kearney have studied the degradation of several phenylcarbamates. Microorganisms, effective in degrading and utilizing these compounds as sole source of carbon, included *Pseudomonas striata* Chester, *Flavobacterium* sp., *Achromobacter* sp., *Arthrobacter* sp., and *Agrobacterium* sp. [198]. In the degradation of CIPC (isopropyl-N-3-chlorophenyl-carbamate) , studied by utilizing an enzyme preparation from *Pseudomonas* sp., 3-chloroaniline was detected. This intermediate was derived from hydrolysis of CIPC; the same enzyme, however, was not able to hydrolyze 3-(*p*-chlorophenyl)-1, 1-dimethylurea (Monuron) [199]. In subsequent studies, Kearney reported purification and properties of this hydrolytic enzyme [200], and studied also the physico-chemical properties of twelve phenylcarbamate herbicides, in an attempt to determine their influence on microbial degradation. Studing the effects of the size of the alcoholic group on the rate of enzyme hydrolysis, Kearney found that by increasing the size of the alcoholic moiety, the rate of hydrolysis was retarded [201]. Whringht and Forey reported that *Penicillium* sp. hydrolyzed 4-chloro-2-butinyl-N-3-chlorophenyl carbamate) (Barban), yielding 3-chloroaniline [202]. Chloroanilines, derived from the degradation of chlorophenylcarbamates, are then degraded by several species of bacteria and fungi [203-205].

2,3-Dihydro-2,2-dimethyl-7-benzofuranyl methylcarbamate (Carbofuran), extensively used to control corn rootworm, has shown to be slowly and incompletely degraded by soil microorganisms or fungi [206-209]. Recently, it has been reported that an *Achromobacter* sp. WM 111, isolated by soil enrichment cultures, is capable of hydrolyzing Carbofuran at an exceptionally rapid rate. The organism utilizes the resulting methylamine as sole source of nitrogen, and produces 2,3-dihydro-2,2-dimethyl-7-benzofuranol [210]. This microorganism catalyzes the degradation of other N-methylcarbamate insecticides (Carbaryl, Aldicarb, Baygon), but is ineffective in the degradation of acylanilide or urea pesticides. These results indicated that the hydrolytic enzyme of this strain is specific for phenol-carbamate ester linkages. In another report, the same authors by studying plasmid and chromosomal DNA of *Achromobacter* found that the strain harboured a plasmid which encoded for hydrolase activity [211]. Head et al. isolated, from soil, a bacterial strain, named MS2d which hydrolyzed Carbofuran, degrading phenolic moiety. In addition to the observation that phenol degradation is mediated by a plasmid, initial evidence that a Carbofuran hydrolase gene is present on a second plasmid has also been given [212].

Phenylureas

The *decomposition* of phenylurea herbicides by *Bacillus sphaericus* has been reported by Wallnvfer, [213], and Wallnvfer and Bader partially purified the enzyme responsible for hydrolytic cleavage [214]. Engelhardt et al. studied the degradation of (N-3,4-dichlorophenyl)-N-methoxy-N-methyl urea (Linuron), and some other herbicides and fungicides with the same organism [215]. This microorganism degraded low levels of Linuron when it grew in the medium containing glucose, yeast extract, and asparagine, producing N,O-dimethylhydroxylamine, CO_2, and 3,4-dichloroaniline [216]. The amidase hydrolyzing Linuron was partially purified. This enzyme, named arylacylamidase, hydrolyzed also acylanilide herbicides and was also induced by different acylanilide herbicides, acylanilide fungicides and by phenylcarbamate herbicide (Propham); however, the maximum enzymatic activity was revealable with Linuron as inducer [217]. Diflubenzuron (1-(4-chlorophenyl)-3-(2,6-difluoro-benzoyl)urea) was hydrolyzed in soil to 4-chlorophenylurea and 2,6-difluorobenzoate [218]. 4-Chlorophenylurea was presumably converted into 4-chloroaniline and bound to soil, while 2,6-difluorobenzoate mineralized to CO_2 [219].

Phenylurea herbicides are also degraded under anaerobic conditions. N-(3,4-Dichlorophenyil)-N'-dimethylurea (Diuron) and Linuron have shown to be dechlorinated in anaerobic sediments with elimination of the chlorine atom in the *para*-position [220, 221].

Anilides

Anilide herbicides constitute a group of compounds whose degradation occurs mainly through hydrolysis with formation of an acidic moiety, easily consumed by microorganisms.

The herbicides N-(3,4-dichloro-phenyl) propionamide (Propanil), N-(3,4-dichlorophenyl) methylacryl amide (Dicryl), and N-(3,4-dichlorophenyl)-2-methylpentanamide (Karsil) were transformed into 3,4-dichloroaniline [222-225]. The acylanilide hydrolases have been isolated from strains of *Fusarium* and *Bacillus* [226-228]. N-alkylacylanilides are, instead, compounds less easily hydrolyzable. The two most widely used are 2-chloro-2',6'- diethyl-N-(methoxymethyl)-acetanilide (Alachlor) and 2-chloro-2'-ethyl-6'-methyl-N-(1-methyl-2-methoxyethyl)-acetanilide (Metolachor). For these compounds, hydrolysis with arylacylamidases has not been reported as an important mechanism of degradation [229-232], and they are very resistant to mineralisation. Novick et al. reported on the production of N-isopropylaniline from 2-chloro-N-isopropylacetanilide (Propachlor) by a microbial consortium [233].

Propachlor was also metabolyzed by two microorganisms named DAK 3 and MAB 2 assigned respectively to the genera *Moraxella* and *Xantobacter*. DAK3 strain degraded Propachlor with the formation of catechol and 2-chloro-N-isopropyl-acetamide, which was released in the medium. MAB2 could grow on this metabolite (Figure 14). In this case, the metabolic pathway does not occur through a hydrolytic step [234].

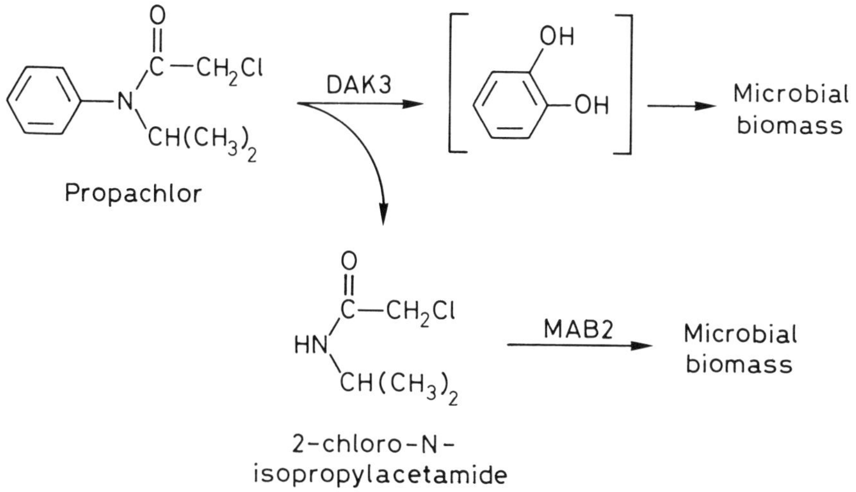

Propachlor

DAK3

Microbial biomass

MAB2

Microbial biomass

2-chloro-N-isopropylacetamide

Figure 14. Proposed pathway of Propachlor degradation by strains DAK3 and MAB2 of *Moraxella* and *Xanthobacter*, respectively [234].

HALOGENATED ANILINES

Chlorinated anilines are common metabolites of the microbial degradation of various phenylurea, acylanilide and phenylcarbamates herbicides (Figure 13). Basicity and oxidability are the features of these compounds. They can react with oxygen and ozone present in air, with soil components [235-237], with lignin of plants [238], and with the microbial enzymatic systems.

Frequently, they form *condensation* products. 3,4-Dichloroaniline may be transformed into 3,3',4,4'-tetrachloro-azobenzene [239-246] and different monochloro- and dichloroanilines are oxidised to dichloro- and tetrachloroazobenzenes [242,243]. An aniline oxidase and a peroxidase from the fungus *Geotrichum candidum* could dimerize different anilines except nitro anilines [244]. Condensation may transform anilines into nitroso benzenes [245]. Acetylation is an other reaction involved in the metabolism of halogenated anilines as *detoxification* mechanism. 4-Chloroaniline was acetylated to 4-chloroacetanilide by bacteria [246, 247]; 4-amino-3,5-dichloroacetanilide was formed in soil by acetylation of 2,6-dichloro-1,4-diphenylendiamine, derived from the degradation of fungicide, 2,6-dichloro-4-nitroaniline [248,249]. Chlorinated anilines are not easily metabolized by microorganisms; however, cometabolic degradation of chloroanilines by several bacteria has been reported [250-256]. Both free and humus-bound chloroanilines were degraded at slow rates [235], but their mineralization was greatly enhanced by the presence of aniline [252]. A *Rhodococcus* strain converted 3-chloroaniline in presence of a growth substrated [254]. Zeyer and Kearney [257] and Zeyer et al. [258] isolated a *Moraxella* sp. strain, capable of utilizing chloroanilines and also 4-bromo- and 4-fluoroaniline, but not 4-iodoaniline as sole source of carbon, nitrogen, and energy. 4-Chloroaniline was metabolized by inducible oxygenase to 4-chlorocatechol, which was further degraded via a modified *ortho*-cleavage pathway. The aniline oxygenase exhibited a broad substrate specificity.

Some *Pseudomonas acidovorans* strains, with high degradative capacity toward, aniline, 3- and 4-chloroanilines to the corresponding chlorocatechols, have been isolated [259]. This initial attack seems to be the limiting step in the rate of degradative process. Among the strains of *Pseudomonas* isolated, a *Pseudomonas acidovorans* CA2b also degraded slowly 2-chloroaniline. The three major reactions involved in the degradation of chloroanilines are summarised in Figure 15.

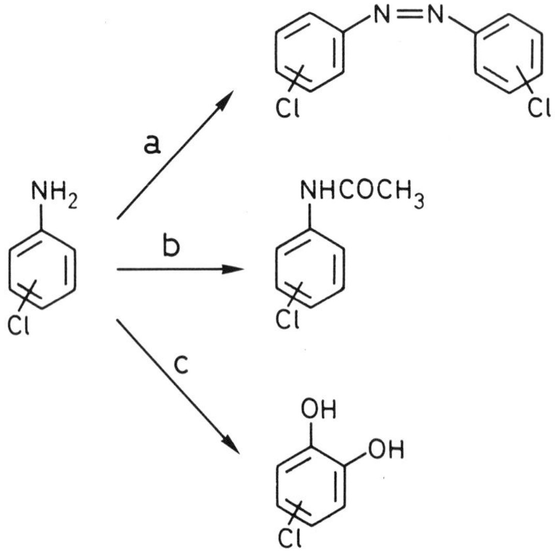

Figure 15. Major aerobic biotransformation reactions of halogenated anilines: a, dimerization; b, acetylation; c, oxygenation.

The anaerobic fate of chloroanilines has not been widely studied up to now. The reductive dehalogenation of chlorinated anilines has been investigated in methanogenic but non sulphate-reducing sediments by Kuhn and Suflita and by Striujs and Rogers [260,261]. Methanogenic aquifer microbiota reductively dehalogenated di-, tri- and tetrachloroaniline by sequential halogen removal to the corresponding monochloroanilines which persisted for up to 8 months [262].

CONCLUSIONS

Xenobiotic compounds have been used extensively in agriculture as herbicides and insecticides and in the manufacturing industry as surfactants, dyes, drugs, solvents and so on. Aliphatic and aromatic organic nitrogen compounds represent an important fraction of these chemicals. Even if many of the nitrogenous compounds are highly toxic and often recalcitrant to microbial attack, the microorganisms exposed to these synthetic chemicals have developed the ability to utilize some of them.

Laboratory studies have established the capacity of many microorganisms to degrade or to transform nitrogenous compounds under aerobic or anaerobic conditions. The results obtained from these investigations which involved pure cultures, or cell-free extracts of microorganisms are important, because they may be predictive of their environmental fate. However, these studies do not reproduce the conditions found in nature where these organisms are exposed to a mixture of compounds and interact with other microbial communities.

For every compound, which has proven to be biodegradable, the load of environmental pollutants is reduced. The assessment of biodegradability opens the way for the development of microbiological methods for the clean-up of soils and waters, contaminated with synthetic compounds. As *bioremediation* has its basis in the physiology and ecology of microorganisms, these methods have to be developed according to the capabilities of these microorganisms to ensure an optimal performance in those habitats. Moreover, the development of genetic manipulation techniques gives us the possibility to construct new strains with the desired "capabilities" for the degradation of xenobiotics. The employment of these strains could enhance the possibilities to *decontaminate* polluted environments.

REFERENCES

1 Parris GE. Resid Rev 1980; 76: 1-30.
2 Kearney PC, Kaufman DD. Herbicides: Chemistry, Degradation and Mode of Action. New York: Marcel Dekker, Inc., 1975.

3 McBride KE, Kenny JV, Stalker DN. Appl Environ Microbiol 1986; 52: 325-330.
4 Giardina MC, Giardi MJ, Filacchioni G. Agric Biol Chem 1982; 46: 1439-1445.
5 McCormick NG, Feeherry FE, Levinsons HS. Appl Environ Microbiol 1982; 31: 949-958.
6 Society of Dyers and Colourists and American Association of Textile Chemists and Colourists (ed) 1956. Colour Index 2nd ed. Society of Dyers and Colourists, Bradford, Yorkshire, England and the American Association of Textile Chemist and Colourists, Lowell Technological Institute, Lowell, Mass.
7 Mason R, Sweeney SC. A literature survey oriented towards adverse environmental effects resultant from the use of azocompounds, brominated hydrocarbons, EDTA, formaldehyde resins and *o*-nitrochlorobenzene. Washington D.C. Environmental Protection Agency 1976; 560/2-76-005.
8 Zoeteman BCJ, Harmsen K, Linders JBHJ, Morra CFH, Slooff W. Chemosphere 1980; 9: 231-249.
9 Piet GJ, Smeenk JGMM. In: Ward CH, Giger W, Carty PLMc, eds. Groundwater Quality. New York: John Whiley & Sons Inc, 1985; 122-144.
10 Golab T, Althaus WA, Wooten HL. J Agric Food Chem 1979; 27: 163-179.
11 Leuenberger C, Czuczwa J, Tremp J, Giger W. Chemosphere 1988; 17: 511-515.
12 Veys CA. Ann Occup Hyg 1972; 15: 11-21.
13 Zavon MR. Arch Environ Health 1973; 27: 1-8.
14 Diachenko GW. Environ Sci Technol 1979; 13: 324-339.

15 Yurawecz MP. GLC and mass spectrometric determination of monochloronitrobenzene residues in Mississippi river fish. Presented AOAC Meeting, Oct 17, 1978; Washington, D.C.
16 Carpenter DF, McCormick NG, Cornell JM, Kaplan AM. Appl Environ Microbiol 1978; 35: 944-954.
17 Sylvestre M, Massh R, Messil F, Fauteux J, Bisaillon G, Beauted R. Appl Environ Microbiol 1982; 44: 871-877.
18 Grosjean D. Environ Sci Technol 1985; 19: 968-974.
19 Kinouchi T, Ohnishi Y. Appl Environ Microbiol 1983; 46: 596-604.
20 Wirth W, Hecht G, Gloxhuber C. Toxikologie-Fibel. Stuttgart: Thieme, 1971
21 Venulet J, Van Etten RL. In: Fener H, ed. The Chemistry of the Nitro and Nitrosogroups, Vol. 5 PATAI Part 2. London: Wiley, 1970; 202-287.
22 Tereda H. Biochim Biophys Acta 1981; 639: 225-242.
23 Zeyer J. Org Micropollut Aquat Environ 1986; 305-311.

24 Cartwright NJ, Cain RB. Biochem J 1959; 71: 248-261.
25 Corbett MD, Corbett BR. Appl Environ Microbiol 1981; 41: 942-949.
26 Schackmann A, Muller R. Appl Microbiol Biotechnol 1991; 34: 809-813.
27 Latorre J, Reineke W, Knackmuss HJ. Arch Microbiol 1984; 140: 159-165.
28 Reber H, Helm V, Karanth NGK. Europ J Appl Microbiol Biotechnol 1979; 7: 181-189.
29 Villanueva JR. J Biol Chem 1964; 239: 773-776.
30 Hoff T, Lin SY, Bollag JM. Appl Environ Microbiol 1985; 49: 1040-1045.
31 Kearney PC, Plimmer JR, Guardia FB. J Agric Food Chem 1969; 17: 1418-1419.
32 Zeyer J, Kearney PC. J Agric Food Chem 1984; 32: 238-242.
33 Spain JC, Wyss O, Gibson TD. Biochem Biophys Res Commun 1979; 88: 634-641.
34 Williams RT, Ziegenfuss PS, Marks PJ. Field demonstration composting of explosives contaminated sediments at the Louisiana Army Ammunition (LAAP). Final Report. 1988; Contract DAAK-11-85-DOO7; Report AMXTH-IR-TE-88242.
35 Pereera WE, Short DL, Manigold DB, Ross PK. Bull Environ Contam Toxicol 1979; 21: 554-562.
36 Sax N. Dangerous Properties of Industrial Materials. New York: Rheinhold Publ Corp, 1957.
37 Mathews ER, Hex EH, Taylor RL, Newell JF. Laboratory studies pertaining to the treatment of TNT wastes. 126th Meeting, American Chemical Society, 1954, New York.
38 Won WD, Di Salvo LH, NG J. Appl Environ Microbiol 1976; 31: 576-580.
39 Won WD, Heckley RJ, Gleover DJ, Hoffsommer JC. Appl Microbiol 1974; 27: 513-516.
40 Channon HJ, Mills GT, Williams RT. Biochem J 1944; 38: 70-85.
41 Dale HH. Med Res Counc (G.B.) Spec Rep Ser 1921; 58: 53-61.
42 Klausmeier RE, Osmon JL, Walls DR. Dev Ind Microbiol 1974; 15: 309-311.
43 Parrish IW. Appl Environ Microbiol 1977; 34: 232-233.
44 Fernando T, Bumpus JA, Aust SD. Appl Environ Microbiol 1990; 56: 1666-1671.
45 U.S. Environmental Protection Agency. 1986. Health and environmental effects profile for nitrotoluenes (*o, m, p*). Report EPA/600/x-86/143. Environmental Criteria and Assessment Office, Cincinnati, Ohio.
46 Liy D, Thomson K, Anderson AC. Appl Environ Microbiol 1984; 47: 1295-1298.

47 Spanggord RJ, Spain JC, Nishino SF, Mortelmans KE. Appl Environ Microbiol 1991; 57: 3200-3205.
48 Hallas LE, Alexander M. Appl Environ Microbiol 1983; 45: 1234-1241.
49 Delgado A, Wubbolts MG, Abril MA, Ramos JL. Appl Environ Microbiol 1991; 58: 415-417.
50 Harayama S, Rekik M, Wubbolts MG, Rose K, Leppik RA, Timmis KN. J Bacteriol 1989; 171: 5048-5055.
51 Robertson JB, Spain JC, Haddock JD, Gibson DT. Appl Environ Microbiol 1992; 58: 2643-2648.
52 Merck & Co Inc. The Merck Index. An Encyclopedia of Chemicals, Drugs and Biologicals. 10th ed. Rahway, New York: Merck & Co, 1983; 945.
53 Chandler CD, Kohlbeck JA, Rolleter JA. J Chromatogr 1972; 64, 127-128.
54 Fewson CA. In: Leisenger T, Cook AM, Hütter R, Nuesh J, eds. Microbial Degradation of Xenobiotics and Recalcitrant Compounds. London: Academic Press Ltd, 1981; 141-149.
55 Haigler BE, Spain JC. Appl Environ Microbiol 1991; 57 3156-3162.
56 Dickel O, Knackmuss HJ. Arch Microbiol 1991; 157: 76-79.
57 Keith LH, Telliard WA. Environ Sci Technol 1979; 13: 416-423.
58 Loomis WF, Lipmann F. J Biol Chem 1948; 173: 807-808.
59 Simon EW. Biol Rev 1953; 28: 453-479.
60 Maier-Brode H. Herbizide und ihre Rückstände. Stuttgart: Eugen Ulmer, 1971.
61 Ivashina SA, Chmill VD. Agrokhimica (USSR) 1987; 1: 94-97.
62 Environmental Protection Agency. Federal Register 1986; 51: 36634-36661.
63 Raymond DGM, Alexander M. Biochem Physiol 1971; 1: 123-130.
64 Spain JC, Van Veld PA, Monti CA, Pritchard PH, Cripe CR. Appl Environ Microbiol 1984; 48: 944-950.
65 Spain CJ, Gibson DT. Appl Environ Microbiol 1991; 57: 812-819.
66 Zeyer J, Kocher HP, Timmis NK. Appl Environ Microbiol 1986; 52: 334-339.
67 Zeyer J, Kocher MP. J Bacteriol 1988; 170: 1789-1794.
68 Schmidt SK, Scow KM, Alexander M. Appl Environ Microbiol 1987; 53: 2617-2623.
69 Ecker S, Widmann T, Lenke H, Dickel O, Fischer P, Bruhn C, Knackmuss JH. Arch Microbiol 1994 (in press).
70 Bruhn C, Lenke H, Knackmuss JH. Appl Environ Microbiol 1987; 53: 208-210.
71 Lenke H, Pieper DH, Bruhn C, Knackmuss JH. Appl Environ Microbiol 1992; 58: 2928-2932.
72 Madhosingh C. Can J Microbiol 1961; 7: 533-567.

73 Wyman JF, Guard HE, Won WD, Quay JH. Appl Environ Microbiol 1978; 37: 222-226.
74 Lenke H, Knackmuss JH. Appl Environ Microbiol 1992; 58: 2933-2937.
75 Durham NN. Can J Microbiol 1958; 4: 141-148.
76 Cartwright NJ, Cain RB. Biochem J 1959; 73: 305-314.
77 Cain RB. J Gen Microbiol 1966; 42: 197-217.
78 Cain RB. J Gen Microbiol 1966; 42: 219-235.
79 Ke YH, Gee LL, Durham NN. J Bacteriol 1958; 77: 593-598.
80 Shackmann A, Muller R. Appl Microbiol Biotechnol 1991; 34: 809-813.
81 Groenewegen PEJ, Breeuwer P, Van Helvort JMLM, Laupenhoff AAM, De Vries FP, De Bont JAM. J Gen Microbiol 1992; 138: 1599-1605.
82 Groenewegen PEJ, De Bont JAM. Arc Microbiol 1992; 158: 381-386.
83 Andreoni V, Villa M, Bernasconi S. XI Meeting SIMGBM, Gubbio, 1992; 4-7 ottobre.
84 Doy CH, Gibson F. Biochim Biophys Acta 1961; 50: 495-592.
85 Yanofsky C. J Biol Chem 1959; 223: 171-184.
86 Fujoka M, Wada H. Biochim Biophys Acta 1968; 158: 70-78.
87 Kamath AV, Vaidyanathan CS. Appl Environ Microbiol 1990; 56: 275-280.
88 Madsen EL, Bollag J-M. Arch Microbiol 1989; 151: 71-76.
89 Neumüller O, Otto-Albrecht P. Rvmpps Chemie Lexikon, 8th ed. Vol 1. Stuttgart: Franckh sche Verlagshandlung W. Keller & Co, 1979.
90 Anderson JJ, Dagley S. J Bacteriol 1981; 146: 291-297.
91 Cain RB. Antonie van Leuwenhoek 1968; 34: 417-432.
92 Braun K, Gibson DT. Appl Environ Microbiol 1984; 48: 102-107.
93 Schnell S, Schink B. Arch Microbiol 1992; 158: 328-334.
94 Klepacka M, Rutkowski A. Acta Aliment Pol 1982; 8: 3-10.
95 Hashimoto S, Kameoka H. J Food Sc 1985; 50: 847-852.
96 Lockwood GB, Afsharypuor S. J Chrom 1986; 356: 438-440.
97 Bergstrom J, Bergstrom G. Nord J Bot 1989; 9: 363-366.
98 Smith PA. Open-Chain Nitrogen Compounds Vol 1. New York: WA Benjamin Inc, 1965.
99 Strobel GA. J Biol Chem 1966; 241: 2618-2621.
100 Uematsu T, Suhadolnik RJ. Arch Biochem Biophys 1974; 162: 614-619.
101 Ashton FM, Crafts AS. Mode of Action of Herbicides. New York: John Wiley & Sons Inc, 1973; 236-255.
102 Henahan YF, Yames D, Idol Yr. Chem Eng News 1971; 49: 16-18.
103 Salame M, Nemphos SP. In: Proceedings of the Symposium on Environmental Impact of Nitrile Barrier Containers. Lopac: A Case Study. St Louis: Monsanto Co, 1973; 21-28.

104 De Meester CF, Poncelet M, Roberfroid M, Menvier M. Toxicology 1978; 11: 19-27.
105 Orusev T, Poporski P. God Zbornik Med Fak Skopje 1973; 19: 187-192.
106 Sakurai H, Kusumoto M. J Sci Labor 1972; 48: 273-282.
107 Thiery A, Maestracci M, Arnaud A, Galzy P. Zentralbl Mikrobiol 1986; 141: 575-582.
108 Nawaz MS, Chapatwala KD, Wolfram JH. Appl Environ Microbiol 1986; 55: 2267-2274.
109 Collins PA, Knowles CJ. J Gen Microbiol 1983; 129: 711-718.
110 Hook RH, Robinson WG. J Biol Chem 1964; 239: 4263-4267.
111 Robinson WG, Hook H. J Biol Chem 1964; 239: 4257-4262.
112 Harper DB. Int J Biochem G 1985; 17: 677-683.
113 Harper DB. Biochem J 1977; 165: 309-319.
114 Harper DB. Biochem J 1977; 167: 685-692.
115 Bandyopadhyay AK, Nagasawa T, Asano Y, Fujishiro K, Tani Y, Yamada H. Appl Environ Microbiol 1986; 51: 302-306.
116 Stalker DM, Malyi LD, McBride KE. J Biol Chem 1988; 263: 6310-6314.
117 Kobayashi M, Nagasawa T, Yamada H. J Biochem 1989; 182: 349-356.
118 Nagasawa T, Kobayashi M, Yamada H. Arch Microbiol 1988; 150: 89-94.
119 Manuger J, Nagasawa T, Yamada H. Arch Microbiol 1990; 155: 1-6.
120 Nagasawa T, Manuger J, Yamada H. Eur J Biochem 1990; 194: 765-772.
121 Kobayashi M, Nagasawa T, Yanaka N, Yamada H. Biotechnol Lett 1989; 11: 27-30.
122 Mathew CD, Nagasawa T, Kobayashi H, Yamada H. Appl Environ Microbiol 1988; 54: 1030-1032.
123 Vaughan PA, Cheetam PSJ, Knowles CJK. J Gen Microbiol 1988; 134: 1099-1107.
124 Kobayashi M, Yanaka N, Nagasawa T, Yamada H. J Bacteriol 1990; 172: 4807-4815.
125 Watanabe Y, Satoh Y, Enomoto K. Agr Biol Chem 1987; 51: 3193-3199.
126 Di Geronimo MJ, Antoine AD. Appl Environ Microbiol 1976; 31: 900-906.
127 Bui K, Fradet H, Arnaud A, Galzy P. J Gen Microbiol 1984; 130: 89-93.
128 Asano Y, Fujishiro K, Tani Y, Yamada H. Agric Biol Chem 1982; 46: 1165-1174.
129 Nagasawa T, Ryuno K, Yamada H. Biochem Biophys Res Commun 1986; 139: 1305-1312.

130 Endo T, Watanabe J. FEBS Letters 1989; 243: 61-64.
131 Asano Y, Tachibana M, Tani Y, Yamada H. Agric Biol Chem 1989; 46: 1175-1181.
132 Tani Y, Kurihara M, Nishise H, Yamamoto K. Agric Biol Chem 1989; 53: 3143-3149.
133 Mimura A, Kawano T, Yamaga K. J Ferment Technol 1969; 47: 631-638.
134 Firmin JL, Gray DO. Biochem J 1976; 158: 223-229.
135 Arnaud A, Galzy P, Jallageas JC. Agric Biol Chem 1977; 41: 2183-2191.
136 Yamada H, Asano Y, Hino T, Tani Y. J Ferm Technol 1979; 57: 8-14.
137 Asano Y, Yasuda T, Tani Y, Yamada H. Agric Biol Chem 1982; 46: 1183-1189.
138 Yamada H, Ryuno K, Nagasawa T, Enomoto K, Watanabe I. Agric Biol Chem 1986; 50: 2859-2865.
139 Nagasawa T, Nanba H, Ryuno K, Takeuchi K, Yamada H. Eur J Biochem 1987; 162: 691-698.
140 Watanabe J, Satoh Y, Enomoto K, Seki S, Sakashita K. Agric Biol Chem 1987; 51: 3201-3206.
141 Ryuno K, Nagasawa T, Yamada H. Agric Biol Chem 1988; 52: 1813-1816.
142 Nawaz MS, Franklin W, Campbell WL, Heinze TM, Cerniglia CE. Arch Microbiol 1991; 156: 231-238.
143 Nagasawa T, Mathew CD, Manger J, Yamada H. Appl Environ Microbiol 1988; 54: 1766-1769.
144 Nagasawa T, Takeuchi K, Yamada H. Biochem Biophys Res Commun 1988; 155: 1008-1016.
145 Mayaux JF, Cerbeland E, Soubrier F, Francher D, Pitri D. J Bacteriol 1990; 172: 6764-6773.
146 Yamamoto K, Ueno Y, Otsubo K, Kawakami K, Komatsu KJ. Appl Environ Microbiol 1990; 56: 3125-3129.
147 Kakeya H, Sakai N, Sugai T, Ohta H. Tetrahedron Lett 1991; 32: 1343-1346.
148 Yamamoto K, Oishi K, Fujimatsu I, Komatsu Y. Appl Environ Microbiol 1991; 57: 3028-3032.
149 Layh N, Stolz A, Fvrster S, Effenberger F, Knackmuss HJ. Arch Microbiol 1992; 158: 405-411.
150 Hsu JC, Camper ND. Can J Microbiol 1976; 22: 537-543.
151 Verloop A. Resid Rev 1972; 43: 55-103.
152 Smith AE, Cullimore DR. Can J Microbiol 1974; 20: 773-776.
153 Golovleva LA, Pertsova RN, Kunc F, Volkounova M. Folia Microbiol 1988; 33: 491-499.
154 Vokounova M, Vacek O, Kunc F. Folia Microbiol 1992; 37: 122-127.

155 Carlson JR, Yokoyama MT, Dickson EO. Science 1972; 176: 298-299.
156 Hammond AC, Carlson JR, Breeze RC. Vet Rec 1980; 107: 344-346.
157 Ochiai M, Wakabayashi K, Sugimura T, Nagao M. Mutat Res 1986; 172: 189-197.
158 Vance WA, Okamoto HS, Wang YY. Mutat Res 1986; 173: 169-176.
159 Sakamoto Y, Uchida M, Ichihara K. Med J Osaka Univ 1953; 477-486.
160 Fujioka M, Wada H. Biochim Biophys Acta 1968; 158: 70-78.
161 Ensley BD, Ratzkin BJ, Osslund TD, Simon MJ, Wackett LP, Gibson DT. Science 1983; 222: 167-169.
162 Claus G, Kutzner HJ. System Appl Microbiol 1983; 4: 169-180.
163 Wang Y, Suidan MT, Pfeffer GT. Appl Environ Microbiol 1984; 48: 1058-1060.
164 Berry DF, Madsen EL, Bollag J-M. Appl Environ Microbiol 1987; 53: 180-182.
165 Madsen EL, Francis AJ, Bollag J-M. Appl Environ Microbiol 1988; 54: 74-78.
166 Gu J, Berry DF. Appl Environ Microbiol 1991; 57, 2622-2627.
167 Madsen EL, Bollag J-M. Arch Microbiol 1989; 151: 71-76.
168 Bak F, Widdel F. Arch Microbiol 1986; 146: 170-176.
169 Smith AE, Walker A. Can J Soil Sci 1989; 69: 587-595.
170 Geller A. Arch Environ Contam Toxicol 1980; 9: 289-305.
171 Esser HO, Dupuis G, Ebert E, Vogel C, Marco GJ. In: P.C. Kearney PC, Kaufman DD, eds. Herbicides: Chemistry, Degradation and Mode of Action. New York: Marcel Dekker Inc, 1975; 129-208.
172 Harris CI, Kaufman DD, Sheets TJ, Nash RG, Kearney PC. Adv Pest Control Research 1968; 8: 1-55.
173 Stratton GW. Arch Environ Contam Toxicol 1984; 13: 35-42.
174 Kaufman DD, Kearney PC, Sheets TJ. J Agric Food Chem 1965; 13: 238-242.
175 Kearney PC, Kaufman DD, Sheets TJ. J Agric Food Chem 1965; 13: 369-373.
176 Kaufman DD, Blake Y. Soil Biol Biochem 1970; 2: 73-80.
177 Cook AM, Hütter R. J Agric Food Chem 1984; 32: 581-585.
178 Cook AM, Hütter R. FEMS Microbiol Lett 1986; 34: 335-338.
179 Grossenbacher H, Horn C, Cook AM, Hütter R. Appl Environ Microbiol 1984; 48: 451-453.
180 Giardina MC, Giardi MT, Filacchioni G. Agric Biol Chem 1980; 44: 2067-2072.
181 Giardi MT, Giardina MC, Filacchioni G. Agric Biol Chem 1985; 49: 1551-1558.
182 Behki RM, Khan SU. J Agric Food Chem 1986; 34: 746-749.
183 Armstrong DE, Chesters G, Harris RF. Soil Sci Soc Am Proc 1967; 31: 61-66.

184 Skipper HD, Gilmour CM, Furtick WR. Soil Sci Soc Am Proc 1967; 31: 653-656.
185 Tafuri F, Patumi M, Marucchini C, Businelli, M. Pestic Sci 1982; 13: 665-669.
186 Pape BE, Zabik MJ. J Agric Food Chem 1970; 18: 202-207.
187 Jessee YA, Benoit RE, Hendricks AC, Allen GC, Neal JL. Appl Environ Microbiol 1983; 45: 97-102.
188 Cook AM, Hütter R. J Agric Food Chem 1981; 29: 1135-1143.
189 Jutzi K, Cook AM, Hütter R. Biochem J 1982; 208: 679-684.
190 Cook AM, Grossenbacher H, Hütter R. Biochem J 1984; 222: 315-320.
191 Beilstein P, Hütter R. Experientia 1980; 36: 1457.
192 Cook AM, Beilstein P, Grossenbacher H, Hütter R. Biochem J 1985; 231: 25-30.
193 Sud RK, Sud AK, Gupta KG. Arch Mikrobiol 1972; 81: 353.
194 Harris CR, Chapman RA, Harris C, Tu CM. J Environ Sci Health 1984; B19: 1-11.
195 Pai V, Bloomfield SF, Gorrod JW. Mutat Res 151; 201-207.
196 Haggblom MM. FEMS Microbiol Rev 1992; 103: 29-71.
197 Johnson LM, Talbot HW Jr. Experientia 1983; 39: 1236-1246.
198 Kaufman DD, Kearney PC. Appl Microbiol 1965; 13: 443-446.
199 Kearney PC, Kaufman DD. Science 1965; 417: 740-741.
200 Kearney PC. J Agric Food Chem 1965; 13: 561-564.
201 Kearney PC. J Agric Food Chem 1967; 15: 568-571.
202 Whrigh SJL, Forey A. Soil Biol Biochem 1972; 4: 207-213.
203 Kaufman DD, Blake J. Soil Biol Biochem 1973; 5: 297-308.
204 McClure. Weed Sci 1974; 22: 323-329.
205 Vega D, Bastide J, Coste CV Soil Biol Biochem 1985; 17: 541-545.
206 Felsot AS, Maddox JV, Bruce W. Bull Environ Contam Toxicol 1981; 26: 781-788.
207 Venkateswarlu K, Sethunathan N. Bull Environ Contam Toxicol 1984; 33: 556-560.
208 Ramanand K, Sharmila M, Sethunathan N. Appl Environ Microbiol 1988; 54: 2129-2133.
209 Chaudhry GR, Ali AN. Appl Environ Microbiol 1988; 54: 1414-1419.
210 Karns JS, Mulbry WW, Nelson JO, Kearney PC. Pest Biochem Physiol 1986; 25: 211-217.
211 Karns JS, Mulbry WW, Kearney PC. In: Augustine PL, Danforth HD, Bakst MR, eds. Biotechnology for Solving Agricultural Problems. BARC Symposium 10. Boston: Martinus Nijhoff Publishers, 1986; 339-354.
212 Head IM, Cain RB, Suett DL. Arch Microbiol 1992; 158: 302-308.

213 Wallnvfer PR. Weed Res 1969; 9: 333-339.
214 Wallnvfer PR, Bader J. Appl Microbiol 1970; 19: 714-717.
215 Engelhardt G, Wallnvfer PR, Plapp R. Appl Microbiol 1971; 22: 284-288.
216 Engelhardt G, Wallnvfer PR, Plapp R. Appl Microbiol 1972; 23: 664-666.
217 Engelhardt G, Wallnvfer PR, Plapp R. Appl Microbiol 1973; 26: 709-718.
218 Nimmo WB, de Wilde PC, Verloop A. Pestic Sci 1984; 15: 574-585.
219 Nimmo WB, Willems AGM, Joustra KD, Verloop A. Pestic Sci 1986; 17: 403-411.
220 Attaway HH, Camper ND, Paynter MJB. Pestic Biochem Physiol 1982; 17: 96-101.
221 Stepp TD, Camper ND, Paynter MJB. Pestic Biochem Physiol 1985; 23: 256-260.
222 Bartha R. J Agric Food Chem 1968; 16: 602-604.
223 Bartha R. J Agric Food Chem 1971; 19: 385-387.
224 Chisaka H, Kearney PC. J Agric Food Chem 1970; 18: 854-858.
225 Sharabi NED, Bordeleau LM. Appl Microbiol 1969; 18: 369-375.
226 Lanzilotta RP, Pramer D. Appl Microbiol 1970; 19: 301-306.
227 Lanzilotta RP, Pramer D. Appl Microbiol 1970; 19: 307-313.
228 Blake J, Kaufman DD. Pestic Biochem Physiol 1975; 5: 305-313.
229 Krause A, Hancock WG, Minard RD, Freyer AJ, Honeycutt RC, LeBaron HM, Paulson DL, Liu SY, Bollag J-M. J Agric Food Chem 1985; 33: 584-589.
230 Liu SY, Zheng Z, Zhang R, Bollag J-M. Appl Environ Microbiol 1989; 55: 733-740.
231 McGahen LL, Tiedje JM. J Agric Food Chem 1978; 26: 414-419.
232 Saxena A, Zhang R, Bollag J-M. Appl Environ Microbiol 1987; 53: 390-396.
233 Novick NJ, Mukherjee R, Alexander M. J Agric Food Chem 1986; 34: 721-725.
234 Villareal DT, Turco RF, Konopkca A. Appl Environ Microbiol 1991; 57: 2135-2140.
235 Hsu TS, Bartha R. J Agric Food Chem 1976; 24: 118-122.
236 Bollag J-M, Blattman P, Laanio T. J Agric Food Chem 1978; 26: 1302-1306.
237 Bollag J-M, Minard RD, Liu SY. Environ Sci Technol 1983; 17: 72-80.
238 Balba HM, Still GG, Mansager ER. Presented Div Pest Chem Amer Chem Soc, Anaheim, CA, 1978.
239 Bartha R, Pramer D. Science 1967; 156: 1617-1618.
240 Bartha R. Science 1969; 166: 1299-1300.
241 Linke HAB. Naturwissenschaften 1970; 57: 307-308.

242 Bartha R, Linke HAB, Pramer D. Science 1968; 161: 582-583.
243 Kearney PC, Plimmer JR. J Agric Food Chem 1972; 20: 584-585.
244 Bordeleau LM, Bartha R. Can J Microbiol 1972; 18: 1873-1882.
245 Simmous KE, Minard RD, Bollag J-M. Environ Sci Technol 1987; 21: 999-1003.
246 Engelhardt G, Wallnvfer P, Fuchsbichler G, Baumeister W. Chemosphere 1977; 2/3: 85-92.
247 Bollag J-M, Russel S. Microb Ecol 1976; 3: 65-73.
248 Van Alfen NK, Kosuge T. J Agric Food Chem 1974; 22: 221-224.
249 Van Alfen NK, Kosuge T. J Agric Food Chem 1976; 24: 584-588.
250 Surovtseva EG, Volnova AI, Shatskaya TY. Mikrobiologiya 1980; 49: 351-354.
251 You IS, Bartha R. J Agric Food Chem 1982; 30: 274-277.
252 You IS, Bartha R. Appl Environ Microbiol 1982; 44: 678-681.
253 Kaminski U, Janke D, Prauser H, Fritsche W. Z Allg Mikrobiol 1983; 23: 235-246.
254 Janke D, Baskunov BP, Nefedova MY, Zyakun AM, Golovleva LA. Z Allg Mikrobiol 1984; 24: 253-259.
255 Janke D, Schukat B, Prauser H. J Basic Microbiol 1986; 26: 341-350.
256 Janke D, Ihn W. Arch Microbiol 1989; 152: 347-352.
257 Zeyer J, Kearney PC. Pestic Biochem Physiol 1982; 17: 224-231.
258 Zeyer J, Vasserfallen A, Timmis KN. Appl Environ Microbial 1985; 50: 447-453.
259 Loidi M, Hinteregger C, Ferschl A, Streichsbeir F. Arch Microbiol 1990; 155: 56-61.
260 Kuhn EP, Suflita JM. Environ Sci Technol 1989; 23: 848-852.
261 Struijs J, Rogers JE. Appl Environ Microbiol 1989; 55: 2527-2531.
262 Kuhn EP, Townsend GT, Suflita JM. Appl Environ Microbiol 1990; 56: 2630-2637.

Biotransformations: Microbial Degradation of Health Risk Compounds
Ved Pal Singh, editor
© 1995 Elsevier Science B.V. All rights reserved.

Synthesis and degradation of dimethyl nitrosamine in the natural environment and in humans

Tadashi Yoshinari

Wadsworth Center for Laboratories and Research,
New York State Department of Health, and School of Public Health,
State University of New York at Albany, Empire State Plaza,
P.O. Box 509, Albany, NY 12201-0509, U.S.A.

INTRODUCTION

Dimethyl nitrosamine (DMNA) is a simplest form of N-nitrosamines. DMNA and other nitrosamines are carcinogenic [1,2] as well as teratogenic and mutagenic [3]. Since Magee and Barnes [4] discovered the carcinogenicity of DMNA with experimental animals, there have been numerous studies on DMNA and other nitrosamines in various aspects: mechanism of induction of carcinogenesis, factors promoting or suppressing production, mechanism of formation, and degradation in different environments, and distribution in food and other environments.

Because of the enormous volume of reports of studies on nitrosamines, many of the recent reviews are focused on a narrow topic; endogenous N-nitrosation [5], bacterial mutagenesis by nitrosamines [6], occurrence of nitrosamines in foodstuffs, and dietary exposure to nitrosamines [7,8,9], environmental exposure to preformed nitroso compounds [10], and the levels of both N-nitroso compounds and their precursors in the human environment [11]. While bacterially mediated N-nitrosation has been reviewed as a part of endogenous N-nitrosation [5], there have been no comprehensive reviews on the production and degradation of DMNA in various environments, that are closely associated with microbial processes.

In this chapter, the processes of DMNA formation in various environments and the degradation of DMNA and other nitrosamines by various biogenic processes are reviewed. Endogenous formation and degradation of DMNA in humans are also included in this review, since they are involved with microbial processes in various degrees and are also considered to be the most important reactions that induce human cancer.

CHEMISTRY AND PRECURSORS OF DMNA

Chemistry of DMNA

DMNA and other nitrosamines can be formed chemically from nitrite and secondary amines (Figure 1). The mechanism of N-nitrosation is,

$$\frac{R}{R'}{>}NH + HO\text{-}NO \longrightarrow \frac{R}{R'}{>}N\text{-}NO + H_2O$$

Figure 1. Formation of nitrosamines through N-nitrosation of secondary amines.

however, much more complex [12, 13] than the reaction scheme shown in Figure 1.

DMNA can be degraded by chemical or microbial processes or microsomal P-450 in liver microsomes. In all cases, the decomposition is either by denitrosation or demethylation (Figure 2). Denitrosation, which produces nitrite, can occur by both chemical and microbial reactions, while demethylation occurs by microbial and microsomal reactions.

Sources of amines and nitrate

Amines and nitrate are the precursors of DMNA. The following is a brief summary of the sources of these compounds.

Amines are derived from food ingestion. Most amines, formed in the lower gastrointestinal tract, enter an enterogastric cycle prior to renal excretion. Mono-, di- and trimethylamines in human gastric fluid resemble those present in human saliva and blood [14]. Dimethylamine (DMA) and piperidine (both ~1 μg/ml) are the most commonly occurring

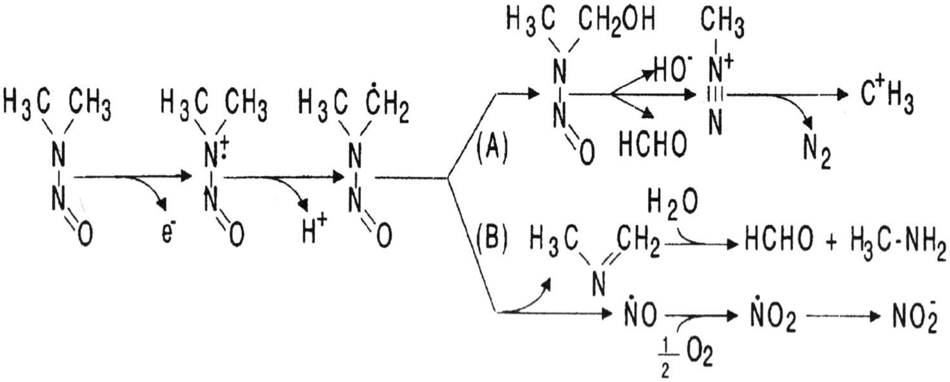

Figure 2. Possible mechanisms of the DMNA metabolism (modified from the schemes proposed by Wade et al. [80]) by demethylation (A) and denitrosation (B).

nitrosatable secondary amines in human saliva, gastric juice, blood, urine, and faeces [15]. In a typical dietary intake of 7.5 mg/day of total secondary amines, dimethylamine constitutes as much as 59% [16].

Nitrate, being a precursor of DMNA and other nitrosamines, is a significant risk factor in human cancer [17,18,19]. Amongst all, nitrate of dietary and drinking-water origin [20] is a major source of endogenous nitrosation, and is actively absorbed in the small intestine, circulated in the bloodstream, and secreted into saliva [21]. Salivary nitrate is reduced to nitrite by bacteria, containing nitrate reductase enzymes, while oral nitrate reduction represents 80% of an individual's nitrite exposure [22]. These bacteria colonize human saliva at levels averaging 10^7 CFU/ml [23]. The nitrite level of human saliva is correlated with the amount of dietary nitrate [24,25]. It should be noted that the life of nitrite in the bloodstream is very short, primarily owing to scavenging by oxyhemoglobin [26,27]. Nitrite in saliva transported to the stomach will become an effective nitrosating agent in the stomach, where the pH is low and secondary and tertiary amines are present. In fact, the existence of a quantitative relationship between oral nitrate-reducing activity and the endogenous formation of N-nitrosoamino acids in humans has been demonstrated [28]. The level of oral nitrate reduction appears to be the major factor affecting the gastric nitrosation [28]. Another important source of endogenous nitrate in humans is biosynthesis from L-arginine [29-32]. The range of daily endogenous biosynthesis of nitrate is 1-1.2 mmol.person^{-1} [33,34]. Determining the incorporation of $^{15}NH_3$, Wagner et al. [35] found that nitrate biosynthesis was enhanced by endotoxin treatment. Stuehr and Marletta [36] found synthesis of nitrite and nitrate in murine macrophage cell lines. While nitrite, nitrate, nitric oxide, and N-nitrosating agents are known to be synthesized by some mammalian cells [37], it was found recently that nitric oxide synthase activity can be induced by *toxic shock* syndrome toxin 1 in a macrophage-monocyte cell line [38].

FORMATION OF DMNA

The formation of DMNA and other nitrosamines by microbial reactions may also be represented by the reaction scheme shown in Figure 1. Much of the efforts, to find the distribution, formation, and degradation of DMNA in different natural environments and by various microbes, were made in the 1970s. The main reason, that there are not many reports on any of those aspects in recent years, may be because the impact is not as great as the other routes of carcinogenesis of humans, as is discussed in later section.

Formation in the natural environment

DMNA can be formed by biochemical processes in various natural environments. Factors, affecting the formation and stability of nitrosamines in different environments, were initially investigated by Alexander and his associates. For example, Ayanaba et al. [39] studied the possible microbial contribution to nitrosamine formation in sewage and soil. Subsequently, transformation of methylamines and formation of DMNA in samples of treated sewage and lake water [40], stability of nitrosamines in samples of lake water, soil, and sewage [41], and factors affecting DMNA formation in soils and water [42] were examined.

Kobayashi et al. [43] found that DMNA can be formed in a polluted environment. They also confirmed that DMNA is formed in artificial wastewater in the range of pH 4-9 in the presence of nitrite. However, there is little evidence to support the view that the production of DMNA and other nitrosamines by microbial actions is an important process in natural environments [44].

From their studies with cultures of several microorganisms, Mills and Alexander [45] found that *Pseudomonas stutzeri* was able to catalyze nitrosation, only when in the growing phase. Although the species was not identified, Ishiwata et al. [46,47] found that bacteria in human saliva formed DMNA from nitrate and dimethylamine.

Microbial formation

There are a large number of reports on DMNA formation from sodium nitrite and dimethylamine by enteric bacteria [48-53], and by resting cells of *Escherichia coli* B [54].

Using a sensitive fluorimetric method to determine nitrosamines, Ralt et al. [55] showed that nitrosation by *E. coli* can be induced under anaerobic conditions by nitrite and nitrate. Ji and Hollocher [56] found that the nitrosation reaction by *E. coli* was carried out first by the production of NO from nitrite, followed by O_2-dependent chemical nitrosation. The nitrosation is a chemical reaction, that proceeds with N_2O_3 and N_2O_4 derived from NO.

Endogenous formation of DMNA

Endogenous nitrosation is a process that is involved with the formation of DMNA and other nitrosamines in human body, with or without the involvement of microbial activity. The reaction can take place in different parts of the human body. Since a strong correlation appears between the exposure of humans to endogenously formed N-nitroso compounds and the induction of cancer [57], extensive studies have been in progress in recent years.

Endogenous N-nitrosation may result from cell-mediated reactions, involving alveolar and peritoneal macrophages, bacteria, and yeasts [5,58,59]. In earlier investigation, dimethylamine was used to determine the rate of nitrosation of amine [52,54]. In contrast, a series of studies were made with morpholine by Calmes et al. [60-62] because the rate of nitrosation of morpholine was about 30 times faster than that of dimethylamine. While the maximum rate of DMNA formation by resting cells of *Escherichia coli* was observed at pH 8.0 [54], Calmels et al. [60] found several species of microorganisms that are capable of nitrosation catalysis at varied rates at pH 7.25. They also studied kinetics of nitrosamine formation from secondary amines by *E. coli* strains. In this paper they examined substrate specificity of three *E. coli* strains to catalyze nitrosation of various amines.

On the basis of their kinetic studies for nitrosation from secondary amines by *E. coli* A10, Calmels et al. [61] concluded that bacterial nitrosation is an enzyme-mediated reaction, closely associated with *molybdenoenzymes,* such as the nitrate reductase/formate hydrogenase system. Subsequent reports [55,62] confirmed the involvement of nitrate reductase in nitrosation. The comparison of relative activity between chemical and bacterial nitrosamine formation appropriate to gastric juice was made [58]. Nitrosation of amines can be stimulated by the presence of macrophages [63]. DMNA can be formed at neutral pH (achlorhydric gastric juice and infected organs, such as bladder) with assistance from microbial activity (nitrosating enzymes). Bacterially catalyzed N-nitrosation reactions proceed much more rapidly at neutral pH than the chemical reaction [58].

Among the species and different strains of bacteria, Leach et al. [58] found that the most rapid catalysis was associated with those bacteria, capable of reducing nitrate and nitrite by the process of denitrification. Their study was based on the use of *Pseudomonas aeruginosa* BM1030, an isolate from achlorhydric gastric juice. This is in contrast to the conclusion by Licht and Deen [64], who have developed a mathematical model to estimate the rate of formation of nitrosamine in human stomach. The calculated amount of gastric formation of DMNA was ~0.02 nmol/day. As the amount is a factor of ~10^2 to 10^3 lower than published estimates of dietary exposure to preformed DMNA, gastric formation of DMNA does not pose a serious additional health risk so long as the stomach is acidic. However, in achlorhydric stomach it may be a serious threat of carcinogenesis due to increased production of nitrosamines. Increased levels of nitrosamines in blood of patients, with chronic renal failure [65], in urine of patients with infection of the bladder due to urinary diversions [10], and in the urinary tract of paraplegic patients [9] indicate that DMNA was formed endogenously by a bacterially mediated in vivo formation of N-nitroso compounds.

Calmels et al. [66] isolated an enzyme catalyzing nitrosamine formation in *Pseudomonas aeruginosa* and *Neisseria mucosae* at neutral pH. Their results suggest that carcinogenic N-nitroso compounds may be formed endogenously with nitrosating enzymes that are provided by some bacteria.

As vitamin E acts as nitrite scavenger, it serves as inhibitor of the formation of N-nitrosamine [67].

DEGRADATION OF DMNA

There are two types of microbial degradation of DMNA (Pathways A and B in Figure 2). One is denitrosation, which produces nitrite and amine. This is a reverse of the process of chemical formation of DMNA (Figure 1), and is presumed to be a main process for removal of DMNA from the environment. The other is demethylation, which produces aldehyde and methyl amine. This reaction is considered as a central paradigm for initiation of carcinogenesis.

Natural environment

Tate and Alexander [68] studied resistance of nitrosamines to microbial attack in the environment. Degradation of nitrosamines in lake water and sewage [69] and in the marine environment [70] has been reported. Kaplan and Kaplan [71] reported biodegradation of DMNA in aqueous and soil systems. In both systems, the rate of degradation was slow, but linearly correlated with the concentration of DMNA (Figure 3). On the

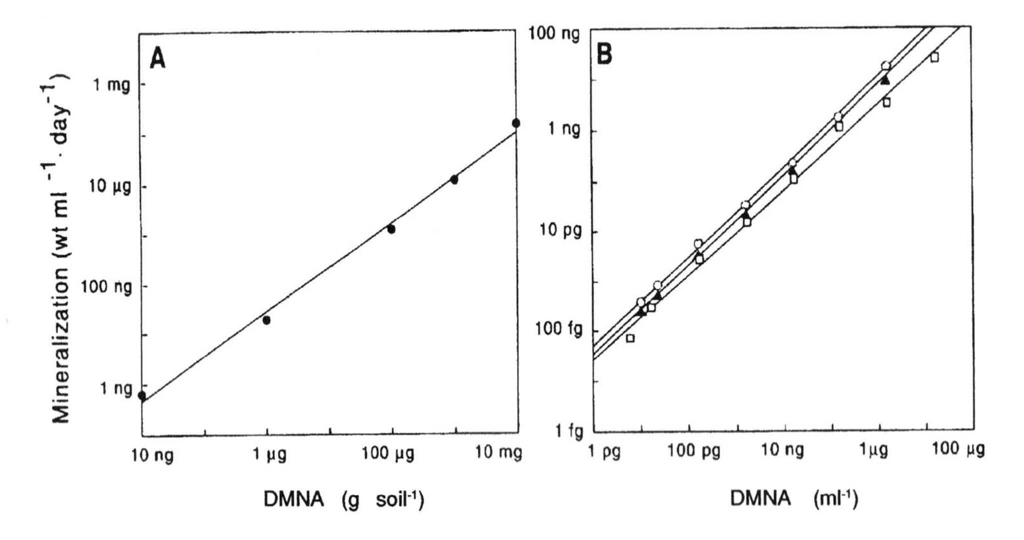

Figure 3. Initial rates of mineralization over a range of concentrations of DMNA in (A) soil batch culture and (B) aqueous batch culture. Symbols: ○, lake water with salts; □, lake water; ▲, lakewater with salts and glucose [71].

basis of their finding, that the substrate concentration reduction curves generated with DMNA were not sigmoidal, the possibility of the DMNA metabolism by the biomass being growth-related was ruled out. Although the characteristics of the microbes are not known, they identified two metabolic intermediates, formaldehyde and methylamine, from the DMNA degradation.

Pure culture studies

Five out of 44 species of bacteria, molds, and yeasts were found to form nitrite from DMNA [72]. Using *Rhizopus oryzae*, *Streptococcus cremoris,* and *Saccharomyces rouxii*, Harada and Yamada [73] examined the degradation of five different N-nitrosamines, including DMNA. While N-nitrosodipropylamine (DPNA) was most easily degradable, DMNA was least degraded among five different nitrosamines. Their results with whole-cell cultures suggested that the enzymes responsible for the degradation of N-nitrosodipropylamine were inducible. Using whole cells in growth phase and cell-free extract of *Rhizopus oryzae*, Harada [74] studied further the microbial degradation of DMNA and DPNA. The degradation activities in the cell-free extract were highest at 30°C and pH 8.0, under anaerobic conditions. The degradation of DMNA and DPNA was concentration-dependent with cell-free extract, while that with whole cells in the growth phase showed a maximum at ~0.1 mM. A series of these studies did not investigate the metabolic pathways of DMNA. Kobayashi et al. [43] found that photosynthetic bacterium, *Rhodopseudomonas capsulata*, is capable of metabolizing DMNA.

Various enteric bacteria metabolized DMNA through denitrosation, by which dimethylamine and nitrite were formed, and dimethylamine was not further metabolized [75].

The other type of the degradation of DMNA is by demethylation. It was found that a methanotroph, *Methylosinus trichosporium* OB3b (MT OB3b), degraded DMNA in the presence of methane, presumably by the catalytic action of NAD(P)H-dependent methane monooxygenase (MMO) [76]. Tracer studies with ^{14}C-labelled DMNA revealed that MT OB3b was capable of both assimilating DMNA-carbon into the cell and respiring it as CO_2 (Figure 4). The rates of CO_2 production (V_{CO_2}) from and cellular uptake (V_P) of DMNA were linearly correlated with the DMNA concentrations of 0.03-10 mM, which corresponded to approximately 3 per cent of the added DMNA metabolized in 24 h. These rates were two to three orders of magnitude less than that of the uptake of methane (V_{CH_4}). V_{CH_4} was suppressed only when the concentrations of DMNA exceeded 0.3 mM (Figure 4). In the presence of 0.1 mM DMNA, V_P and V_{CO_2} were essentially the same with or without the presence of methane in the first 8 h of incubation, but declined sharply thereafter only when methane was absent. These observations suggested that the metabolism

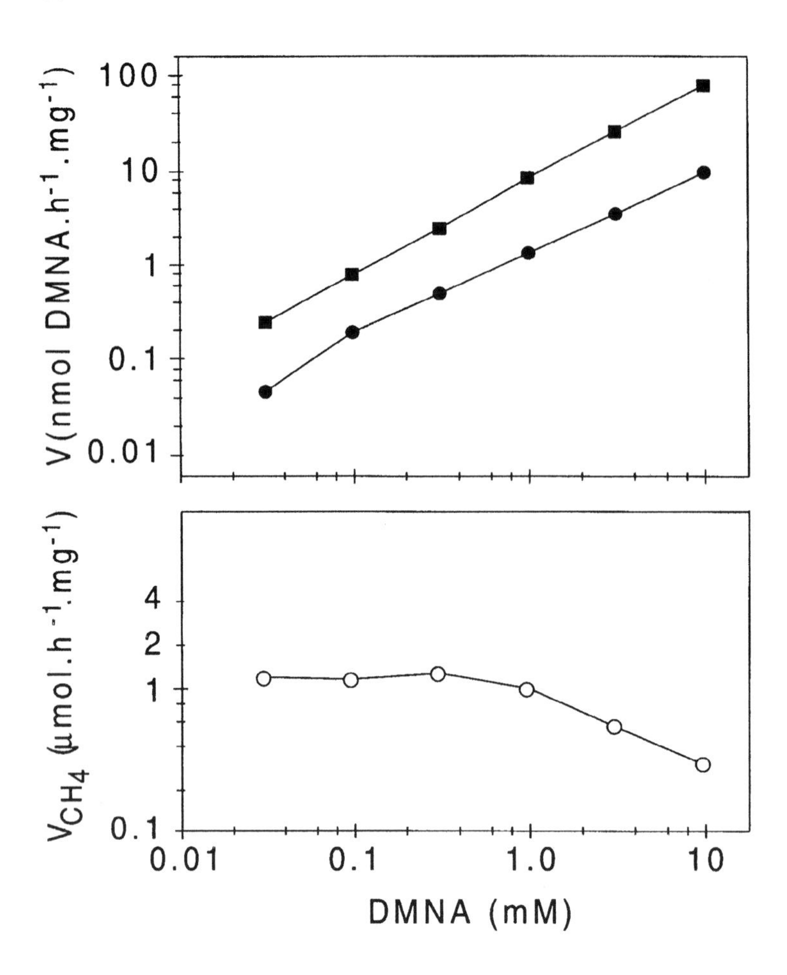

Figure 4. Metabolism of DMNA by *Methylosinus trichosporium* OB3b (MT OB3b) in the presence of different concentrations of DMNA. Average rates at 0-24 h of incubation are given [76]. The rates of cellular uptake, CO_2 production, and methane metabolism are represented by V_P (●), V_{CO_2} (■) and V_{CH_4} (○), respectively.

of DMNA was carried out by methane monooxygenase (MMO), and that NADH, a cofactor for MMO, may be provided from the oxidation of the stored energy in the cells, when methane is not available.

The sequence of the reaction of DMNA by MT OB3b has not been clearly understood. However, in view of the finding that the DMNA-carbon was incorporated to cell material and respired as CO_2, Yoshinari and Shafer [76] postulated that MT OB3b produced formaldehyde through

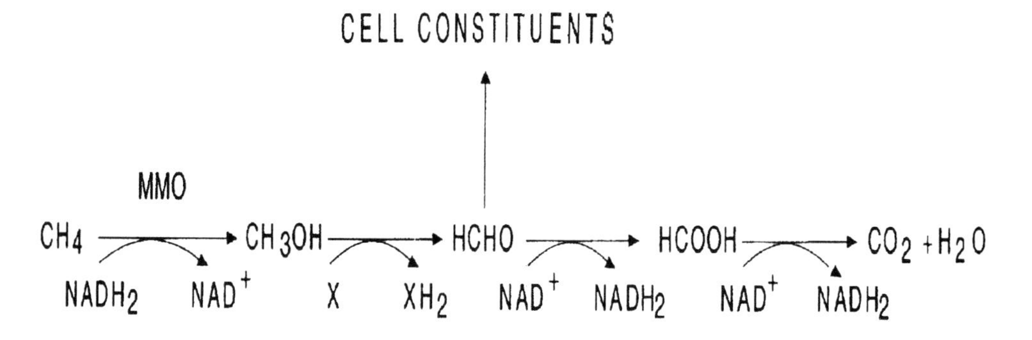

Figure 5. Schematic mathane oxidation pathway by methanotrophs. Initial step for the metabolism of methane is catalyzed by methane monooxygenase (MMO).

hydroxylation of DMNA by MMO, as is the case with the oxidation of methane (Figure 5). They suggested that the initial step for the formation of formaldehyde could be quite similar to the reaction sequence carried out by a liver microsomal cytochrome P-450-dependent mixed-function oxidase (MFO). Both MMO and MFO require cofactor NAD(P)H and oxygen for their activity. With MFO, DMNA is activated to α-hydroxylnitrosamine. Subsequently, it is decomposed to form spontaneously formaldehyde and methyldiazonium (Pathway A in Figure 2). Most of the unstable reactive intermediate, CH_2N_2, which can interact with DNA and other nucleophiles for carcinogenicity and mutagenicity, is suspected to react with water and form ROH and N_2. To test further whether MT OB3b metabolizes DMNA by this sequence, it would be of interest to determine the production of $^{15}N_2$ from the ^{15}N-labelled DMNA, by using the method of Kroeger-Koepke et al. [77].

Endogenous degradation of DMNA

A large body of information regarding metabolism of DMNA by mammalian cells is documented. Mixed-function cytochrome P-450 isozymes (MFO) in mammalian liver microsomes are involved in the metabolism of DMNA [1,6,78]. The same cytochrome P-450 isozymes are reported to be involved in both nitrite formation (denitrosation) and α-hydroxylation (demethylation) of nitrosamines [79-81].

Denitrosation

Denitrosation results in the detoxification of DMNA. Keefer et al. [82] found concurrent generation of methylamine and nitrite during denitrosation of DMNA by rat liver microsomes. Amelizad et al. [83] studied the effect of antibodies against cytochrome P-450 on demethylation and denitrosation of DMNA and N-nitrosomethylaniline. Hydroxy derivatives, aldehydes, and nitrite were formed from N-nitrosomethyl-N-amylamine by rat liver microsomes and by purified cytochrome P-450 IIB1 [84]. Recently, an increased oxidation of DMNA in pericentral microsomes, after pyrazole induction of cytochrome P-450 2El, has been reported [85].

Heur et al. [86] studied the Fenton degradation as a nonenzymic model for microsomal denitrosation of DMNA.

Demethylation

This reaction is the central paradigm for initiation of carcinogenesis. DMNA is metabolically activated to generate the ultimate carcinogenic form. Initial step in this biotransformation is believed to be enzymatic hydroxylation by MFO, followed by spontaneous cleavage of the carbon-nitrogen bond, which releases aldehyde and methyldiazonium (Figure 2). Most of the unstable intermediate, CH_2N_2, that is suspected to interact with DNA and other nucleophiles for carcinogenicity and mutagenicity, seems to react with water and form methanol and N_2. Methyl diazohydroxide appears to be the main cause of carcinogenesis and mutagenesis [1,6,78].

Jensen et al. [87] provided evidence that the in vitro methylation of DNA was carried out by microsomally-activated DMNA and that the methylation was correlated with formaldehyde production. Sagelsdorff et al. [88] showed DNA methylation in rat liver by daminozide, 1, 1-dimethylhydrazine, and DMNA. Studies with other nitroso compounds also confirmed the formation of hydroxy derivatives, aldehydes, and nitrite by rat liver microsomes and by purified cytochrome P-450 IIB1 [84]. The sequence of specific methylation of DNA by N-nitroso compounds shares a common intermediate, methyl diazonium ion [89].

CONCLUSIONS

Dimethyl nitrosamine (DMNA) is a potent carcinogenic compound. Unlike the xenobiotic compounds, that are known to be carcinogenic, it is not the product of industrial processes. Instead, DMNA is the product of chemical and microbial processes in the natural environment and in the human body system. A major source of DMNA for humans is endogenous production rather than ingestion from the food. DMNA can be decomposed chemically and biochemically. The product derived from the enzymatic degradation of DMNA is responsible for methylating the nucleotides, that trigger the process of carcinogenesis.

Endogenously formed DMNA in humans through chemically and/or bacterially mediated reactions, but not of natural environmental origin, appears to be mainly responsible for causing human cancer.

There are two pathways of degradation of DMNA by microbial enzymes and by microsomal P-450. One is denitrosation and the other demethylation. Denitrosation is a main microbial process for removal of DMNA from the environment, although its degradation through demethylation also takes place by a methanotroph. The reaction catalyzed by microsomal P-450 can couple with the methylation of DNA, a central paradigm for initiation of carcinogenesis.

REFERENCES

1 Lai DY, Arcos JC. Life Sci 1980; 27: 2149-2165.
2 Archer M. Cancer Surveys 1989; 8: 241-250.
3 Preussmann R, Stewart BW. In: Searle G, ed. Chemical Carcinogenesis. 2nd edn, Vol2. ACS Monograph 182, Washington D.C.: American Chemical Society, 1984; 643-828.
4 Magee PN, Barnes JM. Br J Cancer 1956; 10: 114-212.
5 Leach S. In: Hill MJ, ed. Nitrosamines: Toxicology and Microbiology. Chichester, England: Ellis Horwood, 1988; 69-87.
6 Guttenplan JB. Mutat Res 1987; 186: 81-134.
7 Scanlon RA. Cancer Res (Suppl) 1983; 43: 2435s-2440s.
8 Hotchkiss JH. Cancer Surveys 1989; 8: 295-321.
9 Tricker AR, Preussmann R. Mutat Res 1991; 259: 277-289.
10 Tricker AR, Spiegelhalder B, Preussmann R. Cancer Surveys 1989; 8: 251-272.
11 Tricker AR, Preussmann R. In: Hill MJ, ed. Nitrosamines: Toxicology and Microbiology. Chichester, England: Ellis Horwood, 1988; 88-116.
12 Challis BC. In: Gibson GG, Ioannides C, eds. Safety Evaluation of Nitrosatable Drugs and Chemicals, 1981; 16-27.
13 Shuker DE. In: Hill MJ, ed. Nitrosamines: Toxicology and Microbiology. Chichester, England: Ellis Horwood, 1988; 48-68.
14 Zeisel SH, daCosta KA, LaMont JT. Carcinogenesis 1988; 9: 179-181.
15 Tricker AR, Pfundstein B, Kalble T, Preussmann R. Carcinogenesis 1992; 13: 563-568.
16 Pfundstein, Tricker AR, Theobald E, Spiegelhalder B, Preussmann R. Food Chem Toxicol 1991; 29: 733-739.
17 Forman D. Cancer Surveys 1989; 8: 443-458.
18 Walker R. Food Additives and Contaminants 1990; 7: 717-768.
19 Kleinjans JCS, Albering HJ, Marx A, van Maanen JMS, van Agen B, ten Hoor F, Swaen GMH, Mertens PLJM. Environ Health Perspectives 1991; 94: 189-193.

20 Moller H, Landt J, Pedersen E, Jensen P, Autrup H, Jensen OM. Cancer Res 1989; 49: 3117-3121.
21 Hartman, PE. Chemical Mutagenesis 1982; 7: 211-294.
22 Spiegelhalder B, Eisenbrand G, Preussmann R. Food Cosmet Toxicol 1976; 14: 545-548.
23 Sasaki T, Matano KJ. Food Hygien Soc Japan 1980; 21: 123-128.
24 Tannenbaum SR, Weisman M, Fett D. Food Cosmet Toxicol 1976; 14: 549-552.
25 Cingi MI, Cingi C, Cingi E. Bull Environ Contam Toxicol 1991; 48: 83-88.
26 Kosaka H, Imaizumi K, Imai K, Tyuma I. Biochim Biophys Acta 1981; 581: 184-188.
27 Parks NJ, Krohn KA, Mathis CA, Chasko JH, Geiger KR, Gregor ME, Peek NF. Science 1981; 212: 58-61.
28 Shapiro KB, Hotchkiss JH, Roe DA. Food Chem Toxicol 1991; 29: 751-755.
29 Green LC, Tannenbaum SR, Goldman P. Science 1981; 212: 56-58.
30 Green LC, Uiz de Luzuriaga K, Wagner DA, Rand W, Istan N, Young VR, Tannenbaum SR. Proc Natl Acad Sci USA 1981; 78: 7764-7768.
31 Leaf CD, Wishnok JS, Tannenbaum SR. Biochem Biophys Res Commun 1989; 163: 1032-1037.
32 Leaf CD, Wishnok JS, Hurley JP, Rosenblad WD, Fox JG, Tannenbaum SR. Carcinogenisis 1990; 11: 855-858.
33 Wagner DA, Schultz DS, Deen WW, Young VR, Tannenbaum SR. Cancer Res 1983; 43: 1921-1925.
34 Leaf CD, Vecchio AJ, Roe DA, Hotchkiss JH. Carcinogenesis 1987; 8: 791-795.
35 Wagner DA, Young VR, Tannenbaum SR. Proc Natl Acad Sci USA 1983; 80: 4518-4521.
36 Stuehr DJ, Marletta MA. Cancer Res 1987; 47: 5590-5594.
37 Marletta MA. Chem Res Toxicol 1988; 1: 249-257.
38 Zembowicz A, Vane JR. Proc Natl Acad Sci USA 1991; 89: 2051-2055.
39 Ayanaba A, Verstraete W, Alexander M. J Natl Cancer Inst 1973; 50: 811-813.
40 Ayanaba A, Alexander M. J Environ Quality 1974; 3: 83-89.
41 Tate RL, Alexander M. J Natl Cancer Inst 1975; 54: 327-330.
42 Mills AL, Alexander M. J Environ Quality 1976; 5: 437-440.
43 Kobayashi M, Taketomo N, Tchan YT. J Ferment Technol 1977; 55: 615-620.
44 Oliver JE. In: Scanlan RA, Tannenbaum SR, eds. N-nitroso Compounds. ACS Symposium Series 174. Washington D.C.: American Chemical Society, 1981; 349-362.
45 Mills AL, Alexander M. Appl Environ Microbiol 1976; 31: 892-895.

46 Ishiwata H, Tanimura A, Ishidate M. J Food Hygien Soc 1975; 16: 234-239.
47 Ishiwata H, Tanimura A, Ishidate M. J Food Hygien Soc 1976; 17: 59-end.
48 Klubes P, Jondorf WR. Res Commun Chem Pathol Pharmacol 1971; 2: 24-34.
49 Hawksworth GM, Hill MJ. Brit J Cancer 1971; 25: 520-526.
50 Coloe PJ, Hayward NJ. J Med Microbiol 1976; 9: 211-223.
51 Hashimoto S, Yokokura T, Kawai Y, Mutani M. Food Cosmet Toxicol 1976; 14: 553-556.
52 Suzuki S, Mitsuoka T. In: O'Neil IK, von Bostel RC, Long JE, Miller CT, Bartsch H, eds. N-Nitroso Compounds: Occurence, Biological Effects and Relevance to Human Cancer. IARC Scientific Publications, No. 57. Lyon: International Agency for Research on Cancer, 1984; 275-282.
53 O'Donnell CM, Edwards C, Ware J. FEMS Microbiol Lett 1988; 51: 193-198.
54 Kunisaki N, Hayashi M. Appl Environ Microbiol 1979; 37: 279-282.
55 Ralt D, Wishnok JS, Fitts R, Tannenbaum SR. J Bacteriol 1988; 170: 359-364.
56 Ji X-B, Hollocher TC. Appl Environ Microbiol 1988; 54: 1791-1794.
57 Bartsch H, Ohshima H, Shuker DEG, Pignatelli P. Mutat Res 1990; 238: 255-267.
58 Leach SA, Thompson M, Hill M. Carcinogenesis 1987; 8: 1907-1912.
59 Leaf CD, Wishnok JS, Tannenbaum SR. Cancer Surveys 1989; 8: 323-334.
60 Calmels S, Ohshima H, Vincent P, Gounot AM, Bartsch H. Carcinogenesis 1985; 6: 911-915.
61 Calmels S, Ohshima H, Rosenkranz H, McCoy E, Bartsch H. Carcinogenesis 1987; 8: 1085-1087.
62 Calmels S, Ohshima H, Bartsch H. J Gen Microbiol 1988; 134: 221-226.
63 Miwa M, Stuehr DJ, Marletta MA, Wishnok JS, Tannenbaum SR. Carcinogenesis 1987; 8: 955-958.
64 Licht WR, Deen WM. Carcinogenesis 1988; 9: 2227-2237.
65 Dunn SR, Simenhoff ML, Lele PS, Goyal S, Pensabene JW, Fiddler WJ. Natl Cancer Inst 1990; 82: 783-787.
66 Calmels S, Dalla Venezia N, Bartsch H. Biochem Biophy Res Commun 1990; 171: 655-660.
67 Lathia D, Blum A. A review Internatl J Vit Nutr Res 1989; 59: 430-438.
68 Tate RL, Alexander M. J Environ Quality 1976; 5: 131-133 [CA 85: 1021z].

69 Yordy JR, Alexander M. Appl Environ Microbiol 1980; 39: 559-565.
70 Petit L, Aubert J, Dubreuil A, Levrat B. Rev Int Oceanogr Med 1984; 73-74: 41-50.
71 Kaplan DL, Kaplan AM. Environ Microbiol 1985; 50: 1077-1086.
72 Harada K, Yamada K. J Shimonoseki Univ Fish 1978; 26: 293-298.
73 Harada K, Yamada K. Nippon Suisan Gakkaishi 1978; 45: 925-928.
74 Harada K. Bull Jap Soc Sci Fish 1980; 46: 723-726.
75 Rowland IR, Grasso P. Applied Microbiol 1975; 29: 7-12.
76 Yoshinari T, Shafer D. Can J Microbiol 1990; 36: 834-838.
77 Kroeger-Koepke MB, Koepke SR, McClusky GA, Magee PN, Michejda CJ. Proc Natl Acad Sci USA 1981; 78: 6489-6493.
78 Magee PN, Barnes JM. Adv Cancer Res 1967; 10: 163-246.
79 Tu YY, Yang CS. Arch Biochem Biophys 1985; 242: 32-40.
80 Wade D, Yang CS, Metral CJ, Roman JM, Hrabie JA, Riggs CW, Anjo T, Keefer LK, Mico BA. Cancer Res 1987; 47: 3373-3377.
81 Sohn OS, Ishizaki H, Yang CS, Fiala ES. Carcinogenesis 1991; 12: 127-131.
82 Keefer LK, Anjo T, Wade D, Wang T, Yang CS. Cancer Res 1987; 47: 447-452.
83 Amelizad S, Appel KE, Oesch F, Hildebrandt AG. J Cancer Res Clin Oncol 1988; 114: 380-384.
84 Ji C, Mirvish SS, Nickols J, Ishizaki H, Lee MJ, Yang CS. Cancer Res 1989; 49: 5299-5304.
85 Dicker E, Cederbaum AI. Clin Exp Res 1991; 116: 1072-1076.
86 Heur YH, Streeter AJ, Nims RW, Keefer LK. Chem Res Toxicol 1989; 2: 247-253.
87 Jensen DE, Lotlikar PD, Magee PM. Carcinogenesis 1981; 2: 349-354.
88 Sagelsdorff P, Lutz WK, Schlatter C. Fund Appl Toxicol 1988; 11: 723-730.
89 Milligan JR, Hirani-Hojatti S, Catz-Biro L, Archer MC. Chem-Biol Interactions 1989; 72: 175-189.

Biotransformations: Microbial Degradation of Health Risk Compounds
Ved Pal Singh, editor
© 1995 Elsevier Science B.V. All rights reserved.

Aflatoxin biotransformations: biodetoxification aspects

Ved Pal Singh

Department of Botany, University of Delhi, Delhi-110007, India

INTRODUCTION

Since antiquity, man has been witnessing the hazardous effects of *mycotoxins* on human and animal health, as almost all agricultural produce, including most foods and feeds used for human and animal consumption, get contaminated by the *toxigenic* fungi. Mycotoxins are highly poisonous secondary metabolites produced mainly by filamentous fungi [1], which contaminate the foods and feeds at some stage of their production, processing, transportation or storage. Over 200 mycotoxins have already been identified [2], and still many others are yet to be characterized. Of all mycotoxins, aflatoxins constitute the most widely studied, and their toxic, mutagenic as well as carcinogenic effects on both man and animals are well-documented [1,3-12]. Aflatoxins are also known to influence metabolism of higher plants, ferns, algae, fungi, and bacteria [13].

Aflatoxins were first discovered by Sargeant and coworkers in England, in the year 1961, while investigating the causal factors for widespread episode of unexplained mortality encountered in poultry flocks, swine, and cattle due to a disease which later came to be known as the *'Turkey-X'* disease. These toxins, found in mould-infected peanut meals, were responsible for the Turkey-X disease in livestock. The mould was identified at the Royal Botanic Gardens, Kew, as *Aspergillus flavus,* and the corresponding toxin was named after this fungus as *A. flavus* toxin or 'Aflatoxin'. These potentially hazardous secondary metabolites are produced primarily by some strains of *A. flavus, A. versicolor, A. nidulans* and most, if not all, strains of *A. parasiticus* as well as a related species *A. nomius* [14]. *Aflatoxigenic* fungi are more commonly found in tropical and subtropical areas where both temperature and humidity are favourable for growth. Aflatoxins are highly oxygenated, heterocyclic compounds, having dihydrofurano or tetrahydrofurano moieties fused to a substituted coumarine moiety and are synthesized through polyketide pathway, linking both primary metabolism, i.e., biosynthesis of fatty acids and secondary metabolism (polyketide biosynthesis), involving an intermediate (acetyl coenzyme A) between these pathways for aflatoxin biosynthesis.

Aflatoxins have been characterized on the basis of their fluorescence properties. Those which fluoresce blue under UV light are classified as B aflatoxins and those with green fluorescence are classified as G aflatoxins. And on the basis of their relative mobilities (Rf values) on TLC plates, these toxins are further classified as aflatoxins B_1 and B_2 (AFB_1 and AFB_2) and aflatoxins G_1 and G_2 (AFG_1 and AFG_2).

That aflatoxins are acutely toxic to man was demonstrated by an outbreak in India in 1974 of epidemic jaundice, involving severe liver disease, which resulted in the deaths of more than 100 people and serious illness in 400 others [15].

Aflatoxins B_1 and G_1 have been shown to be mutagenic in bacteria, vegetative cells as well as in *Drosophila melanogaster* and *Salmonella typhimurium*. Also, metabolites have been studied for their mutagenic effects in *S. typhimurium* and show large variations in their mutagenic potency [16].

Carcinogenicity of aflatoxins has been demonstrated in various animal species [3,12,17,18]. AFB_1 induces malignant tumours in rats, mice, monkeys, marmosets, ducks, guppies, salmon, trout, and tree shrews. The liver is the target organ for these compounds, but some pulmonary tumours have been observed in treated mice, as well as kidney and intestinal tumours in rats. AFB_1 produces carcinogenic effects in many species, following exposures to low doses (as low as 1 μg/kg in the diet). Aflatoxins must be considered to be a probable causative agent for primary liver cancer in man, which is endemic to Thailand, Kenya, Switzerland, and Mozambique. AFG_1 and AFB_2 are also known to induce liver tumours in some animals [18].

Considering all these aspects of aflatoxin-induced health hazards in man and animals, and because of the widespread occurrence of these toxins in otherwise nutritious natural products, many studies have been carried out to find effective and suitable as well as convenient methods of detoxification of aflatoxins.

PHYSICAL AND CHEMICAL METHODS OF AFLATOXIN DETOXIFICATION

Although physical and chemical methods of detoxification are not the subject matter of the present review, some recent works have been included to have a latest insight into the mechanism of aflatoxin detoxification. Some of the conventional methods have already been mentioned elsewhere [19].

Effects of physical factors, such as light, temperature, pH, etc., on aflatoxin detoxification have been of interest to biotechnologists and food scientists. The reduced levels of aflatoxins in contaminated grains of rice due to green light and in those of ragi as well as in liquid broth due to blue light have been recorded by Shrivastava et al. [20]. In artificially infected corn meal and peanuts with *Aspergillus parasiticus*, chlorine gas has been found to show 75% degradation of AFB_1 [21]. The *mutagenicity* of chlorine-treated copra meals and peanuts spiked with AFB_1 was greatly reduced compared with untreated controls, as determined in *Salmonella typhimurium* strain TA98 in the presence of rat liver

5-9 mix; the decrease in the mutagenic potential has been correlated with reduction in AFB_1 levels. However, no mutagenic compound was generated by such treatment. Similar results have been obtained with ammonia treatment of aflatoxin-contaminated cotton seeds by Jorgensen et al. [22]. Abdel-Rahim et al. [23] have shown that the rate of detoxification of aflatoxins in cotton seeds increased with increasing concentrations of ammonium hyroxide (NH_4OH) up to 0.4%, then decreased at further higher concentrations. The detoxification rate for AFB_1 was higher than that of AFB_2 and AFG_2 at very low concentration (0.1%) of NH_4OH. Aqua-ammonia method has been quite successful in detoxifying the aflatoxins in contaminated poultry feed to a non-detectable level on 3rd day after ammoniation, using 1.5% aqueous ammonia solution sprayed over contaminated feed [24].

Interestingly enough, a combination of physical and chemical (physicochemical) method has been quite effective in AFB_1 degradation. Degradation of aflatoxin B_1 in dried figs by sodium bisulphite with or without heat, UV energy or H_2O_2 has been studied by Altug et al. [25]. There was 28.2% degradation by 1% sodium bisulphite treatment alone, but when H_2O_2 (0.2%) was added, 10 min before bisulphite treatment, 65.5% degradation of AFB_1 was achieved in 72 h. In both cases, maximum degradation occurred during 2nd day of treatment. Heating of the bisulphite-treated samples at 45°C to 65°C caused 68.2% degradation of AFB_1. However, UV radiation degraded 45.7% of AFB_1 in fig samples, when exposed for 30 min, but the rate of degradation was not affected by the addition of bisulphite or H_2O_2.

BIODETOXIFICATION OF AFLATOXINS

Biodetoxification aspects can be further divided into three categories, depending on the type of system involved, which are as follows:

(a) Commodity-dependent detoxification.
(b) Enzymatic detoxification.
(c) Microbial detoxification.

(a) Commodity-Dependent Detoxification

Plants can carry out biotransformations of aflatoxins, as made clear by Howes et al. [26], who studied the metabolism of aflatoxin B_1 in *Petroselinum crispum* (Parsley). On administration of AFB_1 to whole plants, *aflatoxicol* was formed. The cell-free preparations, on the other hand, formed two new aflatoxin B_1 or aflatoxicol A. Even plant products, such as neem-leaf extracts, have been shown to control aflatoxin production in *Aspergillus flavus*-infected cotton bolls [27]. Sometimes, the genetic constitution of the aflatoxin contaminated plant commodity can take care of the levels of aflatoxins in itself [28]. For example, the tannin extracts from some genotypes of peanut cultivars, such as PI 337409 and TX 798736, have been found to reduce the level of aflatoxin production [29]. Also, certain corn genotypes which get contaminated with the toxigenic fungus, such as *Aspergillus flavus,* can regulate the toxin production in the grain [28], and this might be the reason why some genotypes of maize germplasm develop resistance to infection by *A. flavus*, and/or subsequent contamination by AFB_1 [30]. Not only the genotypes, but also the biomolecules of the plants, such as glyceollin of soybean seeds, have been implicated in lowering down the aflatoxin levels in the *A. flavus*-infected viable seeds [31]. On the other hand, Rasic et al. [32] reported that the level of AFB_1, added to a food commodity, i.e., milk, before fermentation at concentrations of 600, 1000 and 1400 μg/kg, was reduced in yoghurts (pH 4.0) by 97, 91 and 90%, respectively. According to them, a decrease of AFB_1 (conc. 1000 μg/kg) in milk, acidified with citric, lactic, and acetic acids (pH 4.0), was 90, 84 and 73%, respectively.

(b) Enzymatic Detoxification

Almost all human and animal species get exposed to aflatoxin doses beyond permissible limits, when they consume the aflatoxin-contaminated foods and feeds. However, there are special mechanisms by which aflatoxins are detoxified within the living systems, so as to eliminate the possibility of health hazards caused due to consumption of aflatoxin-contaminated foods and feeds. It has been demonstrated that biotransformations of aflatoxins are mediated by microsomal enzymes, exhibiting the mixed function hydroxylases of endoplasmic reticulum in higher organisms [33]. The liver microsomal preparations from many animal species have been shown to transform the most *hepatotoxic* and *hepatocarcinogenic* aflatoxin - AFB_1 to a 4-hydroxy derivative (i.e., aflatoxin M_1, AFM_1) and an analogous product. Aflatoxin GM_1 may be formed from AFG_1. Both AFB_1 and AFG_1 have been shown to be metabolized to their respective 2-hydroxy derivatives or hemiacetals by $NADPH_2$-dependent enzyme [33]. Patterson [33] also reported the enzymatic detoxification of aflatoxins in the livers of rabbit, duckling, guinea-pig, chick, and mouse, following a minor route to the formation AFM_1. Human liver microsomes were able to detoxify AFB_1 and convert it to AFQ_1. However, once produced, AFQ_1 was not appreciably oxidised in human liver microsomes and was not very genotoxic. The 3α-hydroxylation of AFB_1 to AFQ_1 is considered to be a potentially significant detoxification pathway [34]. Studies have also shown that mice become resistant to carcinogenic effects of aflatoxin B_1 and that, this is due to expression of an isoenzyme of glutathione S-transferase (GST) with high activity towards AFB_1-8-9-epoxide [3]. Daniels et al. [35] have also demonstrated the biotransformations of potentially hepatotoxic and hepatocarcinogenic

AFB_1 to less toxic metabolites, AFM_1 and AFQ_1 in rabbit lung and liver microsomes. Various products of aflatoxin B_1 metabolism are given in Figure 1.

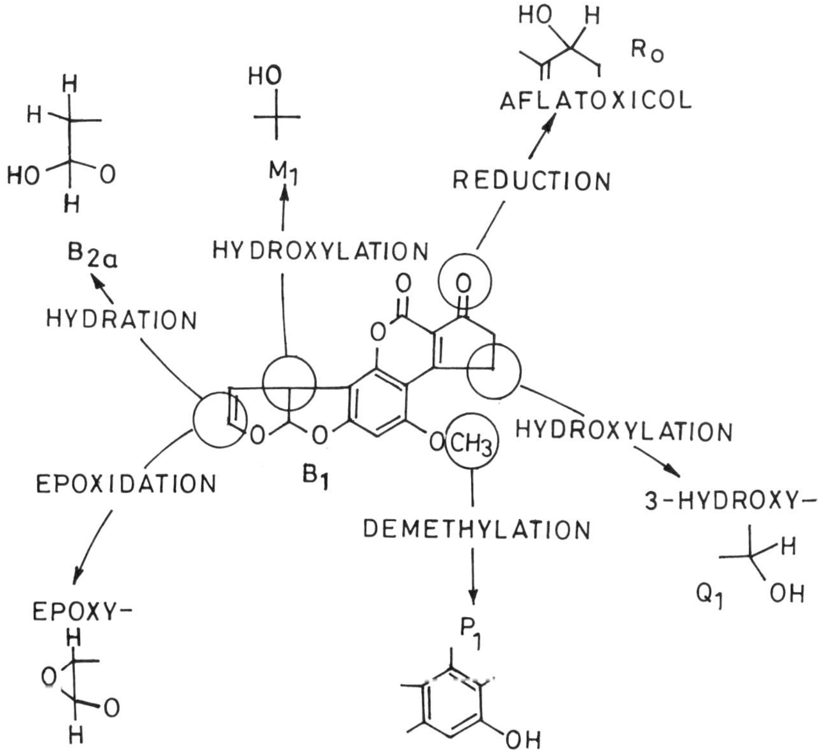

Figure 1. Metabolic biotransformation products of aflatoxin B_1.

Microbial systems show similarities to the animal systems, and Hamid and Smith [36] have clearly demonstrated the involvement of a microsomal enzyme system in aflatoxin degradation in *A. flavus*. The degradation of aflatoxins by cell-free extracts of *A. flavus* was enhanced by NADPH [36], which is consistent with the activity of enzymes in which this cofactor is necessary for enhanced aflatoxin detoxification in eukaryotic systems [33]. NADPH-dependent 17-hydroxy-steroid dehydrogenase has been reported to transform aflatoxin B_1 to aflatoxin R_0. Doyle and Marth [37] have observed maximum aflatoxin degradation by the intramycelial factors of the fungus at physiologically optimal pH and temperature, conducive for the aflatoxin-degrading enzymes.

Also, the enzyme which uses hydrogen peroxide (H_2O_2) as the substrate appears, to play a key role in aflatoxin degradation or detoxification, as H_2O_2 enhances the aflatoxin degradative activity, when it was added to the mycelial proteins of *A. parasiticus* [6]. Thus, peroxidase could be the probable enzyme, which might possibly help in detoxification of aflatoxins. The studies of the author in collaboration with Professor John E. Smith at the University of Strathclyde, Glasgow (U.K.), have indicated the involvement of microsomal peroxidase in aflatoxin detoxification, without precluding the possibility of involvement of *cytochrome* P-450 monooxygenase in aflatoxin detoxification [36]. The role of hepatic *microsomal* P-450 monooxygenase in AFB_1 detoxification in animal systems has already been established. The studies on cytochrome-mediated metabolism of endogenous substrates, such as steroids and fatty acids as well as biotransformation of xenobiotics, have been well-documented [38,39].

(c) Microbial Detoxification

This aspect includes the involvement of microorganisms to detoxify aflatoxins. On the basis of the types of microorganisms involved and the methods through which they detoxify aflatoxins, microbial detoxification approaches can be grouped into the following three categories:

(i) Aflatoxin detoxification by the toxigenic fungus itself.
(ii) Aflatoxin detoxification by the atoxigenic strains of the same fungal species.
(iii) Aflatoxin detoxification by other atoxigenic microorganisms.

(i) Aflatoxin detoxification by the toxigenic fungus itself

Aflatoxin-producing fungi, such as *A. flavus* and *A. parasiticus,* are ubiquitous in nature and contaminate most foods and feeds, rendering them unsuitable for consumption (by producing aflatoxins) and posing potential threat to human and animal health. It would, indeed, be worthwhile to get these toxins detoxified by the producer organism itself by altering growth conditions, including change in pH and temperature, so as to enable it to detoxify its own toxin under such manipulated conditions. Studies have been carried out to demonstrate the ability of *A. flavus* and *A. parasiticus* (both toxigenic) to degrade aflatoxins produced by themselves [6,36,37,40-45]. How this aflatoxin biotransformation is influenced by cultural conditions, such as growth substrates, age of culture, aeration, pH, and temperature, will be considered here briefly.

Effect of growth substrates, age of culture, and aeration on aflatoxin detoxification

The growth substrates (either liquid or solid), used to produce mycelia, can affect the ability of the producer strains of fungi to degrade aflatoxins [42,44]. Marth and Doyle [44] have demonstrated that 9-day-old mycelia of *A. parasiticus* NRRL 2999, grown in potato dextrose, Czapek-Dox, and YES broths, exhibited little or no degradation of aflatoxin B_1, while mycelia grown in glucose salt, Y-M and Moyer's broths were able to show increased levels of aflatoxin detoxification [44]. Shih and Marth [46] reported that aflatoxins B_1, B_2, G_1, and G_2 were degraded by 8- and 16-day-old, but not by 4-day-old mycelium of *A. parasiticus*. The blended mycelia from aerated cultures, grown for 8 to 10 days, have been shown to degrade AFB_1 maximally, while 12-day-old mycelium from the cultures of *A. parasiticus* failed to degrade an appreciable amount of aflatoxin, suggesting thereby that some intracellular substance(s) is/are responsible for aflatoxin degradation/detoxification [40,44]. Hamid and Smith [36] have observed that the intact mycelium of *A. flavus* could detoxify aflatoxins more efficiently than the cell-free extracts. However, the degradative ability of the extracts prepared from the older mycelium was significantly higher than the extracts of younger mycelium, suggesting that the levels of aflatoxin detoxifying enzymes appear to develop within the toxigenic fungus by an obscure mechanism.

Effect of pH and temperature on aflatoxin detoxification

Biologically optimal pH and temperature have been quite instrumental in regulating the secondary metabolism in microorganisms. Doyle and Marth [37] observed that the 9000xg supernatant fraction of 9-day-old mycelium of *A. parasiticus* NRRL 2999 was able to detoxify aflatoxins B_1 and G_1 at pH 5.0 and 6.5. There was some aflatoxin degradation at pH 4.0, but essentially no degradation at pH 2.0 and 3.0. The fact, that such detoxification at pH 5.0 and 6.5 was only biological, was confirmed by the finding that little or no chemical degradation of aflatoxins occurs at pH 5.0 and 6.5 [37,44].

The effect of temperature on detoxification of AFB_1 and AFG_1 was studied by Doyle and March [37] in *A. parasiticus*. They have shown that 9-day-old mycelia caused maximum aflatoxin degradation at 28°C with intermediate rates at 19°C and 36°C, and little degradation at 45°C. In contrast, Faraj [47] has shown that, when toxigenic *A. flavus* was grown at 30°C in either solid or liquid cultures, there was extensive aflatoxin synthesis, followed by the onset of aflatoxin degradation after approximately 5 days. Further incubation at 30°C gave continued aflatoxin detoxification. However, if cultures were transferred to high temperatures viz 40°, 45°, 50°C, there were much increased rates of detoxification. Similar observations have been made by Singh and Smith [45] using the same toxigenic strain of *A. flavus*.

(ii) Aflatoxin detoxification by the atoxigenic strains of the same fungal species

Doyle and March [41] and Marth and Doyle [44] have pioneered such studies by observing aflatoxin detoxification in *A. parasiticus* strains (toxigenic strain NRRL 2999 and atoxigenic strain NRRL 3315). They have compared the two strains and reported that the atoxigenic strain was less efficient in detoxifying the aflatoxins. An atoxigenic strain of *A. flavus* has been found to be most effective in detoxifying the aflatoxins produced by the toxigenic *A. flavus* and *A. parasiticus* strains under co-culture conditions [48]. Nakazato et al. [48] have demonstrated that aflatoxin B_1, produced by the toxigenic *A. flavus* and *A. parasiticus,* is transformed or metabolized by all strains of non-producing *A. flavus*; AF-A and AF-B were the common metabolites. Also, similar studies have already been carried out by Cotty [49] in cotton seeds. Not only that, the aflatoxin produced in *A. flavus* contaminated maize was degraded by the atoxigenic strain of *A. flavus* in solid-state fermentation [50]. This could be a very useful method of biological control of both pre- and post-harvest aflatoxin contamination in agricultural produce, particularly the grains.

(iii) Aflatoxin detoxification by other atoxigenic microorganisms

This approach to aflatoxin detoxification involves the use of microorganisms, other than toxigenic aspergilli, which do not themselves produce aflatoxins but apparently can metabolise them to less toxic or non-toxic molecules. Reduction of most hepatotoxic, hepatocarcinogenic and mutagenic aflatoxin (AFB_1) to a less toxic product, i.e., aflatoxin R_0 (aflatoxicol) has been observed in many organisms [51-53]. Biotransformation of AFB_1 to as yet uncharacterized compounds has also been reported with bacteria, including *Corynebacterium rubrum* and *Lactobacillus* spp., with the fungi - *A. niger, Trichoderma viride, Mucor ambiguos, M. alterans, Helminthosporium sativum, Rhizopus arrhizus, R. oryzae* and *R. stolonifer,* and the protozoan - *Tetrahymena pyriformis* [15,51,54-57]. However, the rate of conversion of AFB_1 to AFR_0 in *Dactylium dendroides, Absidia repens,* and *Mucor grisseo-cyanus* was very slow, taking 3-4 days to achieve only 60% reduction of AFB_1 to AFR_0 [54]. Bol and Smith [58] studied detoxification of aflatoxin B_1 by food grade *Rhizopus* strains. They observed that 87% of *Rhizopus* strains tested were positive to AFB_1 defluorescence on agar media. The isolates from contaminated feedstuffs showed diminished (7.5%) defluorescence capacity. Out of 29 strains tested, 18 were able to eliminate 50-100% aflatoxin B_1 after 5 days incubation at 25°C.

Aflatoxin G_1 has also been found to be degraded by various *Rhizopus* spp. to an intermediate metabolite previously reported in *A. flavus* as aflatoxin B_3 and in *A. parasiticus* as *parasiticol*. Ciegler et al. [59] screened over 100 microorganisms for their ability to either degrade or transform AFB_1 and found that *Flavobacterium aurantiacum* was most effective. *F. aurantiacum* has further been shown to remove aflatoxins B_1 and M_1 from liquid [52,53]. Knol et al. [60] have demonstrated that AFB_1 could be eliminated from peanut meal by *R. oryzae* NRRL 395 in a solid-substrate process, which has been quite effective with major decreases (from 260 to 70 μg/kg) of AFB_1 content in this raw material.

According to Shantha et al. [61], various strains of *A. niger* were able to decrease the levels of aflatoxin concentrations by about 64-99%, when they were grown with the toxigenic *A. flavus* in liquid cultures. The inhibition of AFB_1-induced hepatocarcinogenesis by the *Rhizopus delemer* has been subject matter of the studies of Zhu et al. [12], and hence suggests the effective biodetoxification measure of aflatoxins. *A. niger* has also been shown to be most effective in reducing the levels of respective mycotoxins produced by *A. ochraceus* and *A. flavus* [62]. As well as, *A. oryzae* has been shown to reduce the production of aflatoxins in mixed cultures of this organism and the aflatoxigenic *A. flavus* [63]. Similarly, Choudhary [64] has demonstrated the inhibitory effects of *Fusarium moniliforme, Trichoderma viride,* and *Rhizopus nigricans* on aflatoxin production, when these microbes were co-cultivated with aflatoxin-producing fungus, *A. flavus*.

The detoxification of aflatoxins in solid substrates is very important for both commercial and health reasons. Cuero et al. [65] have demonstrated that there was about 50% reduction in aflatoxin concentration, when *A. flavus* was grown with *A. niger* on maize samples. The decrease in total aflatoxin level was about 40%, 70%, and 75%, when this toxigenic fungus was cultured on maize with *Fusarium graminiarum, A. oryzae* and *Penicillium viridicatum*, respectively. Also, Cuero et al. [66] have implicated chitosan as well as the microbial agents, such as *Bacillus subtilis* and *Trichoderma harzianum* in biological control of aflatoxins in pre-harvest maize. Barrios-Gonzalez et al. [67] have shown, while evaluating the risk of aflatoxin contamination of cassava protein enrichment process with *A. niger*, that the toxigenic *A. parasiticus* can grow and produce aflatoxins under favourable environmental conditions such as suitable temperature, moisture content and nutrition. It was noticed that in mixed cultures, using *A. niger* and differrent amounts of *A. parasiticus*, the operation temperature of protein enrichment process (35°C) drastically reduces the toxin production. Although nitrogen and phosphorus concentrations in the medium were partially inhibitory to aflatoxin biosynthesis, very high reduction could be attained. The best toxicological protection was by the atoxigenic strain itself (*A. niger* No.10). The aflatoxin production was completely inhibited when these two species (*A. niger* and *A. parasiticus*) grew together in solid-state fermentation, thereby suggesting that the microorganisms, other than the toxigenic ones, can detoxify aflatoxins in the consumables (foods and feeds), and hence can present an effective measure of biological control of aflatoxins, thus eliminating the risk to human and animal health.

CONCLUDING REMARKS

Aflatoxins are the serious source of contamination of most foods and feeds, thereby causing potential threat to both human and animal health. There has been a tremendous amount of information available on physical and chemical methods of detoxification of these potentially toxic, carcinogenic and mutagenic secondary metabolites. However, biodetoxification of aflatoxins by the factors already present in the substrates (such as the genetic constitution of the food commodity) infected with the toxigenic fungi, by the toxigenic as well as atoxigenic aspergilli or by various other microorganisms, provide a very useful, novel and safe method for *biological control* of these toxins. More attention should be given to the improved methods of decontamination and detoxification of the contaminated agricultural produce under natural solid-substrate conditions.

ACKNOWLEDGEMENTS

The help rendered by Ms Deepika Mittal and Sandhya Singh as well as Mr Rathendra Raman, Sudhir K. Singh and Hemant K. Singh is gratefully acknowledged. Thanks are also due to Mr Lalit Kumar, Mr Krishan Lal, Mr Satish Kumar Sundan, and Mr S.K. Dass for technical help.

REFERENCES

1 Smith JE, Moss MO. Mycotoxins : Formation, Analysis and Significance. Chichester, New York, Brisbane, Toronto, Singapore: John Wiley & Sons, 1985; pp.148.
2 Cole RJ, Cox RH eds. Handbook of Toxic Fungal Metabolites. New York: Academic Press, 1981; pp.937.
3 Borroz KI, Ramsdell HS, Eaton DL. Toxicol Lett (Amst.) 1991; 58(1): 97-106.
4 Bryden WL, Cumming RB, Lloyd AB. Avian Pathol 1980; 9: 539-550.
5 CAST (Council for Agricultural Science and Technology) 1989; Task Force Report on Mycotoxins : Economic and Health Risk (No.116). Iowa: CAST.
6 Hynh VL, Gerdes RG, Lloyd AB. Aust J Biol Sci 1984; 37: 123-129.
7 Keyl AC, Booth AM. J Am Oil Chem Soc 1971; 48: 599-604.

8 Lynch GP, Covey FT, Smith DF, Weinland BT. J Animal Sci 1972; 35: 65-68.
9 Newberne PM, Butler WH. Cancer Res 1969; 29: 236-250.
10 Newberne PM, Carlton WW, Wogan GN. Pathol Vet 1964; 1: 105-132.
11 Tung TC, Ling KH. J Vitaminol 1968; 14: 48-52.
12 Zhu C, Min-Jie D, Dao-Nian L, Lue-Queen W. Mater Med Pol 1989; 21(2): 87-91.
13 Bilgrami KS, Sinha KK. In : Mukerji KG, Pathak NC, Singh VP, eds. Frontiers in Applied Microbiology. Lucknow: Print House India, 1985; 1: 349-361.
14 Kurtzman CP, Horn BW, Hesseltine CW. Antonie van Leeuwenhoek 1987; 53: 147-158.
15 Singh VP, Mukerji KG. In : Mukerji KG, Singh VP, eds. Concepts in Applied Mirobiology and Biotechnology. New Delhi: Aditya Books Pvt Ltd, 1994; (in press).
16 Wong JJ, Hsieh DPH. Proc Natl Acad Sci USA 1976; 73: 2241-2244.
17 IARC (International Agency for Research on Cancer). IARC Monographs on the Evaluation of Carcinogenic Risk of Chemicals to Man. Lyon: International Agency for Research on Cancer, 1976; 10: 51-72.
18 Purchase IFH ed. 1974. Mycotoxins. Amsterdam: Elsevier.
19 Goldblatt LA, Dollear FG. Pure Appl Chem 1977; 49: 1759-1764.
20 Shrivastava AK, Ranjan KS, Ansari AA. J Food Sci Technol 1991; 28(3): 189-190.
21 Samarajeewa U, Sen AC, Fernando SV, Ahmed EH, Wei CI. Food Chem Toxicol 1991; 29(1): 41-48.
22 Jorgensen KV, Park DL, Rua SN Jr, Price RL. J Food prot 1990; 53(9): 777-778.
23 Abdel-Rahim EA, Naguib KM, Badawi MM, Ibrahim MKK, Guergues SN. Grasas Aceites 1990; 41(2): 144-148.
24 Mahalingam RJ, Govindan S, Punniamurthy N, Balachandran C. Indian Vet J 1990; 67(2): 149-151.
25 Altug T, Yousef AE, Marth EH. J Food Prot 1990; 53(7): 581-582.
26 Howes AW, Dutton MF, Chuturgoon AA. Mycopathol 1991; 113(1): 25-29.
27 Zeringue HJ Jr, Bhatnagar D. J Am Oil Chem Soc 1990; 67(4): 215-216.
28 Costa JL, Da S, Kushalappa AC. Summa Phytopathol 1989; 15(2): 156-162.
29 Azaizeh HA, Pettit RE, Saar BA, Phillips TD. Mycopathol 1990; 110(3): 125-132.
30 Wallin JR, Windstrom NW, Fortnum BA. J Sci Food Agric 1991; 54(2): 235-238.
31 Song D. Acta Microbiol Sin 1991; 31(3): 169-175.

32 Rasic JL, Skrinjar M, Markov S. Mycopathol 1991; 113(2): 117-119.
33 Patterson DSP. Food Cosmet Toxicol 1973; 11: 287-294.
34. Raney KD, Shimada T, Kim D, Groopman JD, Harris TM, Guengerich FP. Chem Res Toxicol 1992; 5(2): 202-210.
35 Daniels JM, Lui L, Stewart RK, Massey TE. Carcinogenesis (Lond.) 1990; 11(5); 823-828.
36 Hamid AB, Smith JE. J Gen Microbiol 1987; 113: 2023-2029.
37 Doyle MP, Marth EH. Eur J Appl Microbiol Biotechnol 1978; 6: 95-100.
38 Käpeli O. Microbiol Rev 1986; 50: 244-258.
39 Ruckpaul K, Rein H, Blanck J. In : Ruckpaul K, Rein H, eds. Frontiers in Biotransformations (Vol.1) : Basic Mechanisms of Regulation of Cytochrome P-450. London, New York, Philadelphia: Taylor & Francis, 1989; 1-65.
40 Doyle MP, Marth EH. J Food Prot 1978; 41: 549-555.
41 Doyle MP, Marth EH. Mycopathol 1978; 63: 145-153.
42 Doyle MP, Marth EH. Mycopathol 1978; 64: 59-62.
43 Doyle MP, Marth EH. Eur J Appl Microbiol Biotechnol 1979; 7: 211-217.
44 Marth EH, Doyle MP. Food Technol 1979; 33: 81-87.
45 Singh VP, Smith JE. Biotechnological implications of high temperature metabolism in microorganisms. In: Proc Summer Conf Soc Appl Bacteriol, Bristol (UK), 1991; 50.
46 Shih CN, Marth EH. Z Lebensm Unters-Forsch 1975; 158; 361-362.
47 Faraj MK. Regulation of Mycotoxin Formation in *Zea mays*. Ph.D. Thesis, University of Strathclyde, Glasgow (UK), 1990.
48 Nakazato M, Morozumi S, Saito K, Fujinuma K, Nishima T, Kasai N. Risei Kagaku 1991; 37(2): 107-116.
49 Cotty PJ. Plant Dis 1990; 74(3): 233-235.
50 Brown RL, Cotty PJ, Cleveland TE. J Food Prot 1991; 54(8): 623-626.
51 Cole RJ, Kirksey JW, Moore JH, Blankenship BP, Diener UL, Davis ND. Appl Microbiol 1972; 24: 248-256.
52 Lillehoj EB, Ciegler A, Hall HH. Can J Microbiol 1967; 13: 624-627.
53 Lillehoj EB, Stubblefield RD, Shamone GM, Shotwell OL. Mycopathol Mycol Appl 1971; 45: 259-264.
54 Detroy RW, Hesseltine CW. Nature 1968; 219: 967.
55 Detroy RW, Hesseltine CW. Can J Biochem 1970; 48: 830-832.
56 Mann R, Rehm HJ. Eur J Appl Microbiol Biotechnol 1976; 2: 297-306.
57 Robertson JA, Teunisson DJ, Boudreaux GJ. J Agric Food Chem 1970; 18: 1090-1091.

58 Bol J, Smith JE. Food Biotechnol 1989; 3: 127-144.
59 Ciegler A, Lillehoj EB, Peterson RE, Hall HH. Appl Microbiol 1966; 14: 934-939.
60 Knol W, Bol J, Huis In T, Veld JHJ In : Zeuthen P, Cheftel JC, Erikson C, Gormley TR, Liko P, Paulus K, eds. Processing and Quality of Food. London, New York: Elsevier Applied Science, 1990; 2: 2.133-2.136.
61 Shantha T, Rati ER, Bhawani Shankar TN. Antonie van Leeuwenhoek 1990; 58(2): 121-128.
62 Paster N, Pushinsky A, Menasherov M, Chet I. J Sci Food Agric 1992; 58(4): 584-592.
63 Sardjono RK, Sudarmadji S. Asian Food J 1992; 7(1): 30-33.
64 Choudhary AK. Lett Appl Microbiol 1992; 14(4): 143-147.
65 Cuero R, Smith JE, Lacey J. J Food Prot 1988; 51: 452-456.
66 Cuero RG, Duffus E, Osuji G, Pettit R. J Agric Sci 1991; 117(2): 165-170.
67 Barrios-Gonzalez J, Rodriguez GM, Tomacini A. J Ferment Bioeng 1990; 70(5): 329-333.

Biotransformations: Microbial Degradation of Health Risk Compounds
Ved Pal Singh, editor
© 1995 Elsevier Science B.V. All rights reserved.

Metabolism and cometabolism of halogenated C-1 and C-2 hydrocarbons

Mukesh K. Jain[a] and Craig S. Criddle[b]

[a]Department of Civil and Environmental Engineering, Michigan State University, East Lansing, MI 48824, U.S.A.

[b]National Science Foundation Center for Microbial Ecology, Michigan State University, East Lansing, MI 48824, U.S.A.

INTRODUCTION

Chlorinated hydrocarbons, containing one or two carbon atoms, constitute a significant fraction of the hazardous substances from industrial, domestic, and agricultural sources. In part, this is due to their high levels of production. Over five million tonnes of 1,2-dichloroethylene (1,2-DCE) are produced annually for use as a solvent and chemical intermediate [1]. Vinyl chloride (VC) is also produced in large amounts (over three million tonnes annually) for the manufacture of polyvinyl chloride [1]. The solvents tetrachloroethylene (PCE), trichloroethylene (TCE), 1,1,1-trichloroethane (TCA), 1,1-dichloroethylene (1,1-DCE), 1,2-dichloroethane (1,2-DCA), and carbon tetrachloride (CT) have a combined annual production of over 6 million tonnes [1]. Since 1970, annual U.S. production of dichloromethane (DCM) has ranged from 212 to 286 million kg, with the principal application being paint removal [2].

One- and two- carbon halogenated compounds tend to be mobile and persistent in soils and ground waters [3]. Among the most commonly detected ground water contaminants are PCE, TCE, TCA, 1,1-DCE, 1,2-DCA, CT, chloroform (CF) and certain chlorofluorocarbons (CFCs). Many of these chemicals are classified as *priority pollutants* by the United States Environmental Protection Agency (USEPA), and are known or suspected *carcinogens* or *mutagens*. Some have potential for ozone depletion. Release to the environment is caused by inadequate disposal techniques, accidents, deliberate agricultural applications, or chlorination of water and wastewater.

Removal of C-1 and C-2 chlorinated aliphatics from water by physicochemical processes such as carbon adsorption or air stripping transfers contaminants from the aqueous phase to a solid or gaseous phase. In contrast, biological processes can destroy contaminants. To be effective, however, conditions that favour growth of a transforming population must be created (*biostimulation*) or the transforming organisms must be added (*bioaugmentation*). In addition, care must be taken to prevent or minimize transformations that yield intermediates or byproducts that are hazardous. In this article, we summarize research on C-1 and C-2 *haloaliphatic compounds* with a focus on the agents of transformation (in the absence of light), growth kinetics, transformation kinetics, and pathways of transformation.

REACTION TYPES

Reactions affecting the environmental fate of halogenated one- and two-carbon compounds can be broadly classified as substitutions, dehydrohalogenations, oxidations, and reductions [4]. These reactions can be either abiotic or biotic. Dehydrohalogenations are typically abiotic, while oxidations in dark environments are mostly biotic. Substitutions and reductions can be either biotic or abiotic. With some notable exceptions, abiotic transformations tend to be slow. Biotic transformations can be rapid when the microorganisms, that synthesize reactive enzymes or cofactors, are present in sufficient numbers.

Several of the less halogenated aliphatic compounds (dichloromethane, 1,2-dichloroethane, etc.) are good electron donors and can serve as growth substrates. Usually, these compounds are susceptible to initial attack by oxidation or hydrolysis. The products of these reactions are typically alcohols or acids, that can be further oxidized by the transforming population to give carbon and energy for growth and maintenance. There is also some evidence that certain highly halogenated aliphatic compounds (such as tetrachloroethylene) can serve as the terminal electron acceptors for growth for some organisms [5]. Usually, though, the highly halogenated aliphatics do not support growth and are transformed only by cometabolism. Cometabolism is defined here as the transformation of a nongrowth substrate by growing cells in the presence of growth substrate, by resting cells in the absence of growth substrate, or by resting cells in the presence of energy substrate [6]. A growth substrate is defined as an electron donor, that supports growth. An energy substrate is defined here as an electron donor, that provides reducing power and energy for the transforming population, but does not, by itself, support growth. Cometabolism results from the lack of specificity of enzymes and cofactors. The products of cometabolic reactions accumulate in pure cultures, but, in a mixed culture, they are typically used by other microorganisms. As a result, cometabolic transformations are key initiatory reactions in pathways, that ultimately result in the complete degradation of many hazardous chemicals [7].

The first known examples of cometabolism were all oxidations, and, as a result, the term "*co-oxidation*" was used to describe them. Subsequently, reductive transformations were discovered, that did not facilitate growth of the transforming organisms, and depend upon the concurrent or previous utilization of a growth or energy substrate. These *"co-reductions"* led to the use of the broader term, cometabolism. It now appears that certain cometabolic reactions are also hydrolytic. Thus, in addition to the well-known examples of co-oxidation, we now recognize the potential for "co-reductions" and *"co-hydrolyses"*. All of the known co-oxidations occur only under obligate aerobic conditions, while most co-reductions occur under anaerobic conditions.

Substitution

The type reaction for substitutions is:

$$RX + N^{-} \text{---->} RNu + X^{-} \qquad (1)$$

In the above reaction, RX is an alkyl halide and Nu is a nucleophile. The most important nucleophile is water. Reactions with water result in replacement of a halogen by -OH (hydrolysis). Halogenated aliphatic compounds undergo hydrolysis in the absence of inorganic or biochemical catalysts. Abiotic hydrolysis reactions are bimolecular, with water as the dominant nucleophile, but because water is present at high concentrations, pseudo-first-order kinetics are observed. Many hydrolysis reactions are potentially faster at higher pH, where the hydroxide ion acts as the nucleophile. However, below pH 11, a pH dependence for substitution reactions is generally not observed [8,9]. The nature of the halogen substituents and the degree of halogenation influences substitution rate. Increased halogenation leads to slower substitution reactions and longer half lives [8,9].

In general, abiotic substitution reactions proceed slowly, but can be greatly accelerated by enzymes. Enzyme-mediated substitutions frequently involve cysteine residues in proteins or peptides, such as glutathione. Biotic and abiotic hydrolysis of halogenated aliphatic compounds yields alcohols by hydroxyl substitution at the halogenated carbon [10]. If these alcohols are themselves halogenated, further hydrolysis to acids or diols can occur. Examples of microbially-mediated hydrolysis reactions, together with responsible enzymes, are provided in Table 1.

Dehydrohalogenations

The type reaction for dehydrohalogenation is:

$$-\underset{H}{\overset{|}{C}}-\underset{X}{\overset{|}{C}}- \longrightarrow \;>C=C< \; + \; HX \qquad (2)$$

Table 1
Examples of hydrolysis and co-hydrolysis

Microorganism	Growth substrate/ nongrowth substrate	Responsible enzyme(s)	Refs
Hydrolysis - metabolism			
Pseudomonas DM1 and DM2	dichloromethane	halidohydrolase	[104]
Hyphomicrobium DM2	dichloromethane dibromomethane	glutathione+ glutathione-S-transferase	[97]
Xanthobacter autotrophicus GJ10 *Ancylobacter aquaticus* AD20 and AD25	1,2-dichloroethane	haloalkane dehalogenase and haloacid dehalogenase	[75,76] [74]
Methylotrophic bacterium sp. strain DM4	dichloromethane dibromomethane	DCM dehalogenase group A	[105]
Arthrobacter sp. strain HA1	1-haloalkanes	dehalogenase	[106]
Methylotrophic bacterium strain DM11	dichloromethane dibromomethane	DCM dehalogenase group B	[98]
Co-hydrolysis			
Clostridium TCAIIB	amino acids/ 1,1,1-trichloroethane	unknown enzymes	[71]
Pseudomonas sp. strain KC	acetate, glycerol/ carbon tetrachloride	unknown iron scavenging agent	[82,83]

For halogenated aliphatics, dehydrohalogenations are abiotic. Polychlorinated alkanes undergo dehydrohalogenation under extreme basic conditions, and at pH 7 [11]. These reactions generally follow bimolecular kinetics, depending on hydroxide ion concentration. At neutral pH conditions, dehydrohalogenation by weaker bases (e.g., water) might be important. The number and kind of halogen substituents have a strong influence on dehydrohalogenation rates. Increased halogenation tends to decrease substitution reaction rates (hydrolysis) and increase dehydrohalogenation rates. Consequently, highly halogenated C-2 alkanes are susceptible to dehydrohalogenation, except, of course, those that are fully halogenated. A few compounds undergo simultaneous dehydrohalogenation and hydrolysis [12]. A well-known example is 1,1,1-TCA, which undergoes simultaneous hydrolysis to acetate and dehydrohalogenation to 1,1-dichloroethylene [13].

Oxidations

Type reactions for aerobic oxidations are:

$$\text{(a)}\quad O_2 + 2H^+ + 2e^- + \;-\overset{|}{\underset{|}{C}}-H \xrightarrow{\text{monooxygenase}} -\overset{|}{\underset{|}{C}}-OH \quad + H_2O \qquad (3)$$

$$\text{(b)}\quad O_2 + 2H^+ + 2e^- + \;{>}C{=}C{<} \xrightarrow{\text{monooxygenase}} {>}\overset{O}{C{-}C}{<} \quad + H_2O$$

$$\text{(c)}\quad O_2 + 2H^+ + 2e^- + \;{>}C{=}C{<} \xrightarrow{\text{dioxygenase}} -\overset{HO}{\underset{|}{C}}-\overset{OH}{\underset{|}{C}}-$$

Aerobic oxidations rely upon the catalytic activity of nonspecific monooxygenases or dioxygenases. As indicated in the type reaction sequences, these enzymes require both reducing power and molecular oxygen. They are widely distributed in nature in many microbial populations, including methanotrophs, nitrifiers, numerous hydrocarbon degraders, and they are even found in higher organisms, including man. Frequently, oxygenase reaction products are not useful to the transforming organisms, so many oxidative transformations are cometabolic. Several examples of co-oxidizing bacteria are provided in Table 2.

Oxygenases catalyze the incorporation of oxygen, derived from molecular oxygen, into the halogenated molecule. As shown in reaction 3a, oxygen may be inserted into the carbon-hydrogen bond creating halogenated alcohols, that spontaneously eliminate HX to give an aldehyde [14,15]. As shown in reaction 3b, oxygen may also be inserted into carbon-carbon double bonds yielding an epoxide [14]. Halogenated aldehydes or acyl chlorides are common intermediates, and are typically oxidized or hydrolyzed to acids or reduced to alcohols [16-18]. Halogenated molecules, that are oxidized include hydrogen-containing alkanes and alkenes. Completely halogenated alkanes and alkenes are resistant to oxidation by oxygenases.

Table 2
Examples of co-oxidation of alkyl halides

Microorganisms	Growth substrate	Nongrowth substrate	Refs
Pseudomonas cepacia strain G4	phenol, toluene and *o*-cresol phenol	TCE	[53,54] [107,108]
Pseudomonas putida F1	toluene	TCE	[55,56]
Strain 46-1	toluene	TCE	[22,59]
Methylosinus trichosporium OB3b	methane methanol formate	TCE chloroalkanes except CT chloroalkenes except PCE	[20,109] [23] [62]
Methylocystis sp. strain M	methane	TCE chloroalkenes except PCE chloroalkanes except CT	[64] [63,110]
Mycobacterium vaccae JOB5	propane	chloroalkenes except PCE	[111]
Nitrosomonas europaea	ammonia	TCE chloroalkenes except PCE chloroalkanes except CT	[50,51, 112]
Xanthobacter strain Py 2	propylene	TCE	[113,114]
Genetically engineered *Escherichia coli*	toluene	TCE	[114]

Nonspecific oxygenases that figure prominently in the process of co-oxidation have great potential for degradation of halogenated aliphatic compounds. Among the most important are the methane monooxygenases (MMOs) found in methanotrophs. These enzymes endow the methanotrophs with the ability to oxidize virtually all of the halogenated aliphatic hydrocarbons, with the exception of those that are completely halogenated. All methanotrophs tested are able to form a particulate type MMO (pMMO) or membrane-bound enzyme, whereas some cultures grown under copper limitation are capable of producing a soluble type of MMO (sMMO), with a broader substrate range than pMMO [19,20]. When induced for sMMO, *Methylosinus trichosporium* OB3b co-oxidized all chlorinated aliphatic hydrocarbons (C1 to C3) except PCE and CT [20-23]. For methanotrophs, formate can serve as an energy substrate, increasing both the rate and extent of TCE cometabolism. Although methanotrophs obtain reducing power from the oxidation of formate to carbon dioxide, they are unable to assimilate formate, and they can not use it as a growth substrate.

Reductions

The type reactions for reductions are:

$$\text{(a)}\quad -\overset{|}{\underset{|}{C}}-X + e^- \longrightarrow -\overset{|}{\underset{|}{C}}\cdot + X^- \qquad (4)$$

$$\text{(b)}\quad -\overset{|}{\underset{|}{C}}\cdot + H^+ + e^- \longrightarrow -\overset{|}{\underset{|}{C}}-H \qquad \text{(hydrogenolysis)}$$

$$\text{(c)}\quad -\overset{|}{\underset{X}{C}}-\overset{|}{\underset{\cdot}{C}}- + e^- \longrightarrow {>}C{=}C{<} + X^- \qquad \text{(di-halo elimination)}$$

$$\text{(d)}\quad {>}C{=}C{<}^{X} + H^+ + 2e^- \longrightarrow {>}C{=}C{<} + X^-$$

Table 3 lists pure cultures, capable of co-reducing halogenated aliphatic compounds. Frequently, these organisms possess transition metal complexes that react with the alkyl halides. Among the most important of these complexes are the cytochromes, corrinoids, factor F430, and vitamin B_{12} [24].

Reductions by transition metal complexes are typically initiated by the transfer of a single electron, loss of a single halide substituent, and the formation of a free radical (reaction 4a). Formation of the free radical is the first and, in most cases, the rate-limiting step in the reduction of halogenated aliphatic compounds. The free radical can undergo a range of reactions, depending upon the nature of the radical and its environment. If the radical abstracts hydrogen from water or

Table 3
Examples of co-reduction of alkyl halides

Microorganism	Growth substrate	Nongrowth substrates	Refs
Clostridium sp. TCAIIB	amino acids	1,1,1-trichloroethane carbon tetrachloride chloroform	[71]
DCB-1	chlorobenzoate	tetrachloroethane	[38]
Methanobacterium thermoautotrophicum	acetate, methanol	1,2-dichloroethane	[37,77]
Methanosarcina mazei *Methanosarcina* sp. strain DCM	acetate, methanol	tetrachloroethane trichloroethane	[36,39]
Methanosarcina sp. strain DCM	methanol	chloroform	[92]
Methanosarcina barkeri	H_2-CO_2	1,2-dichloroethane chloroethane	[78]
Desulfobacterium autotrophicum	lactate	carbon tetrachloride 1,1,1-trichloroethane	[37]
Acetobacterium woodii	fructose	carbon tetrachloride	[81]
Escherichia coli k12 (fermenting)	glucose	carbon tetrachloride	[79]
Escherichia coli k12 (fumarate-respiring)	glycerol	carbon tetrachloride	[79]

from a surrounding organic, the product is similar to the parent compound, with a halogen replacing one of the hydrogen substituents (*hydrogenolysis*-reaction 4b). If the carbon radical is located adjacent to a halogenated carbon, then a second halide can be lost with formation of a double bond (di-halo elimination - reaction 4c). In general, the greater the degree of halogen substitution, the more oxidized a molecule becomes, and the more susceptible it is to reduction by biotic or abiotic electron donors [4].

Alkyl halide mixtures are susceptible to reductive transformation under anaerobic conditions [3,4,25-28]. In particular, Bouwer and McCarty [3,29] and Vogel et al. [4] reported that mixtures of 1,1,1-TCA; 1,2-DCA;

TCE; 1,2 dibromoethane, and PCE are transformed in methanogenic environments. Anaerobic habitats, supporting reductive transformation, include continuous flow methanogenic fixed-film reactors [3,4,30,31], organic sediment-muck [24,27,32-34], anaerobic aquifer microcosms [28], anaerobic sediment from the Rhine river [35], enrichments [26], and pure cultures [36-39].

KINETICS OF GROWTH

If the concentrations of all but one of the substances needed for bacterial growth are present at levels, that exceed growth requirements, then the limiting substrate is termed as the growth-limiting substrate, or simply, the growth substrate. For growth substrates, the rate of substrate utilization is a function of the growth rate of the microorganisms. A widely used model of bacterial growth and decay is the Monod expression as modified by Herbert et al. [40]:

$$\mu = \frac{dX/dt}{X} = Yq_g - b = \frac{YkS}{K_s + S} - b \qquad (5)$$

where:

μ = specific growth rate, day^{-1}
X = organism concentration, mg/l
t = time, days
Y = maximum organism yield, mg cell/mg substrate
q_g = specific rate of utilization of growth substrate, mg growth substrate/mg cell-d
b = decay coefficient, day^{-1}
k = maximum specific rate of substrate utilization, mg growth substrate/mg cell-d
S = growth substrate concentration, mg/l
K_s = half saturation coefficient for the growth substrate, mg growth substrate/l

When S is zero, microorganisms undergo decay, and biomass concentration decreases. A concentration of interest is the concentration of growth substrate at which growth and decay are equal (S_{min}), which is found from equation 1 by letting μ = 0 [41]:

$$S_{min} = K_s \frac{b}{Yk - b} \qquad (6)$$

When the concentration of a growth substrate is above S_{min}, net growth will occur. Under such conditions, the substrate is termed a *primary substrate.* When the concentration is below S_{min}, then the substrate is termed a *secondary substrate.* If a primary substrate is available, or if the collective concentration of secondary substrates permits growth, then the concentration of a secondary substrate can be reduced to a value significantly below its S_{min}. Under such conditions, even low concentrations of a substrate, like those found in many ground waters, can be removed [7].

Values of k, K_s, and hence k' for secondary substrates may differ from those of the primary substrate. In some cases, k' of the secondary substrate is greater than k' of the primary substrate, and in other cases it is much less. When the k' values for the secondary substrate are much less than k' values for the primary substrate, then large amounts of primary substrate may be required to degrade the secondary substrate.

KINETICS OF TRANSFORMATION

Substrate transformation kinetics

Transformations of growth substrate by bacteria can generally be described using saturation kinetics [42]:

$$q_g = \frac{-dS/dT}{X} = \frac{kS}{K_s + S} \qquad (7)$$

Transformations of nongrowth substrate by resting cells can also be described using saturation kinetics:

$$q_c = \frac{-dC/dT}{X} = \frac{k_c C}{K_c + C} \qquad (8)$$

where:

q_c = specific rate of transformation of nongrowth substrate, mg substrate/mg cell-d

k_c = maximum specific rate of nongrowth substrate utilization, mg substrate/mg cell-d

C = concentration of nongrowth substrate, mg/l

K_c = half saturation coefficient for the nongrowth substrate, mg substrate/l.

If the substrate concentration is low ($S \ll K_s$, $C \ll K_c$), the specific transformation rate is directly proportional to the substrate concentration:

$$q_c = k' S \qquad (9)$$

where:
$k' = k/K_c$ = second order rate constant for growth substrate, l/mg-day, or

$$q_c = k'_c C \qquad (10)$$

where:
$k'_c = k_c/K_c$ = second order rate constant for nongrowth substrate, l/mg-day.

At high substrate concentrations ($S \gg K_s$, $C \gg K_c$), the specific rate of substrate transformation is independent of substrate concentration ($q_g = k$, $q_c = k_c$).

Inhibition

Halogenated aliphatic transformations can be complicated by substrate interactions and substrate toxicity. When competitive inhibition occurs, K_s or K_c in equations 7 or 8 can be modified, such that:

$$q_g = \frac{-dS/dT}{X} = \frac{kS}{K_s(1+\frac{I}{K_i}) + S} \qquad (11)$$

or

$$q_c = \frac{-dC/dT}{X} = \frac{k_s C}{K_s(1+\frac{I}{K_i}) + C} \qquad (12)$$

where:
I = concentration of competitive inhibitor, mg/l
K_i = inhibition coefficient of the competitive inhibitor, mg/l

When a growth substrate inhibits its own transformation at high concentrations, Haldane kinetics are often used:

$$q_g = \frac{-dS/dT}{X} = \frac{kS}{K_s + S + \frac{S^2}{K_i}} \qquad (13)$$

where:
K_i = inhibition coefficient of the growth substrate inhibitor, mg/l.

Stoichiometry of cometabolism

New models have recently emerged to describe aspects of cometabolism, that are not described by equations 5 and 8, including the loss of transformation activity with time due to endogenous decay and product toxicity, and the increased rates of cometabolism in the presence of growth substrate. A key concept in these models is the "*transformation capacity*" of cometabolizing cells. This concept was first introduced by Alvarez-Cohen and McCarty [43] who defined a *biomass* transformation capacity $Tc^b = dC/dX$. This definition of Tc^b is appropriate for a batch system with a nonvolatile substrate. More generally, it can be defined as:

$$Tc^b = \frac{-q_c}{\mu} \qquad (14)$$

where:
Tc^b = observed biomass transformation capacity, mg nongrowth substrate/mg cell.

To describe the observed enhancement in the rate of cometabolism in the presence of growth substrate, a modification of equation 8 was recently proposed [6]. This modification introduces a second transformation capacity term, the *growth substrate* transformation capacity Tc^g:

$$q_c = (Tc^g q_g + k_c)\frac{C}{K_c + C} \qquad (15)$$

where:
Tc^g = growth substrate transformation capacity, mg nongrowth substrate/mg growth substrate consumed during growth.

Finally, for cells that are growing, decaying, and simultaneously carrying out cometabolic reactions, equation 1 is an inadequate description of growth. A proposed modification [6] is:

$$\mu = Yq_g - b - \frac{q_c}{Tc^{b*}} \qquad (16)$$

where:
Tc^{b*} = biomass transformation capacity in the absence of endogenous decay, mg nongrowth substrate/mg cell.

Equation 16 indicates that an increase in the rate of cometabolism q_c causes a decrease in the specific growth rate of a cometabolizing population. While this appears to be true for pure cultures, it may not be true for mixed cultures, where organisms capable of using the products of cometabolism for growth are frequently present. Equations 7, 15, and 16 may provide the framework for a complete kinetic description of cometabolism by growing or resting cells, although further experimentation is needed to assess their range of utility [6].

In Tables 4-6, we have compiled kinetic coefficients, as defined by equations 5-14, from literature values for a wide range of experimental systems. Although we attempted to standardize units for purposes of comparison, differences in experimental protocol and methods of reporting, sometimes, made an accurate and complete comparison impossible. As yet, no data is available for the coefficients of equation 15 because this expressions is new, and it has so far only been evaluated using a different experimental system [44]. Clearly, broadly accepted unstructured models of cometabolism and standardized experimental protocols are needed to enable fair comparison of organisms from different sources, and to assist in the design and operation of engineered systems.

Table 4
Kinetics of growth substrate utilization for alkyl halides

Growth-substrate/ organism	Y (mg cell per mg growth substrate)	k (μmole growth substrate per mg cell-day)	K_s (μM)	K_i (μM)	b (day^{-1})	Refs
Oxidation reactions						
Phenol/ *P. cepecia*		671[a]	8.5	454[b]		[107]
Methane/ mixed culture methanotrophs	0.20	107.5[a]	12.5		0.12	[115]
Methane/ mixed culture methanotrophs	0.20	93.8[a]	13.8		0.12	[115]

Table 4 continued

Growth-substrate/ organism	Y (mg cell per mg growth substrate)	k (μmole growth substrate per mg cell-day)	K_s (μM)	K_i (μM)	b (day^{-1})	Refs
Methane/ mixed culture methanotrophs	0.35					[65]
Methane/ *Methylosinus trichosporium* OB3b		523	92	5.7 (TCE)		[109]
Methane/ mixed culture methanotrophs					0.14	[116]
Hydolysis reactions						
DCM/ Methylotroph						[98]
strain DM11	0.076[c]	815	51			
strain DM11	0.035[c]	720	20			
strain DM4	0.089[c]	222	30			
strain DM2						
DCM/ methanogenic enrichment	0.022[d]					[93]
1,2-DCA/ *Ancylobacter aquaticus*						[74]
strain AD20	0.17	167[e]	222			
strain AD25	0.16	149[e]	24			
Xanthobacter autotrophicus						
GJ10	0.24	105[e]	260			
GJ10MR0	0.28	52[e]	51			
GJ10MR1			81			

[a]μmoles/mg protein per day.
[b]From Haldane expression, equation 13.
[c]mg of protein/mg DCM.
[d]Calculated assuming biomass is 53% carbon.
[e]Estimated from k = μmax/Y.

Table 5
Kinetics of nongrowth substrate transformation: co-oxidation of alkyl halides

Growth substrate/ nongrowth substrate(s)	k_c (μ moles nongrowth substrate /mg cell-day)	K_c (μM)	k'_c (l/mg cell-day)	K_i (μM)	Tc[b] (mg nongrowth substrae/ mg cell as VS)	Refs
Toluene/TCE	2.6[a]		0.0325			[24,56, 117]
Toluene/TCE	2.88[a]					
Ammonia/TCE	1.58[a]					[50]
Phenol/TCE	11.5[a]	3.0				[107]
Phenol/TCE	0.29[a]					[118]
Methane[c]						
TCE	28.8					[23]
TCE	216		3.08			[20]
TCE				91.3		[115]
1,1,1-TCA				14.3		
TCE+formate	38.8	55.5			0.073	[2]
TCE-formate	4.6				0.036	
CF	792	34	23.3			[109]
t-1,2-DCE	475	148	3.2			
c-1,2-DCE	259	30	8.6			
TCE	418	145	2.9			
CF+formate	12.6	12.6			0.0015	[66]
CF-formate	2.9	34.3			0.0065	

Table 5 continued

Growth substrate/ nongrowth substrate(s)	k_c (μ moles nongrowth substrate /mg cell-day)	K_c (μM)	k'_c (l/mg cell-day)	K_i (μM)	Tc[b] (mg nongrowth substrae/ mg cell as VS)	Refs
TCE	6.4	12.6			0.042	[66]
CF	2.85	10.9			0.0083	
TCE+formate	18.2				0.017 to 0.028	[116]
TCE-formate	10.6					
TCE+formate	317	138				[17]
TCE				11.6		
TCE+CO				4.2		
TCE+formate			10.2			
TCE-formate			4.6			
TCE+formate	36.5	60	0.61		0.061	[43]
TCE-formate	6.4	5.2	1.23		0.043	
TCE	982[b]	35				[22]
TCA			0.002			[119]
TCE			0.009			
TCE			0.48 to 0.78[b]			[120]
TCE	0.045[a]					[121]

[a]μmoles/mg protein per day.
[b]Whittenbury mineral media.
[c]Data not available.

Table 6
Kinetics of nongrowth substrate transformation: co-reduction and co-hydrolysis of alkyl halides

Growth substrate/ nongrowth substrate	k_c (μmoles/mg cell-day)	K_c (μM)	K'_c (l/mg cell-day)	Tc[b] (mg nongrowth substrate/ mg cells)	Refs
3-Chlorobenzoate					
/PCE	0.00234[a]				[38]
Methanol/PCE	0.00084[a]				
Methanol/PCE	0.00048[a]				
Acetate/					
PCE	0.094				
1,1,1-TCA	0.96				[30]
CT	0.63				
CF	0.21				
Glycerol (fumarate respiring)/CT			0.0025		[79]
Glucose fermenting/CT			0.0041		
Acetate/					
1,1,1-TCA			0.013		
CF			0.02		[71]
CT			0.18		
Co-hydrolysis					
Amino acids/					[122]
1,1,1-TCA	0.28	31	0.009	0.41[b]	
Acetate/CT			0.362 to 9.07	>0.015	[83]

[a]μmole TCE/mg protein per day.
[b]Estimated as mg TCA/mg protein from Tc[b] = k_c/b.

PATHWAYS OF TRANSFORMATION

Pathway considerations are as important as kinetic considerations in assessing the prospects for *detoxification* of halogenated aliphatics. In the following sections, we summarize recent literature on the anaerobic and aerobic degradation of the most significant halogenated C-1 and C-2 contaminants.

Tetrachloroethylene(PCE), trichloroethylene(TCE), and vinyl chloride(VC)

Anaerobic

Reductive pathways of PCE and TCE are illustrated in Figure 1. Transformation of PCE under anaerobic conditions proceeds by sequential hydrogenolysis to TCE, DCE, and VC [3,25-27,34,45]. Several studies report that among DCE isomers, *cis*-1,2-DCE predominates over *trans*-1,2-DCE, while 1,1-DCE is the least significant intermediate [25,27,32, 34]. PCE can be mineralized to VC through reductive dechlorination under methanogenic conditions, and some of the VC produced can be transformed to CO_2 via alcohol and aldehyde [4]. PCE and TCE can also be reductively dechlorinated to VC and ethylene in mixed bacterial culture [5]. Further reduction of ethylene to ethane was observed in a continuous-flow, fixed-film column filled with Rhine-river sediment and granular sludge [35]. Dechlorination of PCE at high concentrations to ethene in the absence of methanogenesis implicates *acetogens* or fermenters as possible agents of transformation, and indicates that PCE may serve as an electron acceptor for growth [46]. PCE can be dechlorinated to DCE without methane formation in mixed anaerobic cultures by fermenting organisms with benzoate as the electron donor [47] and by sulphate-reducing enrichments [48].

PCE is transformed to TCE by pure cultures of methanogens, including *Methanobacterium thermoautotrophicum*, *Methanosarcina mazei* and *Methanosarcina* sp. strain DCM [36-39]. *Methanosarcina* sp. DCM was originally isolated from a chlorophenol-degrading consortium, and showed significantly higher PCE-dechlorination rates than *Methanosarcina mazei*. Methanol-fed cultures of strain DCM transformed PCE to TCE faster than acetate-fed cultures [38,39], and, at all methanol concentrations tested, a constant ratio of the TCE to CH_4 was obtained, suggesting that the electrons required for PCE dechlorination were derived from methanogenesis and that the extent of dechlorination could be enhanced by stimulating methanogenesis [36]. Methanogen pure culture studies have demonstrated only limited PCE degradation beyond TCE, indicating that mixed cultures may be required to achieve complete dechlorination of halogenated alkenes under methanogenic conditions. Mixed-culture studies demonstrated that complete

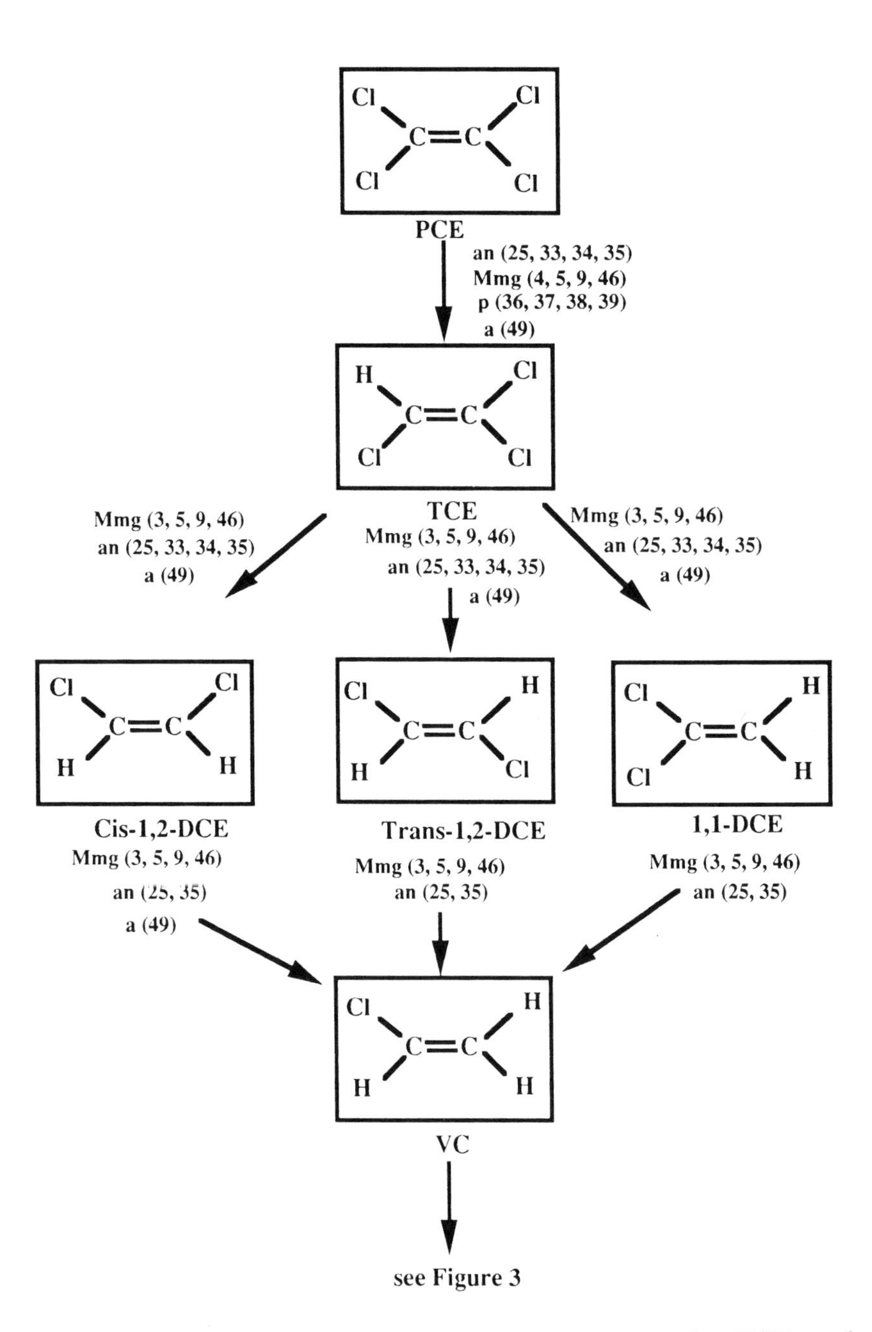

Figure 1. Anaerobic pathways of transformation for PCE and TCE (a=abiotic, an=anoxic, M=mixed culture, mg=methanogens, p=pure culture).

dechlorination of PCE and TCE to ethane is possible, although the final dechlorination step from vinyl chloride to ethene appears to be rate-limiting, with a significant level of vinyl chloride typically persisting. In a recent study [46], anaerobic methanol-PCE enrichment cultures were developed, which were capable of dechlorinating high concentrations of PCE to ethene.

The biochemistry underlying reductive dechlorination was recently examined in vitro using a series of transition-metal coenzymes and other biochemical reductants [49]. Vitamin B_{12} and coenzyme F_{430} sequentially dechlorinated PCE to ethylene, while hematin dechlorinated PCE to vinyl chloride.

Aerobic

As shown in Figure 2, many nonspecific oxygenases attack TCE, but none attack PCE. Among the monooxygenases and organisms, that attack TCE, are the methane monooxygenases (MMOs) of methanotrophs, ammonia monooxygenases of nitrifiers [50-52], propane monooxygenases, certain toluene dioxygenases, and certain phenol monooxygenases. Although several toluene-degrading cultures (*Pseudomonas* strain G4, *Pseudomonas putida* F1, *Pseudomonas putida* B5, and *Pseudomonas putida* PpF1) can transform TCE [53-56], the ability to degrade aromatic compounds does not always correlate with the ability to degrade TCE. The MMO system of the methanotrophs appears to be somewhat more consistent in its ability to degrade TCE, although the rates of oxidation vary substantially. Methanotrophs possess different forms of MMO: the membrane-associated or particulate form (pMMO) oxidizes TCE slowly and is induced at high copper concentrations; the cytoplasmic or soluble MMO (sMMO) oxidizes TCE 100 times faster than the particulate form, and is induced at low copper concentrations. Copper-limited *M. trichosporium* OB3b also oxidizes TCE about 100 times faster than toluene-grown *P. putida* F1 [20, 56].

Methanotrophs oxidize TCE to an epoxide, that spontaneously hydrolyzes, yielding glyoxylate and dichloroacetate under acidic conditions, and carbon monoxide and formate under basic conditions [57-60]. Glyoxylate and dichloroacetate are oxidized to CO_2 by heterotrophic bacteria. Formate and CO are oxidized by methanotrophs to CO_2. Formate is an energy substrate for methanotrophs, but CO inhibits TCE oxidation by exerting a demand for reducing power, and by competitively inhibiting MMO [61]. Trichloroacetaldehyde (chloral) and 2,2,2-trichloroethanol are also produced during the degradation of TCE by sMMO-induced *Methylosinus trichosporium* OB3b [20]. Chloral is apparently produced by chlorine migration during oxidation of TCE by the sMMO of *M. trichosporium* OB3b [22]. Chloral is toxic, and it may be responsible for the product toxicity observed during TCE transformation by methanotrophs. In addition, chloral can be reduced to trichloroethanol or oxidized to trichloroacetate [62]. Trichloroacetate degraded slowly in one methanotrophic mixed culture [63]. Formation of significant levels of trichloroacetate and epoxide degradation products by *Methylocystis* sp. strain M indicates that chlorine migration and epoxide formation proceed in parallel [64]. Degradative pathways for TCE, including percentage estimates for different intermediates, are provided in Figure 2.

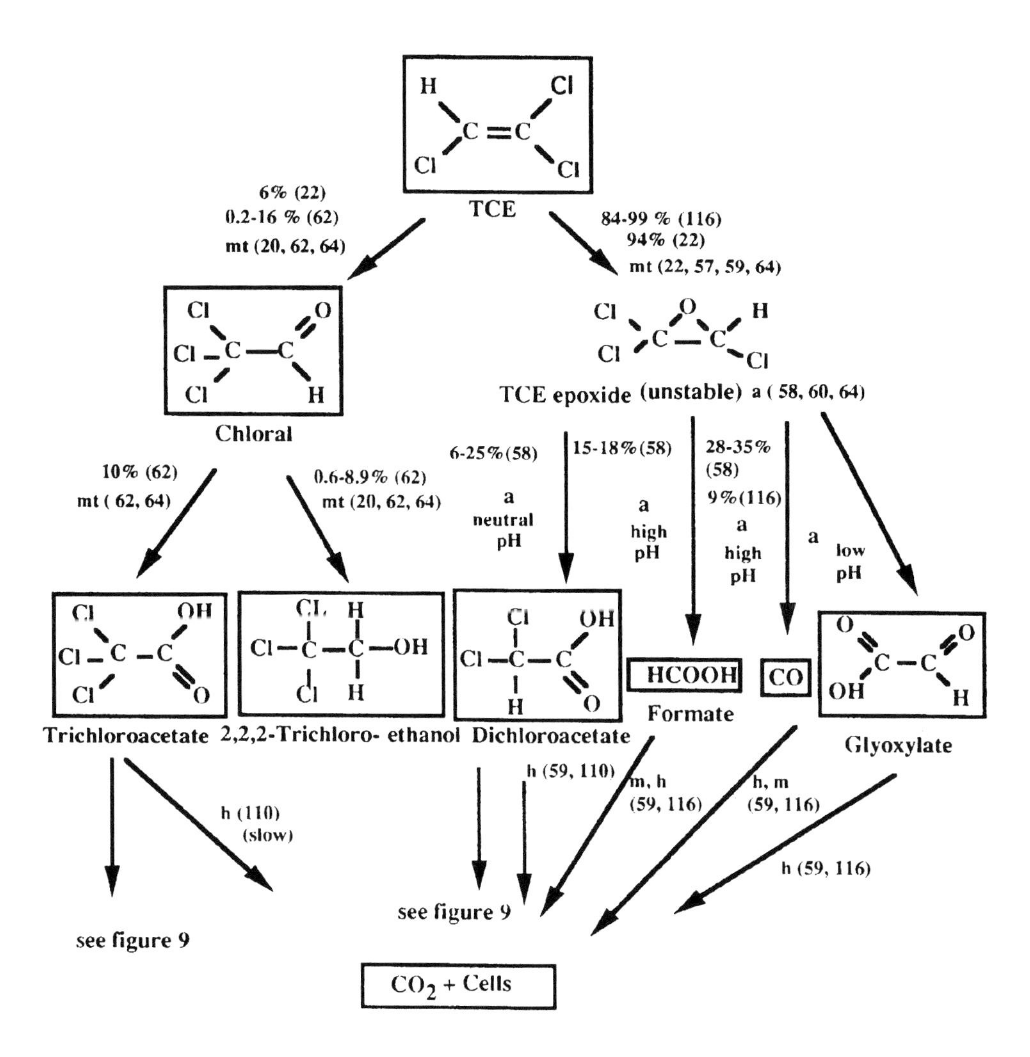

Figure 2. Pathways of transformation for TCE by methanotrophic mixed cultures (a=abiotic, m=mixed culture, mt=methanotrophs, h=heterotrophs).

TCE degradation was evaluated using resting cells of a methanotrophic mixed culture in the presence and absence of formate [43,61,65,66]. Formate is an energy substrate, so that while it does not support growth of methanotrophs, it can provide electrons for MMO transformations. Higher TCE oxidation rates were observed at high TCE concentrations, when formate was provided, but no increase was observed at low TCE concentrations. During methane starvation, storage polymers, found in some methanotrophs, served as endogenous sources of reducing power, and their disappearance correlated with consumption of TCE [65].

As shown in Figure 3, vinyl chloride is broken down in a reaction sequence similar to that of TCE. Vinyl chloride can be oxidized by methane monooxygenase to chloroethylene epoxide, which has a half-life of 1 min. This epoxide rearranges to chloroacetaldehyde and hydrolyzes

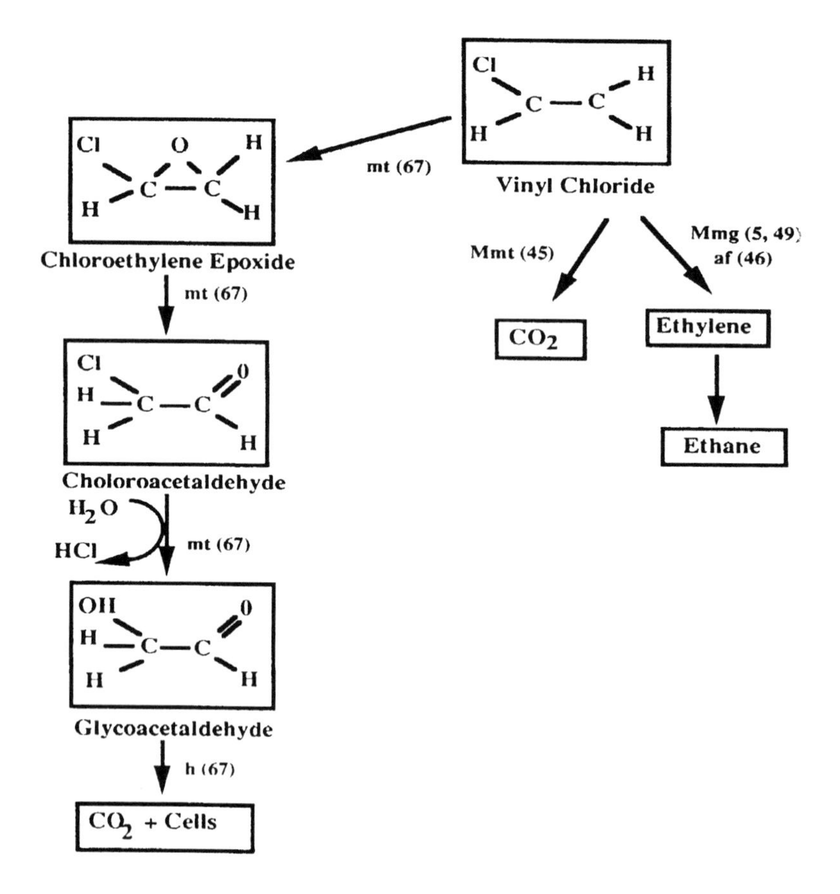

Figure 3. Pathways of transformation for vinyl chloride (af=acetogens or fermenters, M=mixed culture, mt=methanotrophs, mg=methanogens).

to glycoaldehyde, which are further metabolized by heterotrophic bacteria [67]. Vinyl chloride may also be used as a growth substrate [68].

1,1,1-Trichloroethane (TCA)

Anaerobic

As shown in Figure 4, 1,1,1-TCA is transformed by parallel abiotic and biotic pathways. Dehydrohalogenation yields 1,1-DCE [45]. As shown in pathway 1, 1,1-DCE is susceptible to reductive dehalogenation to vinyl chloride. Abiotic or biotic hydrolysis of TCA is illustrated by pathway 3 of Figure 4 [11]. Reductive dechlorination of TCA to 1,1-DCA [27,31,34,69] and chloroethane (CA) is illustrated in pathway 2 [31,34, 69,70]. CA apparently undergoes hydrolysis to ethanol (a "nonvolatile" product), followed by oxidation of the ethanol to CO_2 [31]. In many systems, the above pathways operate simultaneously. For example, *Clostridium* sp. strain TCAIIB, transformed TCA to 1,1-DCA (30-40%), acetic acid (7%), and unidentified products (45%) [71].

Aerobic

TCA can be aerobically oxidized by oxygenase-containing organisms. Degradation of 1,1,1-TCA by the methanotrophic bacterium *Methylosinus trichosporium* OB3b was studied by using cells grown on continuous culture [20]. 1,1,1-TCA was completely converted, and the first compound yielded 2,2,2-trichloroethanol as a chlorinated intermediate. This compound can further be degraded to trichloroacetic acid. A possible pathway of 1,1,1-TCA is illustrated by pathway 4 of Figure 4.

1,2-Dichloroethane (1,2-DCA)

Aerobic

As indicated in Table 3, 1,2-DCA can serve as the sole carbon and energy source for pure cultures [72]. 1,2-DCA is used as a growth substrate by strains of *Xanthobacter autotrophicus* [73] and *Ancylobacter aquaticus* [74]. *Xanthobacter autotrophicus* GJ10 constitutively produces two different dehalogenases. One is specific for halogenated alkanes and the other for halogenated carboxylic acids. Janssen et al. [75,76] proposed that 1,2-DCA is hydrolyzed to 2-chloroethanol, which is subsequently oxidized to 2-chloro-acetaldehyde by an alcohol dehalogenase. An NAD-dependent aldehyde dehalogenase converts chloroacetaldehyde to 2-chloroacetate, which is then converted to glycolate. A complete pathway for 1,2-DCA is given in Figure 5.

Anaerobic

As shown in Figure 5, 1,2-DCA is susceptible to reductive dechlorination in anaerobic environments. Ethylene is produced from 1,2-DCA by

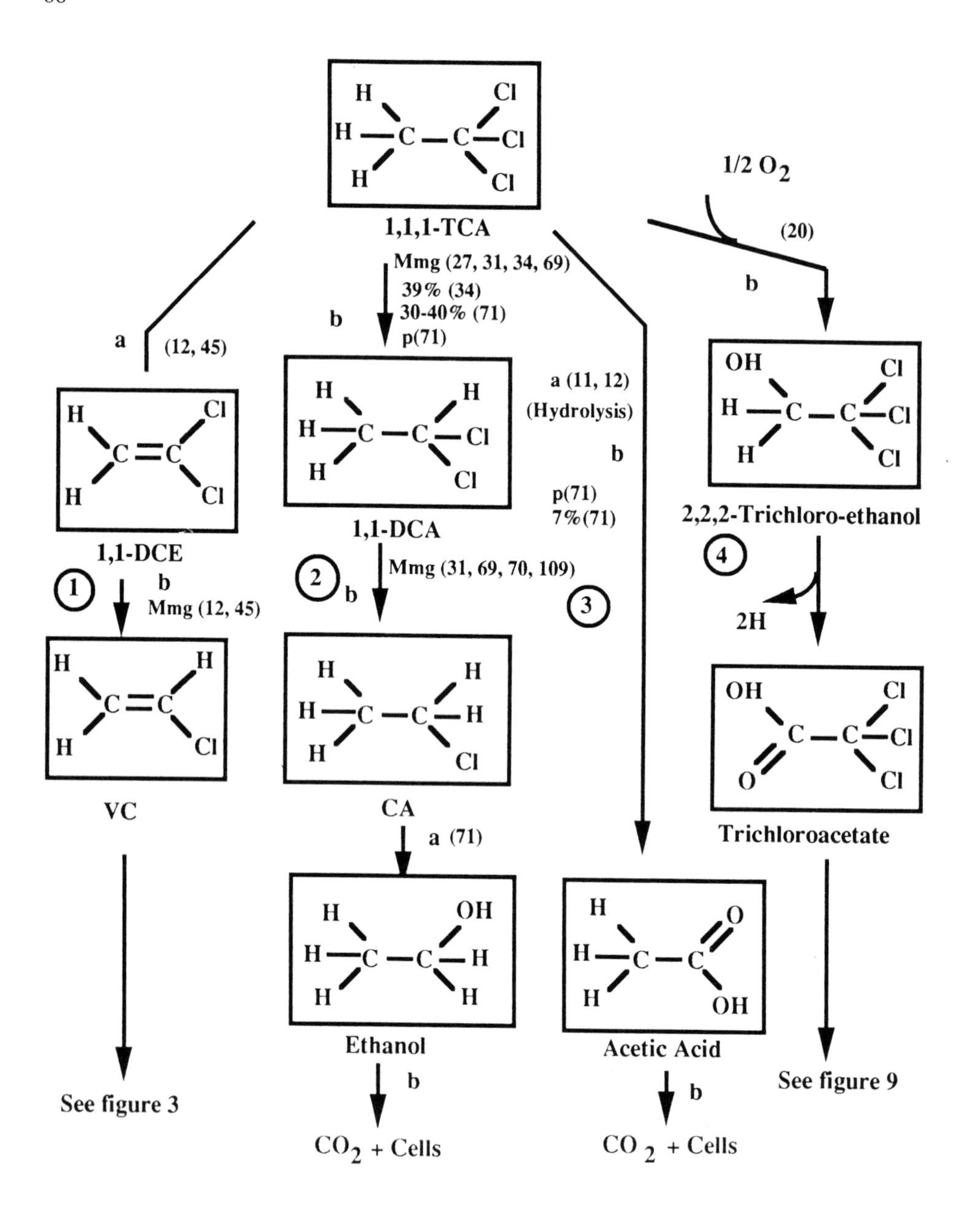

Figure 4. Abiotic and biotic pathways for transformation of 1,1,1-TCA (a=abiotic, b=biotic, p=pure culture, M=mixed culture, mg=methanogens).

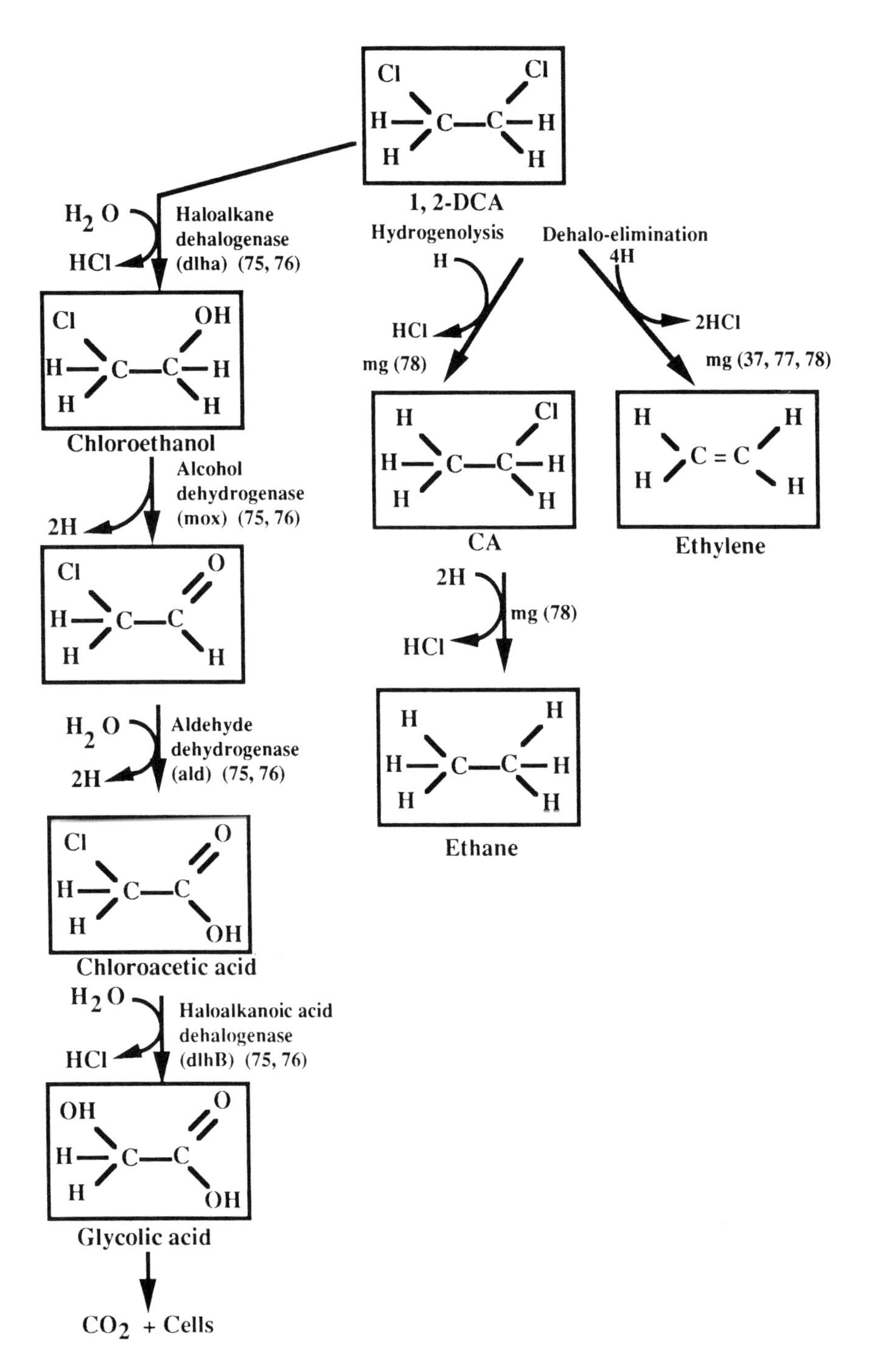

Figure 5. Pathways for transformation for 1,2-DCA (mg=methanogens).

Methanobacterium thermoautotrophicum [37,77]. Hydrogenotrophic and acetoclastic methanogenic bacteria reductively dechlorinate 1,2-DCA via two reaction mechanisms. Ethylene is formed by dihalo-elimination, and chloroethane (CA) is formed by hydrogenolysis. CA is further transformed to ethane via hydrogenolysis [78] or it may undergo hydrolysis to ethanol. Cells of *Methanosarcina barkeri*, when grown on H_2, converted 1,2-DCA and chloroethane at rates that were higher than acetate- or methanol-fed cells [78].

Carbon tetrachloride (CT)

Aerobic

Under fully aerobic conditions, CT persists. However, *E. coli* k12 is capable of transforming CT under low oxygen conditions (~1%) [79].

Anaerobic

As shown in Figure 6, CT is susceptible to transformation by parallel, competing pathways. Transformation of CT to CO_2 occurs in mixed cultures under methanogenic, denitrifying, and sulphate-reducing conditions, but typically, CF is also produced by a parallel pathway [3, 26,30,69,80]. *Desulfobacterium autotrophicum* [37] and *Clostridium* sp. [71] reductively dehalogenated CT to trichloromethane and dichloromethane. *Acetobacterium woodii* transformed CT to trichloromethane and dichloromethane and to CO_2 by parallel pathways [81]. The CO_2 was subsequently converted to acetate by acetogenesis. *Pseudomonas* sp. strain KC rapidly transforms CT to CO_2, and to an unidentified nonvolatile fraction under denitrifying conditions [82]. This organism appears to be unique in that it does not produce CF from CT. The transformation is also unusual in that it is regulated by the availability of trace metals, notably iron and cobalt, and it is linked to the mechanism of trace metal scavenging [82,83].

No significant transformation of CT was observed in nitrate-respiring and oxygen-respiring cultures of *E. coli* k12, except at low oxygen levels (~1%), although fumarate-respiring and fermenting cultures slowly transformed CT to CF and to other products. In these cultures, CF persisted, indicating that other products of CT transformation were generated by parallel pathways and were not derived from CF [79]. Carbon disulphide was an unexpected product of CT transformation in *E. coli* studies [79]. A recent study of sulphide reactivity [84] demonstrates that CT reacts with HS^- to produce CS_2, but it is not known if CT undergoes direct nucleophilic substitution with HS^- or S_x^{2-} to form CS_2 or if CT undergoes reduction to form a trichloromethyl radical, which can then react with HS^-, S_x^{2-}, or $S_2O_3^{2-}$ to form CS_2. CS_2 hydrolyzes to CO_2. About 85% of the CT is ultimately transformed to CO_2 in these systems.

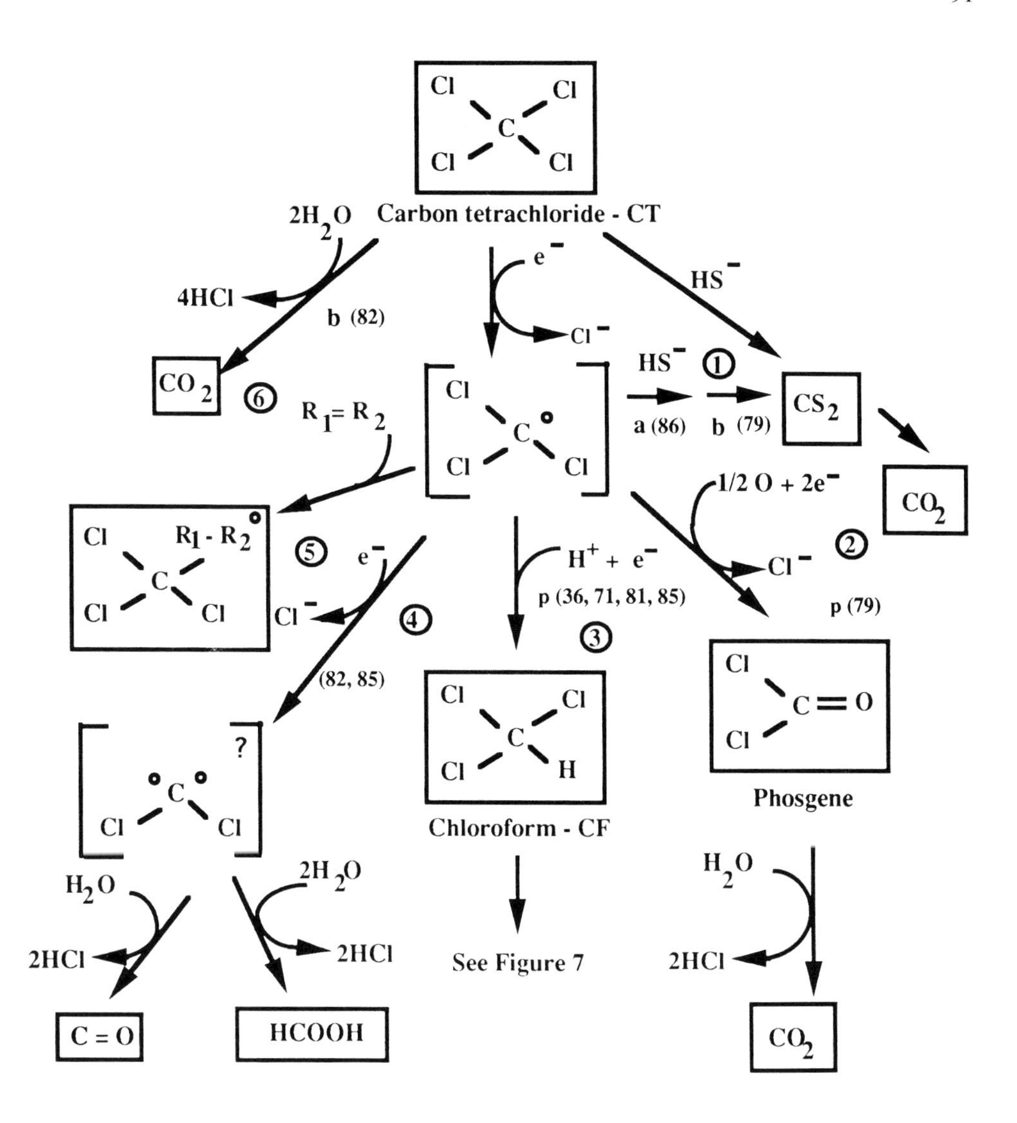

Figure 6. Pathways of transformation for carbon tetrachloride (a=abiotic, b=biotic, p=pure culture).

The observed product distribution is consistent with the general model of competing parallel pathways presented by Criddle and McCarty [85] and by Kriegmann-King and Reinhard [86].

Chloroform (CF)

Aerobic

As shown in Figure 7, methanotrophs can transform CF [15,87-89]. The mechanism of MMO-catalyzed oxidations is similar to that of cytochrome P-450-catalyzed oxidations [22,90]. The intermediate product of P-450 transformation is phosgene, and this compound is probably the intermediate produced by MMO reactions as well [15]. Although phosgene is rapidly hydrolyzed to carbon dioxide and chloride, it also binds to

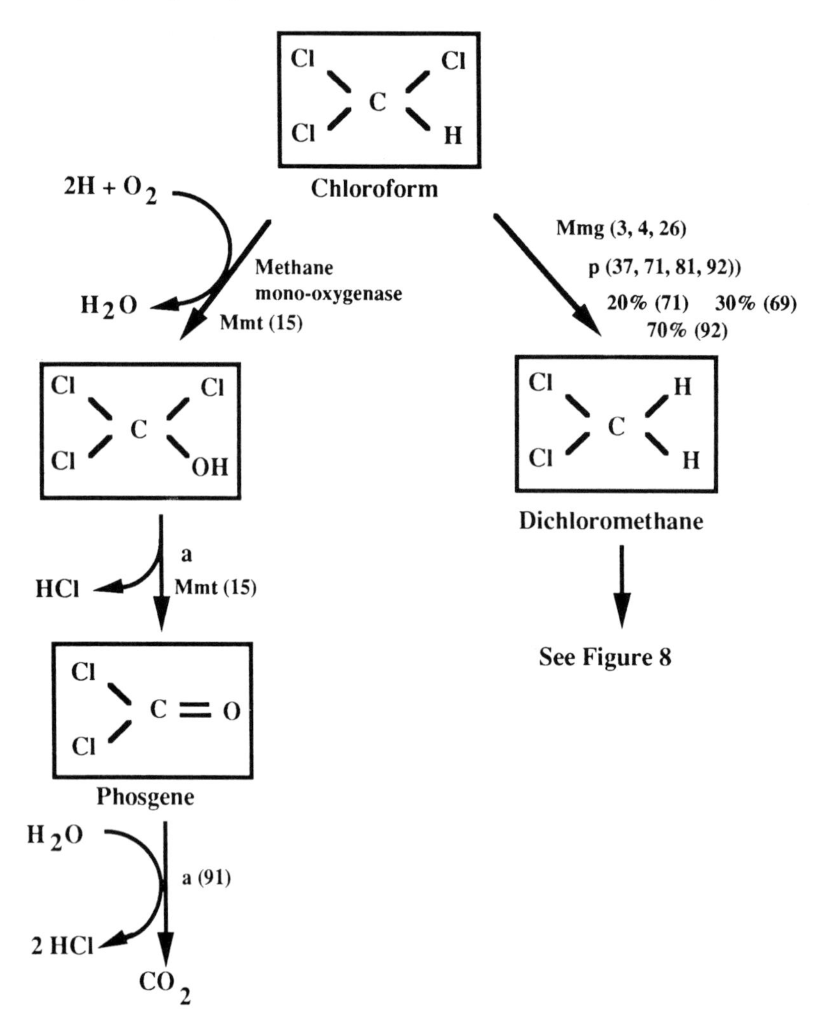

Figure 7. Pathways of transformation for chloroform (a=abiotic, M=mixed culture, mg=methanogens, mt=methanotrophs).

proteins, presumably accounting for its toxicity [91]. Cell-free MMO extracts also oxidize CF [19]. A recent study demonstrated that formate addition increases CF transformation rate (0.35 day^{-1} for resting cell and 1.5 day^{-1} for formate-fed cells) and transformation capacity (= 0.0065 mg CF per mg cells for starving cells, and = 0.015 mg of CF per mg cells for formate-fed cells) [66] . These observations indicate that energy substrate alters biomass transformation capacity. However, even in the presence of an energy substrate, biomass transformation capacity had an upper limit, suggesting that toxicity was important.

Anaerobic

Chloroform was dechlorinated by pure cultures of *Methanosarcina* sp. strain DCM and *Methanosarcina mazei* S6. The initial dechlorination product of chloroform was dichloromethane, accounting for about 70 percent of added chloroform. The product of dichloromethane, chloromethane was detected consistently at trace levels but could not be accurately quantified. The production of $^{14}CO_2$ from [^{14}C] - chloroform and the absence of $^{14}CH_4$ imply that reaction mechanisms other than reductive dechlorination alone are important [92]. Degradation of dichloromethane by oxidation via phosgene or alcohols was proposed previously, for methanogenic systems [13].

Dichloromethane (DCM)

Aerobic

Pathways for dichloromethane *biodegradation* are given in Figure 8. Use of DCM as a growth substrate has been demonstrated in enrichment cultures [93,94], and by pure cultures of *Pseudomonas* and *Hyphomicrobium* species [21,95,96]. The ability to grow on DCM requires the biosynthesis of a DCM dehalogenase (100). DCM dehalogenase from a *Hyphomicrobium* sp. was purified and characterized [97]. DCM degradation by *Hyphomicrobium* involves nucleophilic substitution by a transferase. The glutathione (GS)-dependent enzyme, glutathione-S-transferase (GS) dechlorinates DCM to S-chloromethyl glutathione. This compound is nonenzymatically converted to formaldehyde and glutathione [96]. The DCM-degrading methylotrophic bacterium, DCM11, grows more rapidly on DCM than previous isolates - strains DM1, DM2, DM4 and GJ21, and this difference was traced to a difference in enzyme structure [98].

Anaerobic

DCM can serve as a growth substrate under anaerobic conditions [93, 99]. The principal transformation pathways are oxidation to CO_2 and fermentation to acetate. CO_2-reducing methanogens use some of the electrons made available from DCM oxidation to form methane - the methyl carbon comes from DCM and the carboxyl carbon from CO_2. *Acetoclastic* methanogens produce methane from the acetic acid, formed by fermentation of DCM. DCM degradation is, thus, a disproportionation: a portion of the DCM is oxidized, making reducing equivalents available for reduction of an equal amount of DCM.

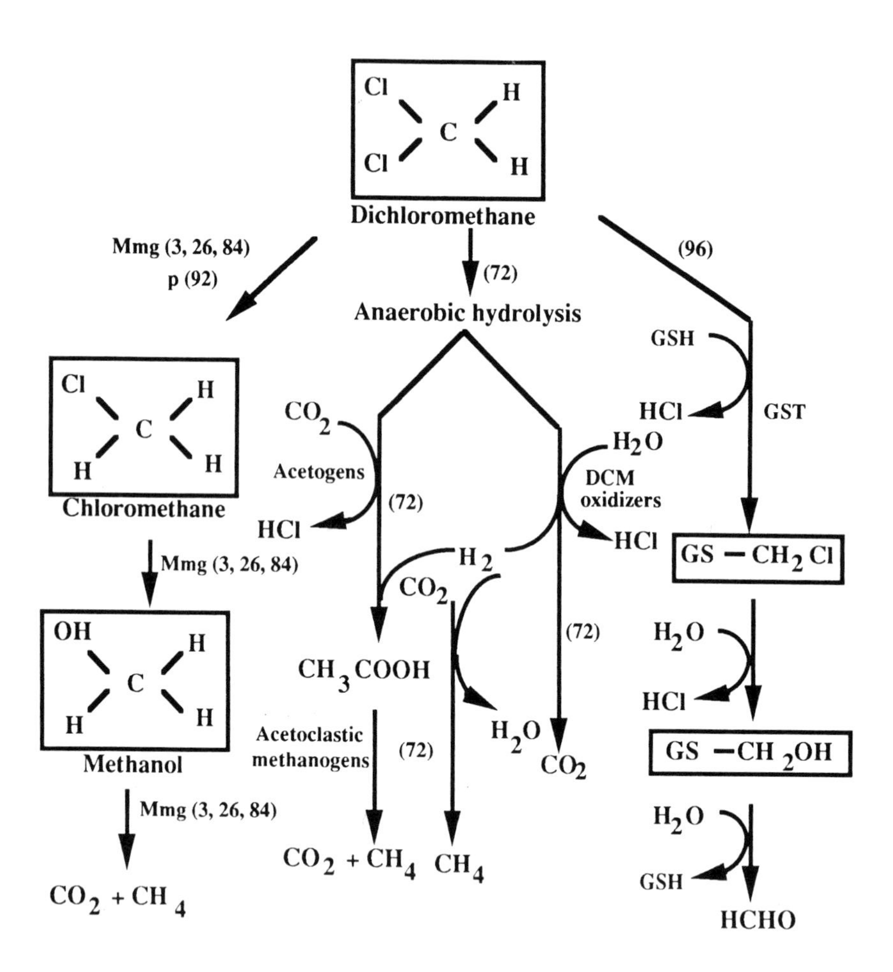

Figure 8. Pathways of transformation for dichloromethane (p=pure culture, M=mixed culture, mg=methanogens, GST=glutathione-S-transferase).

Dichloroacetate (DCAA) and trichloroacetate (TCAA)

Three groups of bacteria capable of decomposing chloro-substituted aliphatic acids have been isolated from soil [100]. In 1960, Jensen described these bacteria as *Pseudomonas dehalogenans, Arthrobacter* sp. and *Agrobacterium* sp. The first group decomposed monochloroacetate (MCAA), but had little effect on DCAA, and none on TCAA. The second group decomposed TCAA when cultivated with a vitamin B_{12}-producing strain of *Streptomyces*. This group had little effect on DCAA, but none on MCAA. The third group decomposed DCAA, but not MCAA and TCAA [101]. These studies demonstrated that an inducible dehalogenase enzyme system removed the halogen from the carbon chain by hydrolysis of the carbon-hydrogen bond. Further study [102] confirmed that growth with halogenated acids induced adaptation to the corresponding oxy-acids which, thus, seem to be normal intermediates. Organisms capable of metabolizing aliphatic oxy-acids are not normally able to decompose the corresponding halogen-substituted acid. The complete pathway of decomposition of dichloroacetic and trichloroacetic acid by this mechanism are shown in Figure 9. *Arthrobacter* sp. dechlorinates TCAA, when provided methionine and vitamin B_{12}, but it does not use TCAA as a carbon or energy source [103].

SUMMARY

Halocarbon biotransformations can generally be classified as substitutions, reductions, or oxidations. Reductions are typically mediated by reduced cofactors present under anaerobic environments; they proceed faster in more reduced environments with highly halogenated compounds; and they typically yield hydrogenolysis reaction products. Oxidations are typically mediated by nonspecific oxygenases; they proceed faster in more oxidative environments with less halogenated compounds; and they typically yield alcohols and organic acid reaction products. Reductions and oxidations of compounds, with more than 2 halogens per carbon atom, are typically cometabolic, requiring prior or concurrent consumption of a growth or energy substrate for sustained transformation. In the absence of growth or energy substrates, cometabolic transformations tend to slow down with time as a result of product toxicity. Halocarbon oxidations and reductions often yield reactive intermediates. These species can subsequently give rise to a spectrum of products via parallel and competing pathways. Some of the intermediates are of human health concern. Hydrolytic substitution reactions occur with compounds that have few halogen substituents and are not sterically hindered. These reactions often yield useful growth substrates, and they typically initiate complete metabolic pathways.

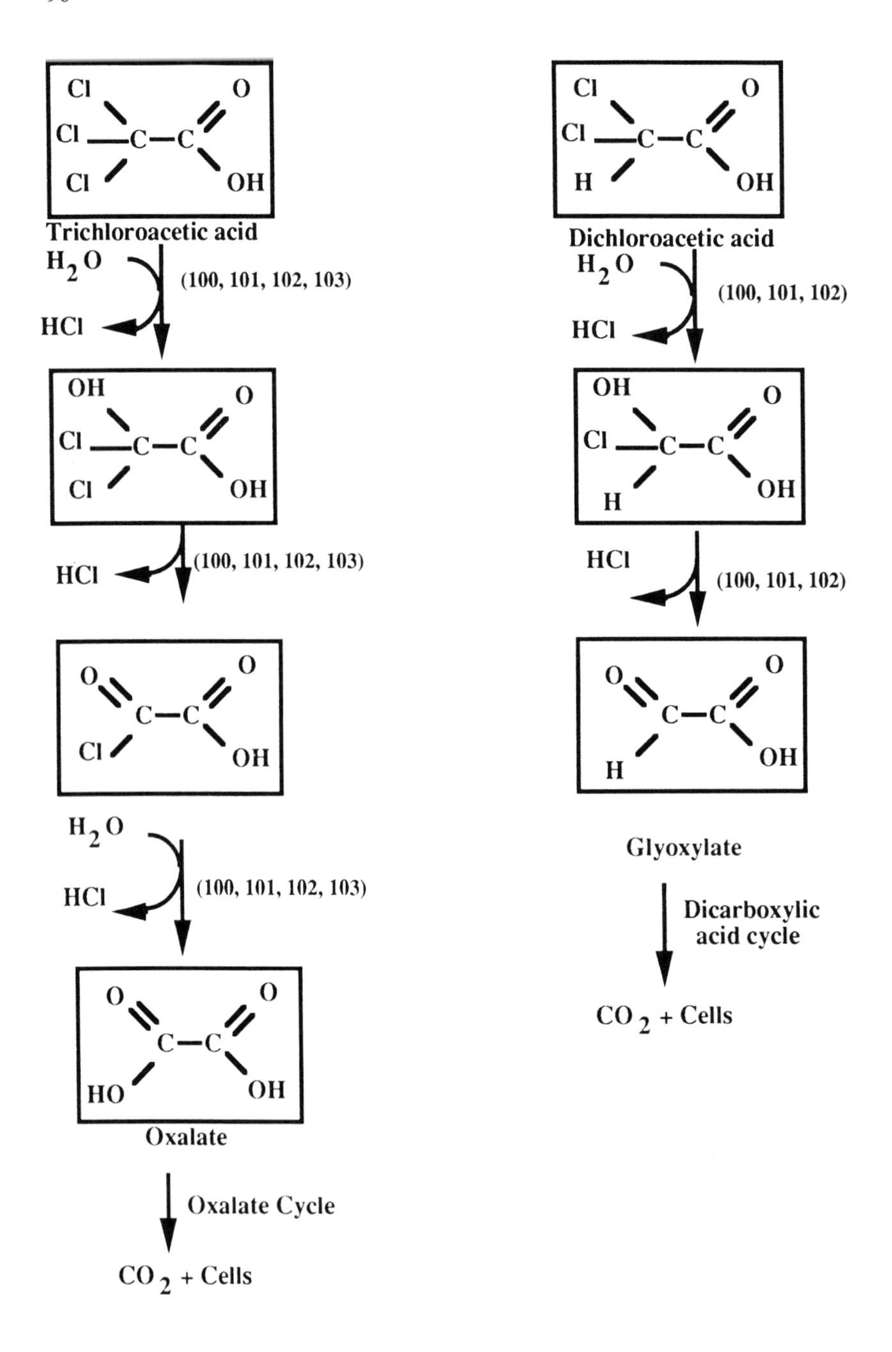

Figure 9. Pathways of transformation for dichloroacetate and trichloroacetate.

REFERENCES

1 Price PS. VOC Degradation, Memo of U.S. Environmental Protection Agency, Office of Water, Washington, D.C. (August 21), 1985.
2 Commission., U.S.I.T. Synthetic organic chemicals, U.S. production and sales, International Trade Commission, Washington, D.C, 1970-1985.
3 Bouwer EJ, McCarty PL. Appl Environ Microbiol 1983; 45: 1286-1294.
4 Vogel TM, Criddle CS, McCarty PL. Environ Sci Technol 1987; 21: 722-736.
5 Freedman DL, Gossett JM. Appl Environ Microbiol 1989; 55: 2144-2151.
6 Criddle CS. Biotechnol Bioeng 1993; 41: 1048-1056.
7 Criddle CS, Alvarez LA, McCarty PL. In: Bear J, Corapcioglu MY, eds. Microbial Processes in Porous Transport Processes in Porous Media. The Netherlands: Kluwer Academic Publishers, 1991; 639-691.
8 Mabey W, Mill TJ. J Phys Chem Ref Data 1978; 7: 383-415.
9 Vogel TM, McCarty PL. Environ Sci Technol 1987; 21: 1208-1213.
10 March J. Advanced Organic Chemistry. 3rd ed. New York: Wiley, 1985.
11 Mabey WR, Barich V, Mill T. Hydrolysis of polychlorinated ethanes. Extended Abstracts, 186th National Meeting of the American Chemical Society, Washington, D.C., 1980.
12 Hagg WR, Mill T, Richardson A. Effect of subsurface sediment on hydrolysis reactions. Extended Abstract, 192nd National Meeting of the American Chemical Society, Anaheim, CA, 1986.
13 Vogel TM, McCarty PL. J Contam Hydrol 1987; 1: 299-308.
14 Kubic VL, Anders MW. Biochem Pharmacol 1978; 27: 2349-2355.
15 Pohl LR, Bhoshan B, Whittaker NF, Krishna G. Biochem Biophys Res Commun 1977; 79: 684-691.
16 Anders MW. In: Jacoby WB, Bend JR, Caldwell J, eds. Metabolic Baisis of Detoxification. New York: Academic Press, 1982; 29-49.
17 Macdonald TL. CRC Crit Rev Toxicol 1983; 11: 85-120.
18 Leibman KC, Ortiz E. Environ Health Prospect 1977; 21: 91-97.
19 Colby JD, Stirling DI, Dalton H. Biochem J 1977; 65: 395-402.
20 Oldenhuis R, Vink RLJM, Janssen DB, Witholt B. Appl Environ Microbiol 1989; 55: 2819-2826.
21 Brusseau GA, Tsien HC, Hanson RS, Wackett LP. Biodegradation 1990; 1: 19-29.
22 Fox BG, Borneman JG, Wackett LP, Lipscomb JD. Biochemistry 1990; 29: 6419-6427.

23 Tsien HC, Brusseau GA, Hanson RS, Wackett LP Appl Environ Microbiol 1989; 55: 3155-3161.
24 Wackett LP. In: Martin AM, ed. Biological Degradation of Wastes. London and New York: Elsevier Applied Science, 1991; 187-205.
25 Barrio-Lage G, Parsons FZ, Nasser RS, Lorenzo PA. Environ Sci Technol 1986; 20: 96-99.
26 Bouwer, EJ, Rittmann BE, McCarty PL. Environ Sci Technol 1981; 15: 596-599.
27 Parsons F, Barrio-Lage G, Rice R. Environ Toxicol Chem 1985; 4: 739-742.
28 Wilson, BH, Smith GB, Rees JF. Environ Sci Technol 1986; 20: 997-1002.
29 Bouwer EJ, McCarty PL. Appl Environ Microbiol 1985; 50: 527-528.
30 Bouwer EJ, Wright JP. J Contam Hydrol 1988; 2: 155-169.
31 Vogel TM, McCarty PL. Appl Environ Microbio 1985; 49: 1980-1983.
32 Barrio-Lage GA, Parsons FZ, Nasser RS. Environ Toxicol Chem 1987; 6: 571-578.
33 Parsons F, Wood PR, DeMarco J. J Am Water Works Assoc 1984; 76: 56-59.
34 Parsons F, Barrio-Lage G. J Am Water Works Assoc 1985; 77: 52-59.
35 DeBruin WP, Kotterman MJJ, Posthumus MA, Schraa G, Zehnder AGB. Appl Environ Microbiol 1991; 58: 1996-2000.
36 Egli C, Scholtz R, Cook AM, Leisinger T. FEMS Microbiol Lett 1987; 43: 257-261.
37 Fathepure BZ, Nengu JP, Boyd SA. Appl Environ Microbiol 1987; 53: 2671-2674.
38 Fathepure BZ, Boyd SA. Appl Environ Microbiol 1988; 54: 2976-2980.
39 Fathepure BZ, Boyd SA. FEMS Microbiol Lett 1988; 49: 149-156.
40 Herbert D, Elsworth RE, Telling RC, J Gen Microbiol 1956; 14: 601-607.
41 Rittman BE, McCarty PL. Biotechnol Bioeng 1980; 22: 2343-2357.
42 Lawrence AW, McCarty PL. Jour. Sanitary Engineering Division, Amer Soc of Civil Engineers 1970; 96: 757-778.
43 Alvarez-Cohen L, McCarty PL. Environ Sci Technol 1991; 25: 1381-1387.
44 Chang M-K, Voice TC, Criddle CS. Biotechnol Bioeng 1993; 41: 1057-1065.
45 Vogel TM, Reinhard M. Environ Sci Technol 1986; 20: 992-997.
46 DiStefano TD, Gossett JM, Zinder SM. Appl Environ Microbiol 1991; 57: 2287-2292.

47 Scholz-Muramatsu H, Szewzyk R, Szewzyk U, Gaiser S. FEMS Microbiol Lett 1990; 66: 81-86.
48 Bagley DM, Gossett JM. Appl Environ Microbiol 1990; 56: 2511-2516.
49 Gantzer CJ, Wackett LP. Environ Sci Technol 1991; 25: 715-722.
50 Arciero D, Vannelli T, Logan M, Hooper AB. Biochem Biophys Res Commun 1989; 159: 640-643.
51 Hyman MR, Murton IB, Arp DJ. Appl Environ Microbiol 1988; 54: 3187-3190.
52 Vannelli T, Logan M, Arciero DA, Hooper AB. Appl Environ Microbiol 1990; 56: 1169-1171.
53 Nelson MJK, Montgomery SO, O'Neill EJ, Pritchard PH. Appl Environ Microbiol 1986; 52: 383-384.
54 Nelson MJK, Montgomery SC, Mahaffey WR, Pritchard PH. Appl Environ Microbiol 1987; 53: 949-954.
55 Nelson MJK, Montgomery SO, Pritchard PH. Appl Environ Microbiol 1988; 54: 604-606.
56 Wackett LP, Gibson DT. Appl Environ Microbiol 1988; 54: 1703-1708.
57 Henry SM, Grbic'-Galic' D. Aerobic degradation of trichloroethylene (TCE) by methylotrophs isolated from a contaminated aquifer. Abstr. Q-64, p. 294. Abstr. 86th Ann Meet Am Soc Microbiol, Washington, D.C., 1986.
58 Henschler D, Hoos WR, Fetz H, Dallmeier E, Metzler M. Biochem Pharmacol 1979; 28: 543-548.
59 Little CD, Palumbo AV, Herbes SE, Lidstrom ME, Tyndall RL, Gilmer PJ. Appl Environ Microbiol 1988; 54: 951.
60 Miller RE, Guengerich FP. Biochemistry 1982; 21: 1090-1097.
61 Henry SM, Grbic'-Galic' D. Appl Environ Microbiol 1991; 57: 1770-1776.
62 Newman LS, Wackett LP. Appl Environ Microbiol 1991; 57: 2399-2402.
63 Uchiyama H, Nakajima T, Yagi O, Tabuchi T. Agric Biol Chem 1989; 53: 2903-2907.
64 Nakajima T, Uchiyama H, Yagi O, Nakahara T. Biosci Biotech Biochem 1992; 56: 486-489.
65 Alvarez-Cohen L, McCarty PL. Appl Environ Microbiol 1991; 5: 228-235.
66 Alvarez-Cohen L, McCarty PL. Appl Environ Microbiol 1991; 57: 1031-1037.
67 Guengerich FP, Crawford W Jr, Watanabe PG. Biochemistry 1979; 18: 5177-5182.
68 Hartmans S, deBont JAM, Tramper J, Luyben K. Biotechnol Lett 1985; 7: 383-388.

69 Gossett JM. Anaerobic degradation of C_1 and C_2 chlorinated hydrocarbons. U.S. Air Force Engineering and Services Center. ESI-TR-85-38. Final report. NTIS. Springfield, VA.
70 Hallen RT, Pyne JW Jr, Molton PM. Transformation of chlorinated ethanes and ethenes by anaerobic microorganisms. Extended Abstracts, 192nd National Meeting of the American Chemical Society, Anaheim, CA., 1986.
71 Gälli R, McCarty PL. Appl Environ Microbiol 1989; 55: 837-844.
72 Stucki G, Krebser U, Leisinger T. Experientia 1983; 39: 1271-1273.
73 Keuning S, Janssen DB, Witholt B. J Bacteriol 1985; 163: 635-639.
74 Van Den Wijngaard AJ, Wind RD, Janseen DB. Appl Environ Microbiol 1993; 59: 2041-2048.
75 Janssen DB, Scheper A, Dijkhuizen L, Witholt B. Appl Environ Microbiol 1985; 49: 673-677.
76 Janssen DB, Keuning S, Witholt B. J Gen Microbiol 1987; 133: 85-92.
77 Belay N, Daniels L. Appl Environ Microbiol 1987; 53: 1604-1610.
78 Holliger C, Schraa G, Stams AJM, Zehnder AJB. Biodegradation 1990; 1: 253-261.
79 Criddle CS, DeWitt JT, McCarty PL. Appl Environ Microbiol 1990; 56: 3247-3254.
80 Bouwer EJ, McCarty PL. Appl Environ Microbiol 1983; 45: 1295-1299.
81 Egli C, Tschan T, Scholtz R, Cook AM, Leisinger T. 1988. Appl Environ Microbiol 1988; 54: 2819-2824.
82 Criddle CS, DeWitt JT, Grbic'-Galic' D, McCarty PL. Appl Environ Microbiol 1990; 56: 3240-3246.
83 Tatara GM, Dybas MJ, Criddle CS. Appl Environ Microbiol 1993; 59: 2126-2131.
84 Kleopfer RD, Easley DM, Haas BB Jr, Deihl TG, Jackson DE, Wurrey CJ. Environ Sci Technol 1985; 19: 277-290.
85 Criddle CS, McCarty PL. Environ Sci Technol 1991; 25: 973-978.
86 Kriegman-King M, Reinhard M. Environ Sci Technol 1992; 26: 2198-2206.
87 Crestiel T, Beaune P, Leroux JP, Lange M, Mansuy D. Chem Biol Interact 1979; 24: 153-165.
88 Strand SE, Shippert L. Appl Environ Microbiol 1986; 52: 203-205.
89 Wilson JT, Wilson BH. Appl Environ Microbiol 1985; 49: 242-243.
90 Fox BG, Froland WA, Dege JG, Lipscomb JD. J Biol Chem 264: 10023-10033.
91 Davidson IWF, Sumner DD, Parker JC. Drug Chem Toxicol 1982; 5: 1-87.

92 Mikesell MD, Boyd SA. Appl Environ Microbiol 1990; 56: 1198-1201.
93 Freedman DL, Gossett JM. Appl Environ Microbiol 1991; 57: 2847-2857.
94 Rittman BE, McCarty PL. Appl Environ Microbiol. 1980; 39: 1225-1226.
95 Gälli R, Leisinger T. Conserv Recycl 1986; 8: 91-100.
96 Stucki G, Gälli R, Ebersold HR, Leisinger T. Arch Microbiol 1981; 130: 366-371.
97 Kohler-Staub D, Leisinger T. J Bacteriol 1985; 162: 676-681.
98 Scholtz R, Wackett LP, Egli C, Cook AM, Leisinger T. J Bacteriol 1988; 170: 5698-5704.
99 Stromeyer SA, Winkelbauer W, Kohler H, Cook ATLAM. Biodegradation 1991; 2: 129-137.
100 Jensen HL. Acta Agri Scand 1963; 13: 404-412.
101 Jensen HL. Acta Agri Scand 1960; 10: 83-103.
102 Jensen HL. Can J Microbiol 1957; 3: 151-164.
103 Lode O. Acta Agri Scand 1967; 17: 140-148.
104 Brunner W, Staub D, Leisinger T. Appl Environ Microbiol 1980; 40: 950-958.
105 Kohler-Staub D, Gälli R, Suter F, Leisinger T. J Gen Microbiol 1986; 132: 2837-2843.
106 Scholtz R, Messi F, Leisinger T, Cook AM. Appl Environ Microbiol 1988; 54: 3034-3038.
107 Folsom BR, Chapman PJ, Pritchard PH. Appl Environ Microbiol 1990; 56: 1279-1285.
108 Folsom BR, Chapman PJ. Appl Environ Microbiol 1991; 57: 1602-1608.
109 Oldenhuis R, Oedzes JY, VanDer Waarde JJ, Janssen DB. Appl Environ Microbiol 1991; 57: 7-14.
110 Uchiyama H, Nakajima T, Yagi O, Nakahara T. Appl Environ Microbiol 1992; 58: 3067-3071.
111 Wackett LP, Brusseau GA, Householder SR, Hanson RS. Appl Environ Microbiol 1989; 55: 2960-2964.
112 Rasche ME, Hyman MR, Arp DJ. Appl Environ Microbiol 1991; 57: 2986-2994.
113. Ensign SA, Hyman MR, Arp DJ. Appl Environ Microbiol 1992; 58: 3038-3046.
114 Zylstra GJ, Wackett LP, Gibson DT. Appl Environ Microbiol 55: 3162-3166.
115 Broholm K. Jensen BK, Christensen TH, Olsen L. Appl Environ Microbiol 1990; 56: 2488-2493.
116 Henry SM, Grbic'-Galic' D. Appl Environ Microbiol 1991; 57: 236-244.

117 Winter RB, Kwang-Mu. Yen, Ensley BD. Bio/Technology 1989; 7: 282-285.
118 Harker AR, Kim Y. Appl Environ Microbiol 1990; 6: 1179-1181.
119 Strand SE, Bjelland MD, Stensel HD. Research Journal WPCF 1990; 62: 124-129.
120 Henry SM, Grbic'-Galic' D. Microb Ecol 1990; 20: 151-169.
121 Fogel MM, Taddeo AR, Fogel S. Appl Environ Microbiol. 1986; 51: 720-724.
122 Gälli R, McCarty PL. Appl Environ Microbiol 1989; 55: 845-851.

Biotransformations: Microbial Degradation of Health Risk Compounds
Ved Pal Singh, editor
© 1995 Elsevier Science B.V. All rights reserved.

Aerobic biodegradation of polycyclic and halogenated aromatic compounds

Erwin Grund[a], Annegret Schmitz[b], Jörg Fiedler[b] and Karl-Heinz Gartemann[b]

[a]GBF - Gesellschaft für Biotechnologische Forschung mBH, Mascheroder Weg 1, D-38124 Braunschweig, Germany

[b]Universität Bielefeld, Fakultät für Biologie, Gentechnologie/Mikrobiologie, Postfach 100131, Universitätsstraße, D-33594 Bielefeld 1, Germany

INTRODUCTION

Polycyclic aromatic hydrocarbons (PAHs) occur as natural constituents and combustion products of fossil fuels and, therefore, are ubiquitous environmental contaminants [1-3]. Furthermore, coal-tar creosote, a chemical which mainly consists of PAHs, has been widely used as a wood preservative for over 150 years, and accidental spillage or improper disposal of creosote has resulted in extensive contamination of soil, surface waters, and groundwater aquifers [4]. PAHs may have toxic, mutagenic, and carcinogenic properties [5,6] and, therefore, the fate of these *xenobiotics* in nature is of environmental concern.

Polyhalogenated dibenzo-*p*-dioxins and the closely related dibenzofurans are unintentionally produced as contaminating by-products during the manufacture of pesticides, incineration of halogen-containing aromatic chemicals and bleaching of paper pulp [7-9]. Halogen-containing aromatic compounds are common constituents of industrial and domestic waste, thus polyhalogenated dibenzo-*p*-dioxins and dibenzofurans are formed in incineration plants. These highly toxic and mutagenic compounds [10,11] have, therefore, become widespread contaminants of the environment. Another interesting point, with respect to contamination of the environment with chlorinated dioxins, is that these compounds may be enzymatically formed from chlorinated phenols [12] by the action of peroxidases.

The third example for polycyclic aromatic compounds of environmental concern is the polychlorinated biphenyls (PCBs). The vast majority of PCBs in the environment are derived from commercial mixtures (*Aroclors*), which contain 60 to 80 different congeners. These mixtures have found widespread industrial use in the past, owing to their physical and chemical stability and their dielectric properties. Inadequate waste disposal has led to their release into the environment, and they have been routinely detected in soil and water samples since the early 1960s.

Their toxicity has caused PCBs to be prohibited in many countries. These compounds are some of the most serious environmental pollutants, and numerous studies have focused on their toxicity, mutagenicity, *bioaccumulation*, environmental fate, and health risks [13-15].

Many of the above mentioned different types of polycyclic aromatic compounds are *recalcitrant,* and bioaccumulation occurs. However, recent work has shown that these chemicals can be degraded oxidatively by some bacterial isolates.

Complete *mineralization* of chlorinated polycyclic aromatic compounds, like dibenzo-*p*-dioxins, dibenzofurans, and biphenyls, requires the presence of two sets of genes, one set for the degradation of the polycyclic aromatic structure and a second set for the degradation of chlorinated monocyclic aromatics. Previous work has shown that more efficient degradation of 4-chlorobiphenyl occurs in a co-culture, which contains a bacterial strain that is able to degrade 4-chlorobenzoate produced by a 4-chlorobiphenyl-degrading strain [16,17]. Therefore, we will present data on the *aerobic dehalogenation* of chlorobenzoates by bacteria. During the degradation of chlorinated dibenzofurans by bacterial strains, chlorinated salicylates will be accumulated, and during the degradation of chlorinated dibenzo-*p*-dioxins, chlorinated phenols will be accumulated. The degradation of these halogenated aromatics has been discussed in another chapter of this book.

NAPHTHALENE

It is well known that microorganisms can degrade naphthalene and other PAHs. Reports exist, for example, on the aerobic degradation of the tricyclic PAHs, phenanthrene and anthracene [18-20], and the tetracyclic PAHs, pyrene [21,22] and fluoranthrene [23,24] by Gram-negative and Gram-positive soil bacteria. Even the aerobic attack of fungi on PAHs has been reported [25-27]. These few examples show that PAHs can be degraded aerobically, and that the biological breakdown of these human health-risk compounds may represent a promising approach for the detoxification of PAH-contaminated areas. But, as was shown recently, parameters, like adsorption of PAHs by soil particles [28], may influence the biodegradability of PAHs. This shows that one of the problems, that arises during the breakdown of PAHs by microorganisms, is the availability of these hydrocarbons for the microorganisms. The mechanisms, by which microorganisms make available the PAHs, are not fully understood todate. Therefore, we will focus on the enzymatic reactions, that occur during the aerobic breakdown of PAHs. The principle, by which the different PAHs are degraded aerobically by bacteria, is always quite similar. Because of this similarity of the degradation routes for different PAHs, it is possible to understand the principles while explaining one of the best understood degradation routes of PAHs which is the oxidative catabolism of naphthalene.

It is well established that bacteria can oxidatively metabolize naphthalene and, therefore, the degradation route of naphthalene and even the genetics of this naphthalene catabolism are well understood. The different degradation routes for naphthalene are presented in Figure 1. In 1964, the metabolic sequence of enzymatic reactions, leading to the degradation of naphthalene, was first presented by Davies and Evans [29]. Later studies have shown that "*cis*-naphthalene dihydrodiol" is the first metabolite in the bacterial metabolism of naphthalene [30]. The enzyme, that catalyzes this reaction, is the naphthalene dioxygenase [31]. The aerobic catabolism of homocyclic aromatic compounds, that do not contain hydroxy groups, is usually initiated by the action of a *dioxygenase*. In general, dioxygenases are enzymes that activate molecular oxygen and introduce two hydroxy-groups. All bacterial dioxygenases, that catalyze the formation of dihydrodiols and that have been analyzed in detail, consist of two or three iron-containing enzyme components, and require NADH or NADPH as electron donors. The naphthalene dioxygenase consists of three protein components, an iron-sulphur flavoprotein, a two-iron, two-sulphur ferredoxin, and an iron-sulphur protein, which are essential for *cis*-naphthalene dihydrodiol formation. The naphthalene dioxygenase also accepts indole as a substrate which leads to the formation of indigo [32]. The second step in the bacterial oxidation of naphthalene is the conversion of *cis*-naphthalene dihydrodiol to 1,2-dihydroxynaphthalene. This reaction is catalyzed by *cis*-naphthalene dihydrodiol dehydrogenase and requires NAD as an electron acceptor [33].

1,2-Dihydroxynaphthalene is enzymatically cleaved by an oxygenase [34], and by the action of an isomerase and an aldolase, salicylaldehyde and pyruvate are formed [29,35,36]. Salicylaldehyde is oxidized to salicylate by the action of a NAD-dependent dehydrogenase [37].

The reactions, described so far, are derived from the most extensively studied catabolic pathway for naphthalene, which is encoded by the NAH7 plasmid of *Pseudomonas putida*. But, the same sequence of reactions seems to take place in all naphthalene-degrading bacteria, which means that salicylate is a common intermediate in naphthalene catabolism. With respect to the enzymes involved in the breakdown of naphthalene, there seem to exist differences between the bacterial species, that can grow on naphthalene-containing media. A recently isolated and described *Rhodococcus* sp., for example, seems to have a 1,2-dihydroxynaphthalene oxygenase that requires NADH [38].

All known bacterial species, that can grow on naphthalene as sole source of carbon and energy, can degrade salicylate as well. Three different catabolic routes are known for the total degradation of salicylate (Figure 1), and all these reaction sequences are realized in naphthalene-

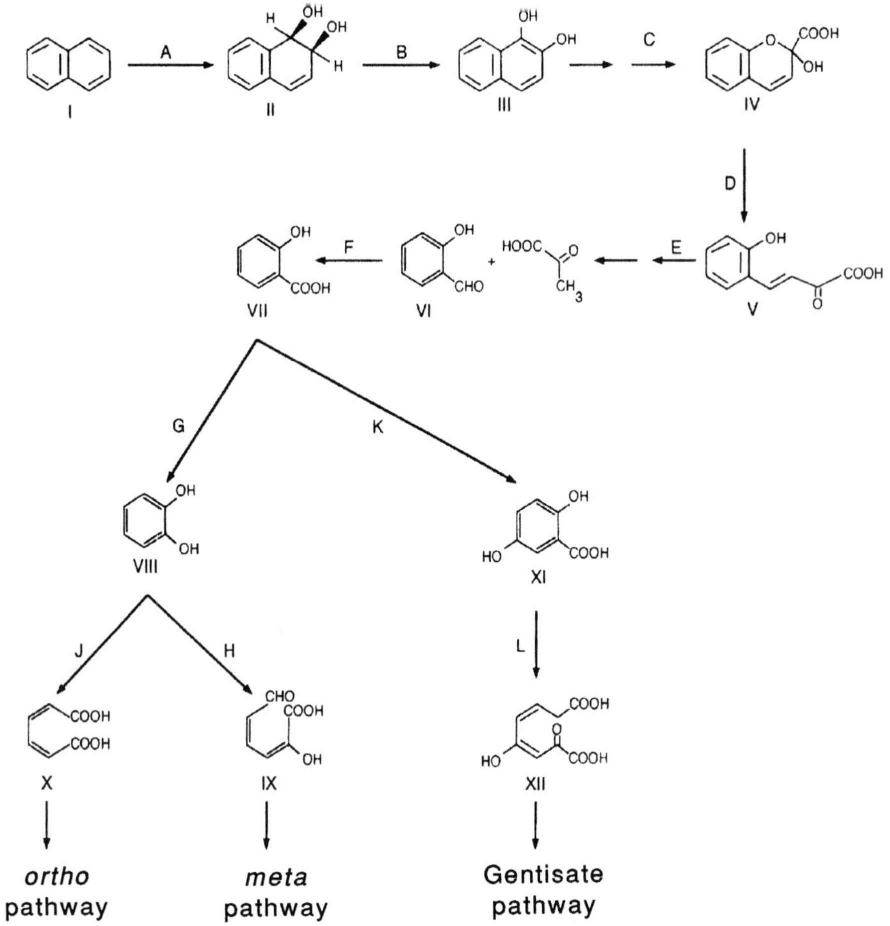

A B C D E F G K J H L

I II III IV V VI VII VIII IX X XI XII

ortho pathway

meta pathway

Gentisate pathway

Figure 1. Catabolic pathways for degradation of naphthalene by bacteria. I, naphthalene; II, *cis*-1,2-dihydro-1,2-dihydroxynaphthalene (naphthalene dihydrodiol); III, 1,2-dihydroxynaphthalene; IV, 2-hydroxy-chromene-2-carboxylate; V, *trans*-*o*-hydroxybenzylidene pyruvate; VI, salicylaldehyde; VII, salicylate; VIII catechol; IX, *cis,cis*-muconate semialdehyde; X, *cis,cis*-muconate; XI, gentisate; XII, maleylpyruvate.
Enzymatic activities: A, naphthalene dioxygenase; B, naphthalene dihydrodiol dehydrogenase; C, 1,2-dihydroxynaphthalene dioxygenase; D, 2-hydroxychromene-2-carboxylate isomerase; E, *o*-hydroxy-benzylidene pyruvate aldolase; F, salicylaldehyde dehydrogenase; G, salicylate 1-hydroxylase; H, catechol 2,3-dioxygenase; J, catechol 1,2-dioxygenase; K, salicylate 5-hydroxylase; L, gentisate 1,2-dioxygenase.

degrading bacteria. In most cases, the salicylate derived from naphthalene is oxidized to catechol by the action of the salicylate 1-hydroxylase. This enzyme has been extensively studied and is one of the model enzymes for flavin-containing monooxygenases [39-41]. In the originally isolated strain of *Pseudomonas* sp., NCIB 9816, the catechol was metabolized by the enzymes of the *ortho*-pathway or by the *meta*-pathway, and it became apparent that a number of variants of NCIB 9816 were in circulation. A partial explanation for these variants was obtained when the *plasmid* of *Pseudomonas* sp. NCIB 9816, which carried the naphthalene genes, was transferred into a plasmid-free *Pseudomonas putida* strain. The plasmid was responsible for the conversion of naphthalene, only as far as catechol, which was then further metabolized by the enzymes of the chromosomally coded *ortho*-pathway. However, it was possible to select spontaneous mutants from the *transconjugant* PaW701 which used a plasmid-coded *meta*-pathway. Examination of the plasmids in some of these mutants showed a deletion of 1.2 to 1.5 kbp between genes for salicylate 1-hydroxylase and catechol 2,3-dioxygenase, which apparently allowed expression of previously silent *meta*-pathway genes [42]. The key enzymes for the *meta*- or *ortho*-pathway of catechol are the ring-opening dioxygenases. The catechol 2,3-dioxygenase (*meta*-pathway) contains Fe^{2+} [43], while the catechol 1,2-dioxygenase (*ortho*-pathway) contains Fe^{3+} [44]. Both enzymes activate oxygen by an unknown mechanism and introduce two hydroxy-groups. In contrast to the above described naphthalene dioxygenase, no reduction equivalents are required and the ring-opening dioxygenases do not contain several protein components.

The third metabolic route for salicylate catabolism is via gentisate [38,45]. The oxidation of salicylate to gentisate is catalyzed by the salicylate 5-hydroxylase. Unlike the salicylate 1-hydroxylase, only a little information is available on the salicylate 5-hydroxylase. First results, obtained with a Gram-negative soil bacterium, indicated that this enzyme is a monooxygenase, requiring NADH [46]. Later studies on a naphthalene-utilizing *Rhodococcus* strain, however, gave totally different results [38]. The salicylate 5-hydroxylase of this Gram-positive soil bacterium requires ATP, CoA and NADPH, and is stimulated by the addition of Mn^{2+} and Zn^{2+}. This requirement for ATP and CoA may be due to the involvement of a salicylate CoA ligase. CoA ligases, which form thioesters of aromatic acids, are known to be involved in the synthesis of phenolic compounds in higher plants, in the anaerobic catabolism of aromatic acids, and in the dehalogenation of 4-chlorobenzoate. The gentisate formed in the reaction, catalyzed by the salicylate 5-hydroxylase is futher oxidized by the gentisate 1,2-dioxygenase, another ring-opening dioxygenase. Unlike the above mentioned catechol dioxygenases, this enzyme does not oxidize the aromatic compound at two vicinyl hydroxy-groups, but oxidizes the gentisic acid between a hydroxy group and a carboxyl group.

Most of the catabolic routes for aromatic compounds are inducible, and the same is true for the naphthalene catabolism. Induction patterns for catabolic routes normally differ from one bacterial species to another. Therefore, the following description of the regulation and organization of the naphthalene catabolic genes (*nah* genes) on plasmid NAH7 can not be generalized. Activation of the *nah* genes from plasmid NAH7 requires both an inducer and the product of a regulatory gene. Salicylate was shown to induce the expression of all genes of plasmid NAH7, coding for the oxidation of naphthalene via salicylate and the *meta*-pathway [47]. Although all *nah* genes of NAH7 are induced by salicylate, the analysis of a series of NAH7 mutations, generated by transposon mutagenesis, revealed the presence of two *nah* operons. The first operon includes genes coding for the conversion of naphthalene to salicylate (upper pathway) and the second operon includes genes coding for the oxidation of salicylate via the catechol *meta*-pathway (lower pathway). It has been proposed that the complete naphthalene catabolic pathway of plasmid NAH7 has evolved from three catabolic modules [47]. The genes of the upper pathway, leading to the formation of salicylate, represent one module. The second module carries the regulatory gene, *nah*R, and the *nah*G gene codes for the salicylate 1-hydroxylase. The third module contributes the *meta*-pathway genes for catechol catabolism. One explanation of how these modules could have been arranged in a catabolic plasmid is the finding that the *nah* genes on plasmid NAH7 are on a defective *transposon* [48]. Transposons and *conjugative plasmids,* that carry sets of catabolic genes for the breakdown of aromatic compounds, may represent the tools by which this genetic information is exchanged between different species of soil bacteria (mainly pseudomonads). This mechanism might also be involved in the rapid adaptation of a population of soil bacteria to the degradation of PAHs, when contamination occurs.

DIBENZO-*P*-DIOXIN

The principal structural difference between the above mentioned PAHs and the dibenzo-*p*-dioxins and the structurally related dibenzofurans is the occurrance of the biaryl ether structure in the latter two compounds. In addition, it is known that the most toxic dibenzo-*p*-dioxins and dibenzofurans are highly halogenated. In general, halogenated aromatic compounds are more resistant to biodegradation than their nonhalogenated analogs. Therefore, todate only little information is available on the *biodegradation* of polyhalogenated dibenzo-*p*-dioxins and dibenzofurans. However, in the last few years, German scientists were successful in isolating bacterial strains that have the ability to grow on nonhalogenated dibenzo-*p*-dioxin and dibenzofuran. These strains, together with strains that dehalogenate the aromatic nucleus of polyhalogenated dibenzo-*p*-

dioxins and dibenzofurans, may represent a promising attempt for the *detoxification* of contaminated areas.

First reports on the cooxidation of dibenzo-*p*-dioxin, its monochloro derivatives, and dibenzofuran by bacterial strains utilizing naphthalene or biphenyl appeared during 1979-1980 [49-51]. Cleavage of the aromatic structure of the dioxin to respective dibenzofuran, however, did not occur. In 1989, both Gram-positive [52] and Gram-negative bacteria [53] were described which could mineralize dibenzofuran, and in 1992 Wittich et al. described, for the first time, a bacterium that mineralizes dibenzo-*p*-dioxin. The strain is a Gram-negative bacterium isolated from Elbe river, belonging to the genus *Sphingomonas* and utilizes, in addition to dibenzo-*p*-dioxin, dibenzofuran and several other polycyclic and monocyclic aromatic compounds.

The degradation route for dibenzo-*p*-dioxin is presented in Figure 2. An oxygenolytic cleavage of the ether bond occurs as the initial enzymatic step in the course of the bacterial attack on dibenzo-*p*-dioxin and dibenzofuran. During the initial step, dibenzo-*p*-dioxin and dibenzofuran are oxidized at carbon atoms 2 and 3, adjacent to the ether structure. This is followed by cleavage of the resulting unstable hemiacetal and spontaneous dehydration. In case of dibenzo-*p*-dioxin metabolism, 2,2',3-

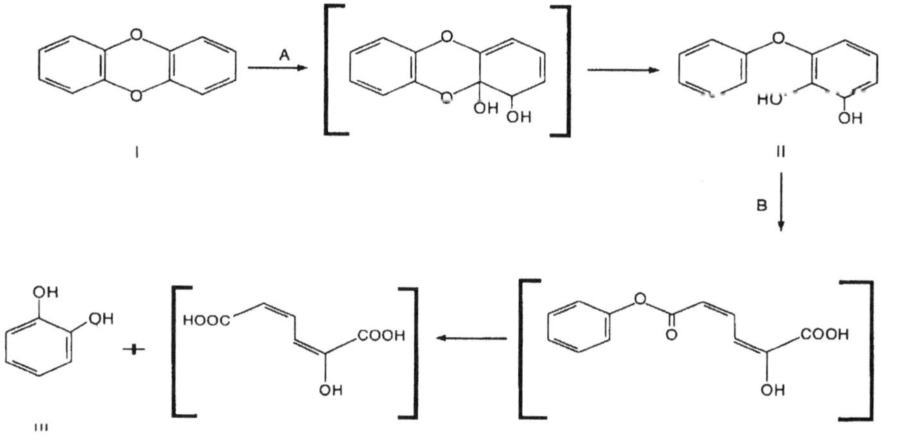

Figure 2. Catabolic pathway for degradation of dibenzo-*p*-dioxin by bacteria. I, dibenzo-*p*-dioxin; II, 2,2',3-trihydroxy diphenylether; III, catechol. Enzymatic activities: A, dibenzo-*p*-dioxin dioxygenase; B, 2,2',3-trihydroxy diphenylether dioxygenase.

trihydroxy diphenylether, and in case of dibenzofuran metabolism 2,2',3-trihydroxy biphenyl [52,54], are the first degradation products of these compounds. The regeneration of NADH seems to be impossible. This unusual enzymatic reaction is the key reaction for the total degradation of both biaryl-ethers by bacteria.

Further metabolism of dibenzofuran leads to salicylate which is metabolized either via catechol or via gentisate, while the 2,2',3-trihydroxy diphenylether, formed during the breakdown of dibenzo-*p*-dioxin, is attacked by a *meta*-cleaving dioxygenase. Whether the *meta*-cleavage occures between C-1 and C-2, yielding 6-(2-hydroxyphenyl)ester of 2-hydroxymuconic acid or between C-3 and C-4, producing 2-hydroxy-3-(2-hydroxyphenoxy)muconic acid semialdehyde remains unclear. The chemical structures of both compounds indicate thermodynamic instability. The ester may be easily saponified, leading to the formation of catechol and 2-hydroxymuconate. The catechol is further metabolized via the *meta*-pathway.

BIPHENYL

Polychlorinated biphenyls (PCBs) are highly recalcitrant, and have toxic effects in animals [55]. A number of chemical families, including the PCBs, polychlorinated dibenzo-*p*-dioxins, and polychlorinated dibenzofurans, have similar toxic effects in animals, and appear to act via the same receptor mechanism [56]. The proposed mechanism of action is initiated by interaction of the chemical with a cytosolic receptor (the Ah receptor), followed by movement of the toxicant-receptor complex to the nucleus, interaction with responsive genetic elements, resulting in altered gene expression, and, ultimately, the manifestations of toxicity.

Metabolic breakdown of PCBs by microorganisms is considered to be one of the major routes of environmental degradation for these widespread pollutants. Environmental monitoring has demonstrated that the PCBs, found in environmental samples, are those containing five or more chlorines [57]. This indicates that the less-chlorinated biphenyls are more rapidly degraded than highly chlorinated biphenyls. The degree of chlorination seems to be the most significant factor affecting biodegradation of PCBs, but other factors, like water solubility, volatility, pH, temperature, oxygen, nutrient supplies, emulsification, and adsorption by various materials all combine to varying extents to affect biodegradability of PCBs.

The ubiquitous occurrence of bacteria, which can utilize biphenyl as a sole source of carbon and energy, has been reported. Several species of Gram-negative and Gram-positive bacteria have been shown to degrade biphenyl and PCBs. The main pathway for biphenyl catabolism is outlined in Figure 3 [58-66].

The first step of the biphenyl catabolic pathway (Figure 3) is catalized by the biphenyl 2,3-dioxygenase. It catalyzes the formation of a 2,3-

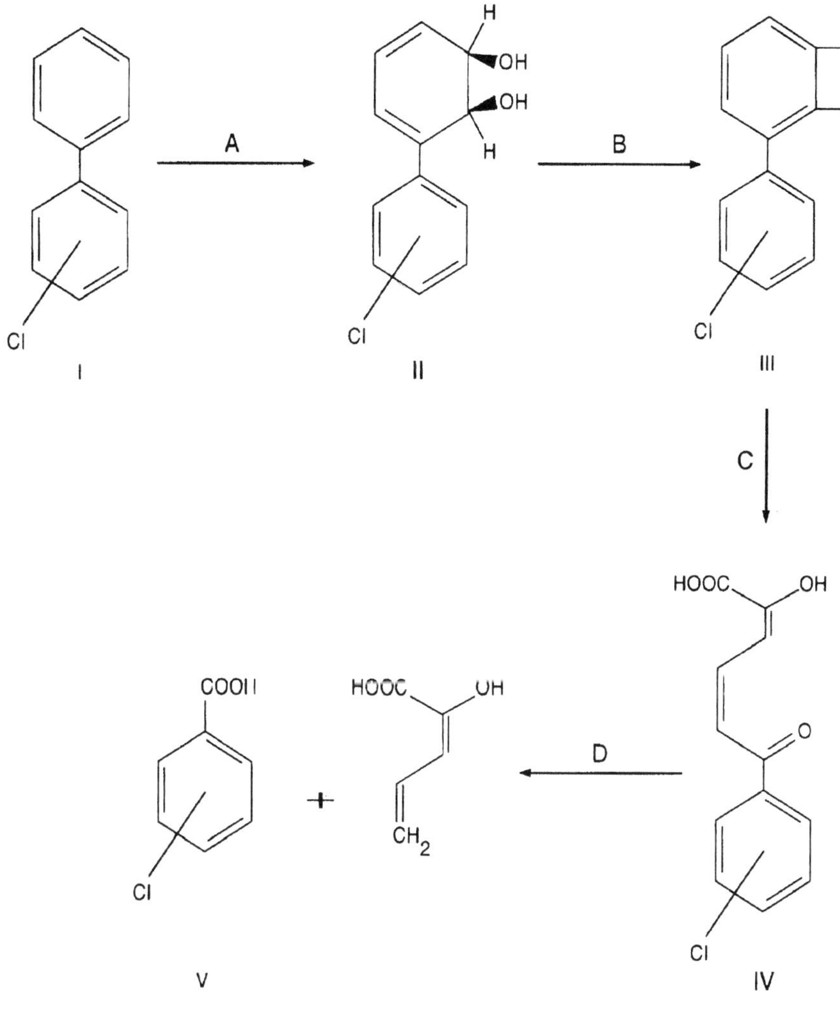

Figure 3. Catabolic pathway for degradation of biphenyl and PCB by bacteria. I, biphenyl; II, 2,3-dihydroxy-4-phenylhexa-2,4-diene (2,3-dihydrodiol); III, 2,3-dihydroxy biphenyl; IV, 2-hydroxy-6-oxo-6-phenylhexa-2,4-dienoic acid (*meta*-cleavage compound); V, benzoic acid.
Enzymatic activities: A, biphenyl dioxygenase; B, 2,3-dihydrodiol dehydrogenase; C, 2,3-dihydroxy biphenyl dioxygenase; D, *meta*-cleavage compound hydrolase.

dihydrodiol, and has been cloned from three different *Pseudomonas* strains, able to degrade PCBs [67-69]. Erickson and Mondello [68] reported that they could identify six open reading frames (ORFs) in the DNA-region, coding for biphenyl 2,3-dioxygenase of *Pseudomonas* sp. LB400. Four of these ORFs show some homology to the components of the toluene dioxygenase from *Pseudomonas putida* F1 [70], and the benzene dioxygenase of another *Pseudomonas putida* isolate [71]. Therefore, Erickson and Mondello [68] propose that the biphenyl 2,3-dioxygenase is a multicomponent enzyme, consisting of a two-subunit iron-sulphur protein, a ferredoxin, and a reductase. The high degree of sequence identity between the genes of the biphenyl 2,3-dioxygenase and the genes of the toluene dioxygenase shows that these pathways are closely related and illustrates how bacteria are developing systems, capable of degrading xenobiotic compounds. With respect to the substrate spectrum, that is accepted by the biphenyl dioxygenases; there seem to exist differences between the several biphenyl, and PCB-degrading bacterial strains. Ahmad et al. [67] showed that the biphenyl 2,3-dioxygenase of *Pseudomonas testosteroni* B-356 accepts nonchlorinated and chlorinated biphenyls, and they report that dehalogenation of biphenyls, chlorinated at positions 2 or 4, can occur during initial attack by the dioxygenase.

The next step of PCB-catabolism is catalyzed by a 2,3-dihydrodiol dehydrogenase and regenerates NADH, and leads to the formation of 2,3-dihydroxybiphenyl. With respect to the reaction mechanism of the fist two steps of PCB degradation, there is a great similarity to other catabolic pathways of aromatics, like benzene, toluene, benzoate, naphthalene, etc.

The 2,3-dihydroxybiphenyl is attacked by a *meta*-cleaving dioxygenase, and 2-hydroxy-6-oxo-6-phenylhexa-2,2-dienoic acid is formed. The 2,3-dihydroxybiphenyl 1,2-dioxygenase of *Pseudomonas alcaligenes* has been cloned and sequenced [72]. Like other *meta*-cleaving dioxygenases, this enzyme contains Fe^{2+}, and in contrast to the above mentioned biphenyl 2,3-dioxygenase, this enzyme contains 8 identical subunits and, thus, it is not a multicomponent enzyme. With respect to these properties, the 2,3-dihydroxybiphenyl dioxygenase resembles very much the catechol 2,3-dioxygenase which is composed of 4 Fe^{2+}-containing identical subunits. On the other hand, Furukawa et al. [72] could not detect any significant sequence homologies between the *xyl*E gene, coding for the catechol 2,3-dioxygenase of the TOL plasmid and the *bph*C gene, coding for the 2,3-dihydroxybiphenyl 1,2-dioxygenase of *Pseudomonas pseudoalcaligenes*. Furthermore, Furukawa et al. [73] demonstrated that the 2,3-dihydroxybiphenyl 1,2-dioxygenases of two biphenyl-degrading *Pseudomonas* species showed no significant sequence homology.

The last step in PCB catabolism, leading to the formation of chlorinated benzoic acids, is catalyzed by a hydrolase. Many of the PCB-degrading bacteria, investigated so far, can not metabolize chlorobenzoates. Furthermore, the PCB degradation seems to be inhibited by the accumulation of chlorobenzoates and other chlorinated metabolites [74]. Therefore, total degradation of chlorobiphenyls occurs only when bacteria, carrying the *bph* gene cluster coding for the above described PCB-degradation, are co-cultured with bacteria able to degrade chlorinated benzoates [75,76], or when genes coding for the degradation of chlorobenzoates are introduced in PCB-degrading bacteria [77,78].

Another promising approach for the detoxification of PCBs is the finding that anaerobic bacteria dechlorinate PCBs reductively [79, 80]. The authors used anaerobic microorganisms from Hudson River sediment and report that, at PCB concentrations of 700 ppm Aroclor, 53 per cent of the total chlorine was removed in 16 weeks, and the proportion of mono- and dichlorobiphenyls increased from 9 to 88 per cent. Dechlorination occurred primarily from the *meta* and *para* positions. These results indicate that reductive dechlorination may be an important environmental fate of PCBs, and suggest that a sequential anaerobic-aerobic biological treatment system for PCBs may be feasible. The proton source for the microbial reductive dechlorination of 2,3,4,5,6-pentachlorobiphenyl has been identified by Nies and Vogel [81]. The authors report that the exact mechanism of the electron transfer for the dechlorination of PCBs is unknown; however, they could show that the source of the hydrogen atom is the proton from water, and that chloride is released from the PCB.

CHLOROBENZOATES

Chlorinated benzoic acids (CBAs) are often dead-end products of the metabolism of PCBs [74,82-84]. Shields et al. [85] reported the complete degradation of 4-chlorobiphenyl by a pure culture, but in all other reports on the biodegradation of chlorobiphenyls, total degradation of the compounds could only be achieved, using mixed cultures or genetically engineered strains. As discussed above, metabolites of chlorobenzoates, that are formed by biphenyl-induced oxygenases, inhibit the degradation of PCBs. Therefore, it is essential to remove all the CBAs accumulated by PCB-degrading bacteria, for efficient biodegradation of these compounds. One approach to solve this problem is the use of CBA-degrading strains in co-cultures with the PCB-degrading strain [17,75, 82,83]. Another possibility is to construct genetically engineered strains, that are able to degrade PCBs and CBAs [75,78,86].

There are two principal ways to indicate how CBAs can be dehalogenated:

a) Dehalogenation is the first step of the pathway, preceding the

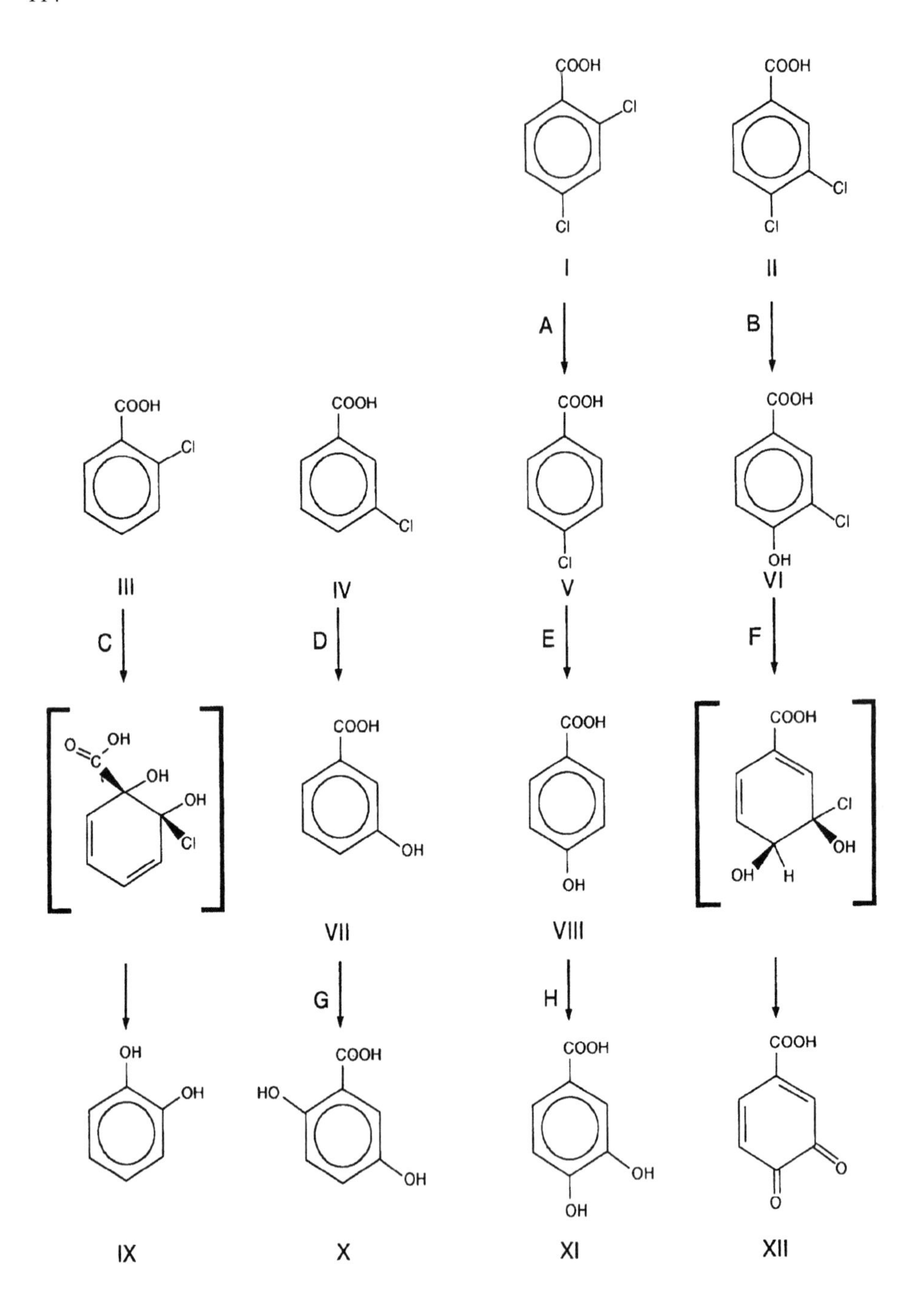

Figure 4. Catabolic pathways for degradation of chlorinated benzoic acids

by bacteria. I, 2,4-dichlorobenzoate; II, 3,4-dichlorobenzoate; III, 2-chlorobenzoate; IV, 3-chlorobenzoate; V, 4-chlorobenzoate; VI, 3-chloro-4-hydroxybenzoate; VII, 3-hydroxybenzoate; VIII, 4-hydroxybenzoate; IX, catechol; X, gentisate; XI, protocatechuate; XII, 4-carboxy-1,2-benzoquinone. Enzymatic acitivities: A, 2,4-dichlorobenzoate dehalogenase (reductive); B, 3,4-dichlorobenzoate dehalogenase; C, 2-halobenzoate 1,2-dioxygenase; D, 3-chlorobenzoate dehalogenase; E, 4-chlorobenzoate dehalogenase (hydrolytic); F, 3-chloro-4-hydroxybenzoate hydratase; G, 3-hydroxybenzoate 6-hydroxylase; H, 4-hydroxybenzoate 3-hydroxylase.

ring cleavage. In most cases, hydroxylated benzoic acids are formed, following the dehalogenation.

b) Dehalogenation occurs after ring cleavage. CBAs are converted to chlorocatechols, which are cleaved by enzymes of a modified *ortho*-pathway. The dehalogenation often occurs spontaneously during further metabolism of the nonaromatic intermediates.

The breakdown of CBAs via chlorocatechols, and subsequent spontaneous dehalogenation of nonaromatic intermediates is widely distributed among Gram-negative bacteria, especially *Pseudomonas, Alcaligenes*, and *Acinetobacter* strains. The genes of this pathway are cloned, and most of the genetic experiments, concerning the construction of recombinant strains with a wider substrate spectrum, including chlorinated biphenyls, are performed with this set of genes. Unfourtunately, in some instances, the induction of *meta*-cleaving dioxygenases can lead to misrouting of metabolites, and suicide inactivation of the catechol 2,3-dioxygenase by 3-chlorocatechol is a well-known phenomonon [87-89]. Other chlorinated metabolites may be either toxic, inhibiting other enzymes of the degradation pathway, or are accumulated [90].

Some of the problems mentioned above can be avoided if bacteria dehalogenate their substrates before ring cleavage. Additionally, this type of dehalogenation represents an initial detoxification, and many other bacterial strains, capable of growth with hydroxylated benzoates, can use the reaction products, allowing a more efficient use of co-cultures. Some Gram-negative and Gram-positive bacteria, that can grow with CBAs as sole carbon source, exhibit this mode of CBA degradation.

Dehalogenation before ring cleavage is known for 4-CBA (91-98), 3-CBA [99], 2-CBA [88,100-102], 2,4-CBA [103,104], and 3,4-CBA [82]. In bacteria, three principal reaction mechanisms exist for dehalogenation of CBAs: Dehalogenation of 2,4-CBA to 4-CBA occurs by a reductive reaction mechanism, the dehalogenation of 2-CBA is catalyzed by an oxygenase (oxidative pathway), and the dehalogenation of 4-CBA and 3-CBA is mediated by a hydrolytic reaction mechanism (Figure 4).

The *reductive dehalogenation* normally occurs in anaerobic bacteria, e.g., dehalogenation of biphenyls (see above), dehalogenation of 3-CBA

[105], although some examples exist for the involvement of the reductive dehalogenation in the aerobic degradation of chlorinated compounds, like 2,4-CBA [103] and pentachlorophenol [106,107].

The *oxidative dehalogenation* by dioxygenases has been reported for the catabolism of 4-chlorophenylacetate, by a *Pseudomonas* species [108-110], and 2-CBA [88,100,101,102,111]. 4-Chlorophenylacetate is converted to 3,4-dihydroxyphenylacetate by the action of 4-chlorophenylacetate 3,4-dioxygenase, while 2-CBA is converted to catechol by a 2-halobenzoate 1,2-dioxygenase. It could be shown that the oxygen of both hydroxy groups is derived from molecular oxygen [111]. Both enzymes belong to the class I type dioxygenases, and require NADH and Fe^{2+} as cofactors. The 2-halobenzoate 1,2-dioxygenase consists of 2 components: an iron-sulphur protein with two different subunits and an iron-sulphur flavoprotein monomer. A close relationship to benzoate 1,2-dioxygenase [112-115] seems to exist. The enzyme exhibits a wide substrate range, being able to oxidize 2-fluro-, 2-iodo-, 2-bromo- and 2-aminobenzoate, in addition. The highly unstable intermediate, 2-chloro-3,5-cyclohexadiene-1,2-diol-1-carboxylic acid may be formed by the action of the dioxygenase, which is spontaneously dehalogenated and decarboxylated to catechol in the course of the rearomatization of the ring structure. The close relationship of the above described 2-halobenzoate 1,2-dioxygenase to the benzoate 1,2-dioxygenase again illustrates how bacteria are developing systems, capable of degrading xenobiotic compounds.

The *hydrolytic dehalogenation* is the best known reaction mechanism, and the enzymes and their corresponding genes are fairly well characterized. First reports on this dehalogenation mechanism appeared in 1984 and the authors could demonstrate that the hydroxy group of *p*-hydroxybenzoate, which is formed during the dehalogenation of 4-CBA, originated from water [94,116]. In *Pseudomonas* sp. CBS3, three enzymes are involved in the hydrolytic dehalogenation of 4-CBA [117], and the reaction requires ATP and Coenzyme A [118,119]. 4-CBA-CoA is an intermediate in this reaction. Each of the three enzymes, involved in hydrolytic dehalogenation of 4-CBA, were subcloned, purified, and characterized separately [120]. In the first reaction, catalyzed by the 4-CBA CoA-ligase, 4-CBA is activated. ATP, CoA, and Mg^{2+} are required for the reaction and AMP and 4-CBA are formed. The 4-CBA CoA-ligase shows homologies to a number of enzymes involved in secondary metabolism and to the enzymes involved in the β-oxidation, suggesting a route for its evolution [121]. All these enzymes activate their carboxylated substrates by adenylation and concomitant formation of a thioester. The second step in the hydrolytic dehalogenation of 4-CBA is catalyzed by the 4-CBA dehalogenase and leads to the formation of 4-hydroxybenzoate-CoA. The 4-CBA dehalogenase of *Pseudomonas* sp. CBS3 is a homotetramer that shows some homologies to enzymes involved in fatty acid degradation.

Finally, the thioester of 4-hydroxybenzoate-CoA is hydrolyzed by a thioesterase. The thioesterase of *Pseudomonas* sp. CBS3 was shown to be a homotetramer with a subunit size of 16 kD.

The 4-CBA dehalogenases of *Arthrobacter* strains, that dehalogenate 4-CBA to 4-hydroxybenzoate, seem to exhibit the same reaction mechanism as described above. Therefore, it is interesting that the 4-CBA dehalogenase genes of two *Arthrobacter* strains have been cloned recently [97,122]. Comparison of the sequence of the DNA region coding for the 4-CBA dehalogenase of *Arthrobacter* sp. strain SU to the corresponding DNA region of *Pseudomonas* sp. CBS3 revealed a high homology of the genes for 4-CBA CoA-ligase and 4-CBA dehalogenase, while the third gene, coding for the thioesterase could not be detected in *Arthrobacter* sp. SU [122]. Whether the third ORF present on the DNA region, coding for the 4-CBA dehalogenase genes of *Arthrobacter* sp. SU is a thioesterase, is unknown to date. This homology between genes involved in the hydrolytic dehalogenation of 4-CBA in Gram-negative bacteria and in Gram-positive bacteria again may indicate that genes coding for the degradation of xenobiotics are present on mobile DNA-elements, like transposons and plasmids. The finding, that the 4-CBA dehalogenase genes of several Gram-positive bacteria are coded on plasmids, is in accordance with this suggestion [122,123].

CONCLUSIONS

The biodegradability of PAHs, dibenzo-*p*-dioxins, dibenzofurans, and PCBs depends on several parameters, like their *bioavailability*, the number of aromatic cycles (PAHs), and the degree of halogenation (dioxins, dibenzofurans, PCBs). It is known that aerobic bacteria can degrade PAHs, the nonhalogenated dibenzo-*p*-dioxin, dibenzofuran, and biphenyl. The enzymes and genes, involved in this aerobic metabolism of xenobiotics, have been characterized and cloned, and it has been demonstrated that many of the enzymes are closely related to the well-known catabolism of monocyclic aromatic compounds. This fact may indicate how the bacterial catabolism of xenobiotics has developed. The bottleneck of the catabolism of polyhalogenated xenobiotics is the accumulation of toxic or inhibitory halogenated degradation products. Suicide inactivation of *meta*-cleaving dioxygenases has been demonstrated. This inactivation of *meta*-cleaving dioxygenases, which often represent key enzymes in the breakdown of PAHs, dibenzo-*p*-dioxin, dibenzofuran, and biphenyl, leads to a severe reduction of the flux of xenobiotics through the catabolism of the degrading strains, and is one of the reasons why dehalogenation, before ring-cleavage, may represent a promising approach for the detoxification of dioxins, dibenzofurans, and biphenyls. The anaerobic reductive dehalogenation of PCBs has been demonstrated. Therefore, a sequential anaerobic-aerobic biological treatment system may be feasable. In this system, the dehalogenation would occur anaerobically, while the breakdown of the less toxic low halogenated dioxins, dibenzofurans, and biphenyls would occur aerobically. Another approach for total biodegradation of the polyhalogenated compounds is the co-culture of dioxin-, dibenzofuran- or biphenyl-degrading bacteria with those, that have evolved the ability to degrade chlorinated monocyclic aromatics.

The degradation of chlorinated monocyclic aromatic compounds, like, CBAs may occur by two different principles:

a) Attack on the halogenated compound by broad substrate range oxygenases and further metabolism of halogenated catechols by a modified *ortho*-pathway : The genes of this catabolism have been cloned, and some scientists were successful in constructing genetically engineered bacteria that can degrade chlorinated biphenyls completely.

b) Dehalogenation of CBAs before ring-cleavage : It is known that some aerobic bacterial strains can dehalogenate CBAs. This dehalogenation can occur reductively, oxidatively, or hydrolytically. The enzymes and genes, involved in the reductive dehalogenation of CBAs by aerobic bacteria, have not been characterized until now. On the other hand, it could be demonstrated that the oxidative dehalogenation is catalyzed by specialized broad substrate range dioxygenases, while the enzymes involved in the hydrolytic dehalogenation of 4-CBA have been characterized, and the genes have been cloned and sequenced. Further cloning of genes, involved in the aerobic dehalogenation of CBAs and characterization of the enzymes, will probably lead to a good characterized set of genes, which will open some more approaches for the construction of genetically engineered strains that can degrade chlorinated dioxins, dibenzofurans, and biphenyls. These strains, and combinations of different xenobiotic-degrading or dehalogenating strains, may be useful in the detoxification of contaminated areas in future.

REFERENCES

1 Hites RA, Laflamme RF, Windsor JG. In: Petrakis L, Weiss FT, eds. Petroleum in the Marine Environment (Advances in Chemistry Series) Washington D.C.: American Chemical Society, 1980; 289-311.

2 Jacob J, Karcher W, Belliando JJ, Wagstaffe PJ. Fesenius Z Anal Chem 1986; 323: 1-10.

3 Johnson AC, Larsen D. Mar Environ Res 1985; 15: 1-16.

4 Goerlitz DF, Troutman DE, Godsy EM, Franks BJ. Environ Sci Technol 1985; 19: 955-961.

5 Bos RP, Theuws JL, Jongeneelen FJ, Henderson PT. Mutat Res 1988; 204: 203-206.
6 Moertelmanns K, Wawarths S, Lawlar T, Speck W, Tainer B, Zeiger E. Environ Mutagen 1986; 8: 1-119.
7 Buser HR. Environ Sci Technol 1986; 20: 404-408.
8 Czuczwa JM, Hites RA. Environ Sci Technol 1984; 18: 444-450.
9 Swanson SE, Rappe C, Malmström J, Kringstad KP. Chemosphere 1988; 17: 681-691.
10 Couture LA, Abbott BD, Birnbaum LS. Teratology 1990; 42: 619-627.
11 Hanson DJ. Contam and Environ 1991; 7-14.
12 Ballschmiter K. Nachr Chem Tech Lab 1991; 39: 988-1000.
13 Hooper SW, Pettigrew CA, Sayler GS. Environ Toxicol Chem 1990; 9: 655-667.
14 Hutzinger O, Veerkamp W. In: Leisinger T, Hutler R, Cook AM, Nuesch J, eds. Microbial Degradation of Xenobiotic and Recalcitrant Compounds. New York: Academic Press Inc, 1981; 3-45.
15 Wilkins JR III, Nickel JT. Rev Environ Health 1984; 4: 269-286.
16 Adriaens P, Kohler HPE, Kohler-Staub D, Focht DD. Appl Environ Microbiol 1989; 55: 887-892.
17 Sylvestre M, Masse R, Ayotte C, Messier F, Fauteux J. Appl Microbiol Biotechnol 1985; 21: 191-195.
18 Evans WC, Fernley HN, Griffiths E. Biochem J 1965; 95: 819-831.
19 Kiyohara H, Nagao K. J Gen Microbiol 1978; 105: 69-75.
20 Guerin WF, Jones GE. Appl Environ Microbiol 1988; 54: 937-944.
21 Heitkamp MA, Franklin W, Cerniglia CE. Appl Environ Microbiol 1988; 54: 2549-2555.
22 Walter U, Beyer M, Klein J, Rehm HJ. Appl Microbiol Biotechnol 1991; 34: 671-676.
23 Mueller JG, Chapman PJ, Blattmann BO, Pritchard PH. Appl Environ Microbiol 1990; 56: 1079-1086.
24 Weissenfels WD, Beyer M, Klein J, Rehm HJ. Appl Microbiol Biotechnol 1991; 34: 528-535.
25 Bumpus JA. Appl Environ Microbiol 1989; 55 : 154-158.
26 Hammel KE, Gai WZ, Green B, Moen MA. Appl Environ Microbiol 1992; 58: 1832-1838.
27 Pothuluri JV, Heflich RH, Fu PP, Cerniglia CE. Appl Environ Microbiol 1992; 58: 937-941.
28 Weissenfels WD, Klewer HJ, Langhoff J. Appl Microbiol Biotechnol 1992; 36: 689-696.
29 Davies JI, Evans WC. Biochem J 1964; 91: 251-261.
30 Jerina DM, Daly JW, Jeffrey AM, Gibson DT. Arch Biochem Biophys 1971; 142: 394-399.
31 Ensley BD, Gibson DT, Laborde A. J Bacteriol 1982; 149: 948-954.

32 Ensley BD, Ratzkin BJ, Osslund TD, Simon MJ, Wackett LP, Gibson DT. Science 1983; 222: 167-169.
33 Patel TR, Gibson DT. J Bacteriol 1974; 119: 879-885.
34 Patel TR, Barnsley EA. J Bacteriol 1980; 143: 668-673.
35 Barnsley EA. Biochem Biophys Res Commun 1976; 72: 1116.
36 Eaton RW, Chapman PJ. J Bacteriol 1992; 174: 7542-7554.
37 Shamsuzzaman KM, Barnsley EA. Biochem Biophys Res Commun 1974; 60: 582-589.
38 Grund E, Denecke B, Eichenlaub R. Appl Environ Microbiol 1992; 58: 1874-1877.
39 Einarsdottir GH, Starkovich MT, Tu ST. Biochemistry 1988; 27: 3277-3285.
40 Katagiri M, Maeno H, Yamamoto S, Hayaishi O. J Biol Chem 1965; 240: 3414-3417.
41 White-Stevens RH, Kamin H. J Biol Chem 1972; 247: 2371-2381.
42 Cane PA, Williams PA. J Gen Microbiol 1986; 132: 2919-2929.
43 Hayaishi O, Katagiri M, Rothberg S. J Biol Chem 1957; 229: 905-920.
44 Fujisawa H, Hayaishi O. J Biol Chem 1968; 243: 2673-2681.
45 Starovoitov II, Nefedova MY, Yakovlev GI, Zyakun AM, Adanin VM. Izv Akad Nauk SSSR Ser Khim 1975; 9: 2091-2092.
46 Buswell JA, Paterson A, Salkinoja-Salonen MS. FEMS Microbiol Lett 1980; 8: 135-137.
47 Yen KM, Serdar CM. Crit Rev Microbiol 1988; 15: 247-268.
48 Tsuda M, Ino T. Mol Gen Genet 1990; 223: 33-39.
49 Cerniglia CE, Morgan JC, Gibson DT. Biochem J 1979; 180: 175-185.
50 Klecka GM, Gibson DT. Biochem J 1979; 180: 639-645.
51 Klecka GM, Gibson DT. Appl Environ Microbiol 1980; 34: 288-296.
52 Engesser KH, Strubel V, Christoglou K, Fischer P, Rast HG. FEMS Microbiol Lett 1989; 65: 205-210.
53 Fortnagel PH, Harms H, Wittich RM, Krohn S, Meyer H, Francke W. Naturwissenschaften 1989; 76 : 222-223.
54 Fortnagel PH, Harms H, Wittich RM, Krohn S, Meyer H, Sinnwell V, Wilkes H, Francke W. Appl Environ Microbiol 1990; 56: 1148-1156.
55 Kamrin MA, Fischer LJ. Environ Health Persp 1991; 91: 157-164.
56 Safe S. CRC Crit Rev Toxicol 1984; 13: 319-394.
57 Koeman JH, Ten Noever De Brauw MC, De Vos RH. Nature (London) 1969; 221: 1126.
58 Ahmed M, Focht DD. Can J Microbiol 1973; 19: 47-52.
59 Ballschmiter K, Unglert KC, Neu HT. Chemosphere 1977; 1: 51-56.
60 Bedard DL, Wagner RE, Brennan MJ, Haberl ML, Brown Jr JF. Appl Environ Microbiol 1987; 53: 1094-1102.

61 Catelani D, Sorlini C, Treccani V. Experientia 1981; 27: 1173.
62 Furukawa K, Matsumura F, Tonomura K. Agric Biol Chem 1978; 42: 542.
63 Furukawa K, Tomizuka N, Kamibayashi A. Appl Environ Microbiol 1979; 38: 301-310.
64 Gibson DT, Roberts RL, Wells MC, Kobal UM. Biochem Biophys Res Commun 1973; 50: 211.
65 Lunt D, Evans C. Biochem J 1970; 118: 54.
66 Ohmori T, Ikai T, Minoda Y, Yamada K. Agric Biol Chem 1973; 37: 680.
67 Ahmad D, Sylvestre M, Sondossi M. Appl Environ Microbiol 1991; 57: 2880-2887.
68 Erickson BD, Mondello FJ. J Bacteriol 1992; 174: 2903-2912.
69 Furukawa K, Miyazaki T. J Bacteriol 1986; 166: 392-398.
70 Zylstra GJ, Gibson DT. J Biol Chem 1989; 264: 14940-14948.
71 Irie S, Doi S, Yorifugi T, Takagi M, Yano K. J Bacteriol 1987; 169: 5174-5179.
72 Furukawa K, Arimura N, Miyazaki T. J Bacteriol 1987; 169: 427-429.
73 Furukawa K, Hayase N, Taira K, Tomizuka N. J Bacteriol 1989; 171: 5467-5472.
74 Sondossi M, Sylvestre M, Ahmad D. Appl Environ Microbiol 1992; 58: 485-495.
75 Havel J, Reineke W. FEMS Microbiol Lett 1991; 78: 163-170.
76 Masse R, Messier F, Peloquin L, Ayotte C, Sylvestre M. Appl Environ Microbiol 1984; 47: 947-951.
77 Fulthorpe RR, Wyndham RC. Appl Environ Microbiol 1992; 58: 314-325.
78 Mokross H, Schmidt E, Reineke W. FEMS Microbiol Lett 1990; 71: 179-186.
79 Brown Jr JF, Bedard DL, Brennan MJ, Carnahan JC, Feng H, Wagner RE. Science 1987; 236: 709-712.
80 Quensen III JF, Tiedje JM, Boyd SA. Science 1988; 242: 752-754.
81 Nies L, Vogel TM. Appl Environ Microbiol 1991; 57: 2771-2774.
82 Adriaens P, Focht DD. Appl Environ Microbiol 1991; 57: 173-179.
83 Havel J, Reineke W. Appl Microbiol Biotechnol 1992; 38: 129-134.
84 Khan A, Walia S. Appl Environ Microbiol 1989; 55: 798-805.
85 Shields MS, Hooper SW, Sayler GS. J Bacteriol 1985; 163: 882-889.
86 Adams RH, Huang CM, Higson FK, Brenner V, Focht DD. Appl Environ Microbiol 1992; 58: 647-654.
87 Bartels I, Knackmuss HJ, Reineke W. Appl Environ Microbiol 1984; 47: 500-508.
88 Fetzner S, Müller R, Lingens F. Biol Chem Hoppe-Seyler 1989; 370: 1173-1182.

89 Schmidt E, Bartels I, Knackmuss HJ. Appl Microbiol Biotechnol 1987; 27: 94-99.
90 Taeger K, Knackmuss HJ, Schmidt E. Appl Microbiol Biotechnol 1988; 28: 603-608.
91 Groenewegen PEJ, Van Den Tweel WJJ, De Bont JAM. Appl Microbiol Biotechnol 1992; 36: 541-547.
92 Klages U, Lingens F. FEMS Microbiol Lett 1979; 6: 201-203.
93 Marks TS, Smith ARW, Quirk AV. Appl Environ Microbiol 1984; 48: 1020-1025.
94 Müller R, Thiele J, Klages U, Lingens F. Biochem Biophys Res Commun 1984; 124: 178-182.
95 Ruisinger S, Klages U, Lingens F. Arch Microbiol 1976; 110: 253-256.
96 Shimao M, Onishi S, Mizumori S, Kato N, Skazawa C. Appl Environ Microbiol 1989; 55: 478-482.
97 Tsoi TV, Zaitsev GM, Plotnikova EG, Kosheleva IA, Boronin AM. FEMS Microbiol Lett 1991; 81: 165-170.
98 Van Den Tweel WJJ, Ter Burg N, Kok JB, De Bont JAM. Appl Microbiol Biotechnol 1986; 25: 169-174.
99 Johnston HW, Briggs GG, Alexander M. Soil Biol Biochem 1972; 4: 187-190.
100 Engesser KH, Schulte P. FEMS Microbiol Lett 1989; 51: 143-147.
101 Sylvestre M, Mailhiot K, Ahmad D, Masse R. Can J Microbiol 1989; 35: 439-443.
102 Zaitsev GM, Karasevich YN. Mikrobiologiya 1984; 53: 75-80.
103 Van Den Tweel WJJ, Kok JB, De Bont JAM. Appl Environ Microbiol 1987; 53: 810-815.
104 Zaitsev GM, Karasevich YN. Mikrobiologiya 1985; 54: 356-359.
105 Dolfing J, Tiedje JM. FEMS Microbial Ecol 1986; 38: 293-298.
106 Apajalahti JHA, Salkinoya-Salonen MS. J Bacteriol 1987; 169: 675-681.
107 Schenk T, Müller R, Mörsberger F, Otto MK, Lingens F. J Bacteriol 1989; 171: 5487-5491.
108 Markus A, Klages U, Krauss S, Lingens F. J Bacteriol 1984; 160: 618-621.
109 Markus A, Krekel D, Lingens F. J Biol Chem 1986; 261: 12883-12888.
110 Schweizer D, Markus A, Seez M, Ruf HH, Lingens F. J Biol Chem 1987; 262: 9340-9346.
111 Fetzner S, Müller R, Lingens F. J Bacteriol 1992; 174: 279-290.
112 Yamaguchi M, Fujisawa H. J Biol Chem 1978; 253: 8848-8853.
113 Yamaguchi M, Fujisawa H. J Biol Chem 1980; 255: 5058-5063.
114 Yamaguchi M, Fujisawa H. J Biol Chem 1981; 256: 6783-6787.
115 Yamaguchi M, Fujisawa H. J Biol Chem 1982; 257: 12497-12502.

116 Marks TS, Wait R, Smith ARW, Quirk AV. Biochem Biophys Res Commun 1984; 124: 669-674.
117 Elsner A, Löffler F, Miyashita K, Müller R, Lingens F. Appl Environ Microbiol 1991; 58: 1385-1387.
118 Löffler F, Müller R, Lingens F. Biochem Biophys Res Commun 1991; 176: 1106-1111.
119 Scholten JD, Chang KH, Babbitt PC, Charest H, Sylvestre M, Dunaway-Mariano D. Science 1991; 253: 182-185.
120 Chang KH, Liang PH, Beck W, Scholten JD, Dunaway-Mariano D. Biochemistry 1992; 31: 5605-5610.
121 Babitt PC, Kenyon GL, Martin BM, Charest H, Sylvestre M, Scholten JD, Chang KH, Liang PH, Dunaway-Mariano D. Biochemistry 1992; 31: 5594-5604.
122 Schmitz A, Gartemann KH, Fiedler J, Grund E, Eichenlaub R. Appl Environ Microbiol 1992; 58: 4068-4071.
123 Zaitsev GM, Tsoi TV, Grishenkov VG, Plotnikova EG, Boronin AM. FEMS Microbiol Lett 1991; 81: 171-176.

Biotransformations: Microbial Degradation of Health Risk Compounds
Ved Pal Singh, editor
© 1995 Elsevier Science B.V. All rights reserved.

Microbial degradation of halogenated aromatics

Manzoor A. Bhat and C.S. Vaidyanathan

Department of Biochemistry and UGC Centre of Advanced Study, Indian Institute of Science, Bangalore 560012, India

INTRODUCTION

Rapid developments over the past three decades, in the fields of industrial and agricultural chemicals, have resulted in the production of enormous quantities of organochlorine compounds. Large amounts of these compounds find use as solvents, lubricants, heat transfer media, insulators, fire retardants, paints, and varnishes. Some find use as pesticides, herbicides and plasticizers. Many of these compounds become widely dispersed in the biosphere either deliberately or accidentally, and have been detected in water, air and even in humans [1]. Another source of halogenated organic compounds in the environment is due to the chlorination procedures used in the treatment of potable water, waste water, and cooling water.

Most of these compounds were absent in the biosphere prior to their synthesis by man, and may be considered as environmentally foreign compounds. Consequently, they persist in the environment, as they are not susceptible to biological transformation. Therefore, massive pollution problems resulting from large scale use of such chemicals have confronted most of the nations worldwide [2]. Public concern over the possible toxic effects of chemicals on humans and their environment has largely focused on a few classes of compounds. Of these, chlorinated compounds have elicited the most interest. In response to such incidence of pollution, the United States Congress in 1976 passed the Toxic Substances Control Act (TSCA), requiring that all new chemical substances intended for large scale use be approved by and registered with the Environmental Protection Agency. Specifically, TSCA, the Federal Insecticide, Fungicide and Rodenticide Act (FIFRA), the Clean Water Act, and several others have called for assessment of the environmental hazard and impact on human health of synthetic halogenated compounds and other toxic environmental pollutants.

The assessment of risks and hazards is usually associated with the determination of two important parameters of the chemicals: their toxicological properties and their fate in the environment. While the long term toxicological properties of halogenated compounds present interesting challenges to toxicologists, the recalcitrance of many halogenated compounds to microbial degradation presents equal challenges to microbiologists, biochemists, and geneticists.

Microorganisms play a very important role in maintaining steady state concentrations of environmental chemicals, and these activities ensure a smooth operation of the carbon cycle in nature. It is generally accepted that microorganisms are able to degrade biosynthetic products. However, man-made organics are structurally different from naturally occurring ones and, being of recent origin, may not be susceptible to microbial degradation, as sufficient time is required to evolve the requisite enzyme systems [3,4]. Even after three decades of uninterrupted growth of information in the field of microbial genetics, the genetic, biochemical, and molecular mechanisms of transformation, and mineralization of halogenated compounds in the biosphere by microbial population are still inadequately understood. Structural plasticity of the genetic machinery of the microbes endows them with a degradative capability, that can be quickly adapted in response to substrate variability. Such expanding catabolic versatilities may reveal principles, not yet encountered in the intensively studied metabolic pathways of *E.coli*, *Salmonella*, and other genera and species. Unlike enterobacteria, which are confined to the intestinal milieu, saprophytic microbes inhabit widely divergent ecosystems, both aerobic and anaerobic, of soil and aquatic environments.

Thus the microbial activity is chiefly responsible for the mineralization of halogenated compounds, which enter into a variety of natural habitats. The toxic, mutagenic, and carcinogenic effects of various chlorinated compounds make studies of these compounds at various levels, extremely important. As excellent recent reviews on the microbial degradation of halogenated aromatics are available [5-8], a brief description of the degradation of some important chloroaromatics, with available enzymological studies, will be presented.

CHLORINATED POLYCYCLIC HYDROCARBONS

Polycyclic aromatic hydrocarbons occur as natural constituents and combustion products of fossil fuels, and are widespread environmental contaminants [9]. These compounds exhibit high chemical and thermal stabilities, and thus have been widely used as insulating liquids for transformers, as grease for vacuum pumps and turbines, and as coolants. Moreover, they have found application in the production of wrapping papers, carbon paper, inks, paints, tyres, and many other products.

Polychlorinated biphenyls (PCBs)

The commercial production of PCBs began in 1929, and today these compounds can be detected throughout the environment. Since PCBs are strongly *hydrophobic*, they tend to accumulate in the fatty tissues of humans and other animals. Chloracne, skin pigmentation, hair damage, defective vision, fatigue, and liver damage are symptoms associated with acute exposure of mammals to PCBs [10]. Considering all this, studies on the biodegradation of PCB are justified by the observation that most of the world's total production of PCBs still contaminates the environment.

Commercial formulations of polychlorinated biphenyls are made up of almost random mixtures of all the 209 possible isomers. The degradation of both commercially available mixtures [11-13] and defined compounds [14-21] have been studied. Only the latter studies, however, can provide information on specific degradative reactions. The most extensive studies on the biodegradability of specific chlorinated biphenyls have been carried out by Furukawa's group and others, who examined the degradation of as many as 36 pure isomers, substituted with, between one to five, chlorine atoms, by an *Alcaligenes* species and an *Acinetobacter* species. Both the number and position of the chlorine atoms determined the extent of the biodegradability. For example, the more highly substituted PCBs were metabolized less extensively than the mono- and dichlorinated biphenyls. Furthermore, the isomers, bearing chlorine substituents on only one ring, were more easily metabolized than those with substituents on both the rings, and isomers doubly substituted in the *o*-position (e.g., 2,2'- or 2,6-substitution) were generally recalcitrant to microbial degradation.

Both aerobic and facultative anaerobic bacteria, capable of utilizing PCBs, have been isolated from the environment [21-23]. Takase et al. [24] isolated a *P. cruciviae* strain that could grow on more than 10 biphenyl-related compounds, including *p*-chlorobiphenyls (*p*-CB). They demonstrated that biphenyl ether was degraded through an *ortho*-cleavage pathway, and that biphenyl was degraded through a *meta*-cleavage pathway. Dmochewitz and Ballschmiter [25] reported an *Aspergillus niger,* that was capable of utilizing mixtures of PCBs, and Eaton [26] showed *mineralization* of PCBs by a ligninolytic fungus.

p-Chlorobiphenyls (*p*-CB)

Furukawa and Chakrabarty [22] isolated an *Acinetobacter* sp., capable of utilizing *p*-CB. Since then, several other microorganisms have been isolated, which degrade *p*-CB. Masse et al. [18] reported two isolates, an *Achromobacter* sp., and a *Bacillus brevis* strain, which were able to utilize *p*-CB. Both strains generated the same metabolites, with 4-chlorobenzoate as the major metabolic product. Sylvestre et al. [16,17] demonstrated a more rapid and complete degradation of *p*-CB by a bacterial culture, containing two different strains. One strain was able to grow on *p*-CB, and transform it into 4-chlorobenzoate, and the other strain degraded 4-chlorobenzoate. Similarly, complete degradation of 4,4-dichlorobiphenyl, by the co-culture of two *Acinetobacter* strains, which separately degraded biphenyl and 4-chlorobenzoate, was reported by Adriaens et al. [27]. The metabolic pathway for the microbial degradation of PCB and *p*-CB is shown in Figure 1.

Barton and Crawford [28] isolated a *Pseudomonas* sp., that was capable of utilizing *p*-CB as the sole source of carbon and energy. The metabolic degradation of *p*-CB by this bacterium is via a different route, and 4-chloroacetophenone may be an end-product.

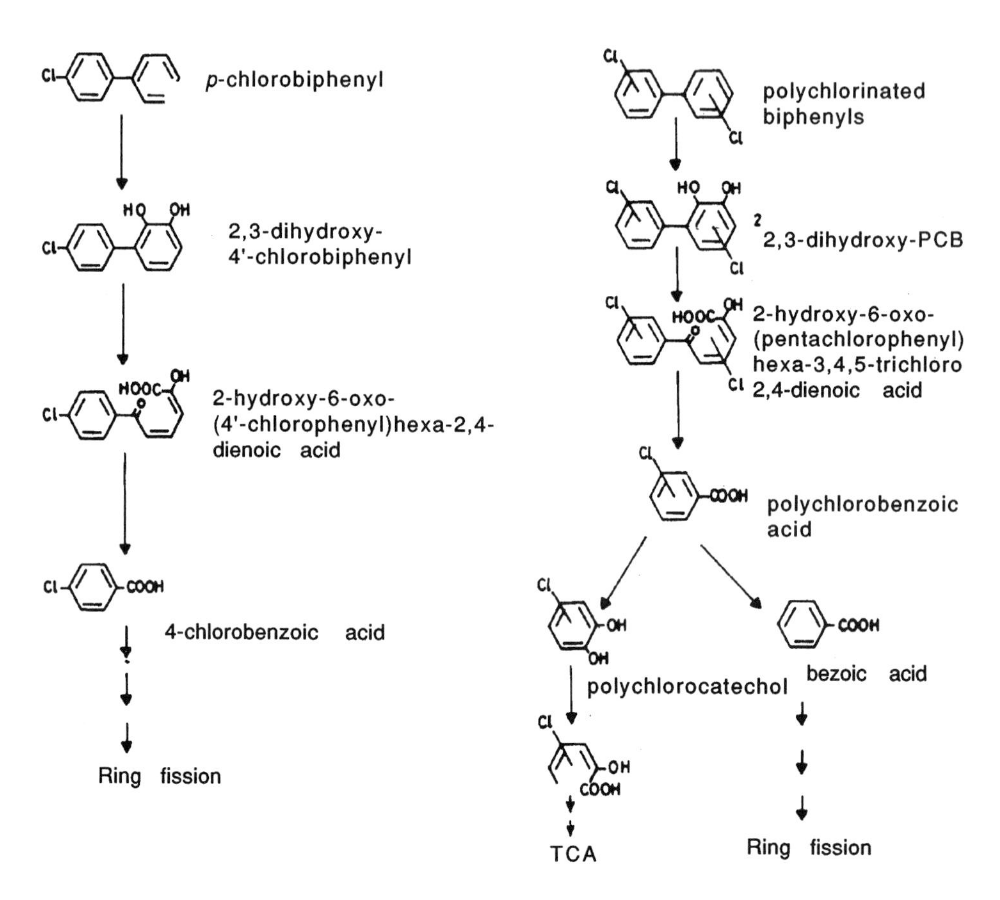

Figure 1. Proposed pathways for the microbial degradation of *p*-chlorobiphenyl (*p*-CB) and polychlorinated biphenyls (PCBs).

1,1,1-Trichloro-2,2-bis(4-chlorobiphenyl) ethane (DDT)

Until recently, 1,1,1-trichloro-2,2-bis(4-chlorobiphenyl) ethane (DDT) was extensively used in the control of insect pests, especially in the fight against the spread of malaria. However, DDT was found to be highly persistent in the environment with a half-life of more than 15 years [29]. Despite the persistence of DDT in the environment, it has been shown to be completely degraded by several different bacteria [30, 31].

The term, cometabolism [31] has been used to describe the microbial transformation of a compound, which cannot serve as a sole source of carbon. In this context, DDT also does not serve as a source of carbon or energy. Alternative sources of carbon and energy, therefore, need to be supplied in these experiments. Lactic acid, as carbon source, and nitrate, as electron acceptor, were added to the medium. Phenylacetic acid formed by bacterial cometabolism of DDT was degraded by the producing organism aerobically. Cometabolism of DDT by a *Hydrogenomonas* sp., to produce 4-chlorophenylacetic acid, was reported by Pfander and Alexander [30]. Bumpus and Aust [32] isolated a lignin-degrading fungus, *Phanerochaete chrysosporium,* which is capable of utilizing DDT. The major metabolic pathway of DDT by this fungus involves oxidation and dechlorination processes, in which ultimately DDT is degraded into carbon dioxide. Various other reports showed that the DDT degradation products could be further degraded by a variety of microorganisms [30,32,33]. This shows that extensive degradation of halogenated hydrocarbons by bacteria, observed in the laboratory does not guarantee that the corresponding degradative reactions occur to a significant extent in nature.

Other polycyclic compounds

Several naphthalene-degrading *Pseudomonas* spp. were able to metabolize monochloro-naphthalene, but the chloronaphthalene did not support growth [34], as an example of cometabolism. The cometabolism of monochloro-naphthalene is dependent on the normal enzymes of naphthalene metabolism, which are induced during the process.

HALOGENATED AROMATIC COMPOUNDS

Chlorinated aromatic compounds are major environmental pollutants, because they are often released in substantial quantities. They are toxic and resistant to degradation, and accumulate in sediment and biota. Although some compounds are degraded only slowly by soil and aquatic microorganisms, others are metabolized relatively quickly. There have been many studies on the ability of soil and aquatic microorganisms to dissimilate chlorinated aromatic hydrocarbons, such as chlorotoluenes [35,36], chlorobenzenes [37-42], chlorobenzoates [43-50], chlorophenols [51-66], chloroacetamide [67], 4-chlorophenylacetate [33,68], and chlorophenoxyacetates [69-83].

Microorganisms are challenged to develop new pathways by altering their own preexisting genetic information as a result of either mutations in single structural and/or regulatory genes, or perhaps recruitment of single silent genes, when they encounter the foreign compounds [84].

Polychlorinated phenols

The chlorinated phenols, in particular pentachlorophenol, have been extensively used as herbicides, fungicides, disinfectants, wood preservatives, and general biocides. These compounds form a large group of xenobiotics that are serious environmental pollutants. In 1983, worldwide production of pentachlorophenol (PCP) was estimated at $5x10^7$ kg [85]. The toxicity of phenols tends to increase with their degree of chlorination, and because few microorganisms can decompose them, the more highly chlorinated phenols tend to accumulate in the environment.

Microorganisms, such as *Arthrobacter* sp. [62], *P. cepacia* [73], *Flavobacterium* sp. [53,55,56,60,63-65] can degrade some, but not all, of the chlorinated phenols. Crawford and his coworkers reported the isolation of *Flavobacterium* sp. from PCP-contaminated dump sites. The initial steps in the catabolism of PCP by the *Flavobacterium* sp. are conversion of PCP to tetrachloro-*p*-hydroquinone and then to trichloroquinone and dichloroquinone. Under anaerobic conditions, the PCP is degraded into tri-, di- and mono-chlorophenol. The benzene ring is then broken to produce methane and carbon dioxide (Figure 2). Recently, cell extracts from an *Arthrobacter* sp. have been shown to dehalogenate PCP [86]. Reductive dechlorination of PCP has been observed in flooded soils. Early studies suggest reductive dechlorination as a degradative pathway for PCP. 2,3,4,6-Tetrachlorophenol, 2,4,5-trichlorophenol, 3,4- and 3,5-dichlorophenol, and 3-chlorophenol were identified in acidic hexane extracts of soil samples, taken from rice fields, which were incubated for 30 days with pentachlorophenol. These results strongly indicated that sequential reductive dechlorination can be carried out by microorganisms under anaerobic conditions.

Suzuki [87] reported the isolation of a *Pseudomonas* sp., capable of utilizing pentachlorophenol as the sole source of carbon and energy. Stanlake and Finn [62] isolated an *Arthrobacter* sp. with the same ability. In neither case was it possible to elucidate the complete pathways for the degradation of pentachlorophenol. However, Suzuki [87] identified tetrachlorocatechol and tetrachlorohydroquinone under aerobic conditions, neither of which was a product of anaerobic degradation. Collectively, these results indicate that aerobic biodegradation of pentachlorophenol proceeds by replacement of a chlorine substituent by a hydroxyl group. However, the reaction mechanism remains unclear. Both oxygenolytic and hydrolytic mechanisms are possible for the cleavage of the carbon-chlorine bond, but no enzymes have been isolated. The subsequent steps in the aerobic degradation of pentachlorophenol also remain to be elucidated.

An interesting application of studies of pentachlorophenol biodegradation has been described by Edgehill and Finn [88], who applied an inoculum of pentachlorophenol-utilizing bacteria grown in a fermentor, to contaminated soil (1 mg of bacteria per gram dry weight of soil). The half-life of pentachlorophenol was, thereby, reduced from two weeks to less than one day. This and other studies strongly suggest that bacterial treatment could provide rapid, cheap, and effective methods for the decontamination of polluted soils.

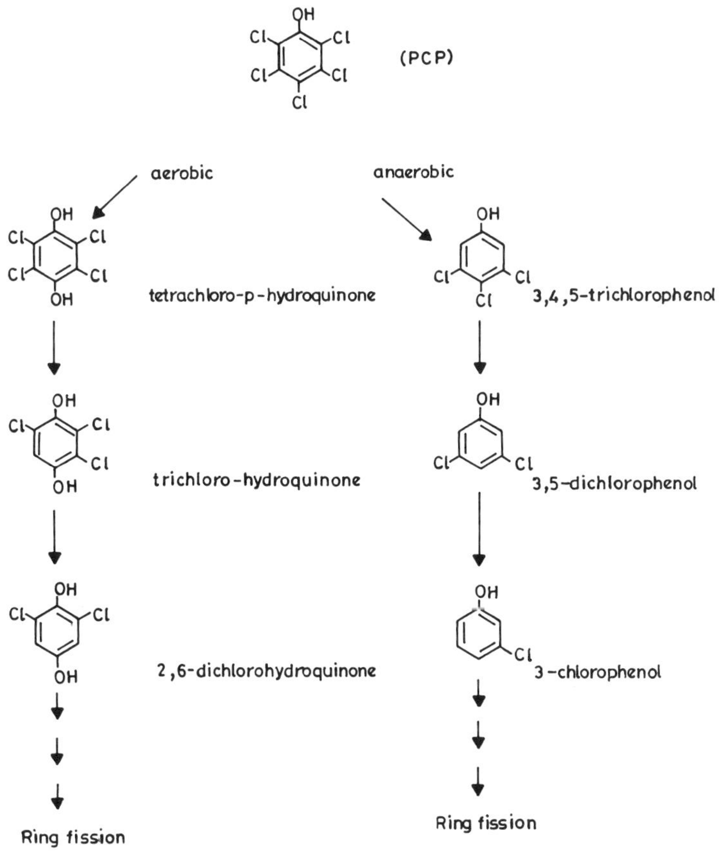

(PCP)

aerobic

anaerobic

tetrachloro-p-hydroquinone

3,4,5-trichlorophenol

trichloro-hydroquinone

3,5-dichlorophenol

2,6-dichlorohydroquinone

3-chlorophenol

Ring fission

Ring fission

Figure 2. Proposed pathways for the microbial degradation of pentachlorophenol (PCP).

Chlorobenzenes

Chlorobenzenes are used extensively as solvents, fumigants, deodorants, pesticides, and synthetic intermediates. Because of their patterns of use, a large fraction of the total production is released into the environment. For example, about 100,000 tons of *p*-dichlorobenzene (*p*-DCB) are released into the atmosphere annually [89]. The toxicity of chlorobenzenes to higher organisms are relatively low but recent studies have indicated that *p*-DCB may be a carcinogen (National Toxicology Program 1987). In addition, detection of chlorobenzenes in ground water has led to increasing concern about the fate and persistence of such compounds in the environment.

Biodegradation of chlorobenzenes has been reported in groundwater [90], fixed [91], and soil columns [92]. However, until recently nothing was known about the pathways used by bacteria for metabolism of chlorobenzenes because strains, able to grow on such compounds in pure culture, had not been isolated. The toxicity of chlorobenzenes accounts for much of the difficulty in isolation of bacteria, able to degrade them. Recently, extended selected enrichment with the substrate provided at low concentrations in the vapour phase has allowed the isolation of bacteria able to grow at the expense of chlorobenzenes. Reineke and Knackmuss [40] isolated a chlorobenzene-degrading strain, designated WR1306, after 9 months of chemostat selection, during which benzene was gradually replaced with chlorobenzene as the growth substrate. The original inoculum had been obtained from a mixture of soil and sewage. Similar approaches yielded an *Alcaligenes* [41] and a *Pseudomonas* sp. [42] able to degrade *p*-DCB, an *Alcaligenes* sp. able to degrade *m*-DCB [93], a *Pseudomonas* sp., that grows on *o*-DCB [94], and a *Pseudomonas* sp. strain P51, able to grow on all the DCB isomers and 1,2,4-trichlorobenzene [95].

In addition to the strains isolated from natural systems, several chlorobenzene-degrading strains have been constructed in the laboratory. For example, a mating between a benzene degrader, *P. putida* F1 [96], and a chlorobenzoate degrader, *Pseudomonas* sp. strain B13 [97], yielded a strain able to grow on chlorobenzene [98]. Extended selective enrichment yielded *P. putida* WR1323 which was able to grow on both chlorobenzene and *p*-DCB [99]. Similarly, Krockel and Focht [100] used a novel chemostat selection technique for construction of a strain of *P. putida* able to grow on both chlorobenzene and *p*-DCB. Chapman [101] used a different approach to construct strains able to degrade chloro- and bromotoluenes by conjugal transfer of the TOL plasmid into a 3-chlorobenzoate-degrading *P. putida*.

Dechlorination of chlorobenzenes has been shown to occur under anaerobic conditions. Sewage sludge completely transformed 190 μM hexachlorobenzene to DCB [102]. The metabolic pathway for biodegradation of *o*-DCB [42], *m*-DCB [93], and *p*-DCB [99] proposes that they all form a common intermediate dichlorocatechol, and then the benzene ring is broken as indicated in Figure 3.

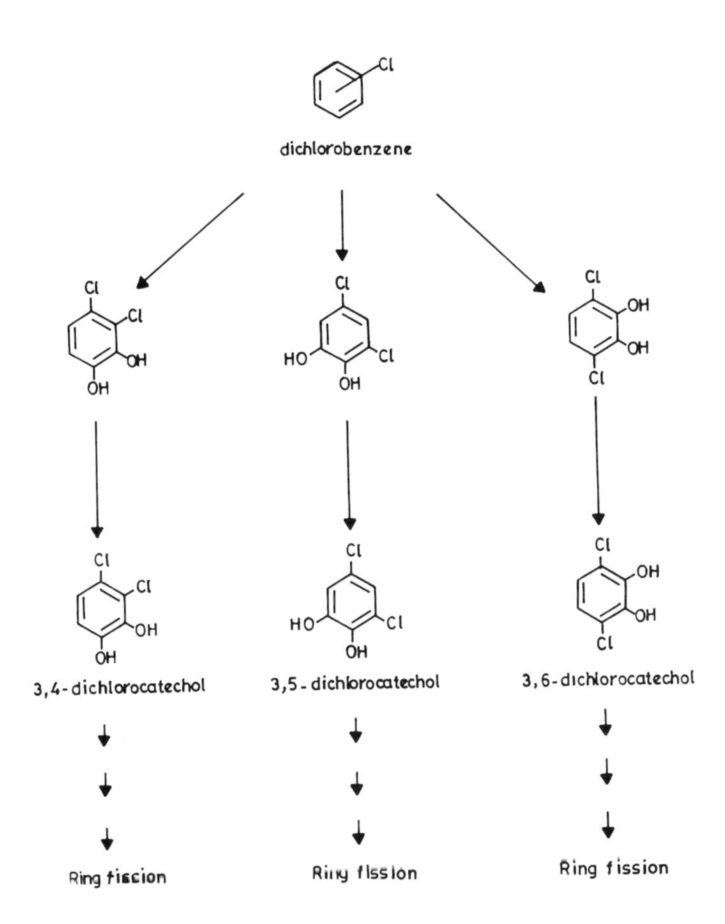

Figure 3. Proposed pathways for the microbial degradation of dichlorobenzenes (DCB).

Chlorobenzoates

Chlorinated benzoic acids have been shown to be intermediates in the biodegradation of several chlorinated aromatic compounds, for example, 4-chlorobenzoic acid is formed in the biodegradation of both polychlorinated biphenyls [103]. Mixed microbial cultures, which have been studied by a number of groups under aerobic conditions, can degrade a wide range of chlorinated benzoic acids, including 2-, 3- and 4-chlorobenzoic acid as well as 3,4-dichloro- and 2,4-dichlorobenzoic acid [104]. By contrast, under anaerobic conditions, only reductive dechlorination of *meta*-substituted benzoic acids has been observed [49,105,106]. Cometabolism of chlorobenzoic acids in the presence of unsubstituted benzoic acids leads to the formation of the corresponding chlorocatechols in some pure cultures of bacteria [107,108]. These cultures were eventually poisoned due to the accumulation of chlorocatechols, which could not be further catabolized, and which polymerized to form toxic products. These results suggest that the enzymes, involved in the conversion of unsubstituted benzoic acid to catechol, can also transform some chlorobenzoic acids in an analogous manner. The critical step of these degradation reactions appears to be the ring cleavage. Conversion of 3-chlorocatechol to the corresponding acyl chloride and irreversible inactivation of a bacterial catechol 2,3-dioxygenase by the product (i.e., "*suicide inactivation*") has been reported by Bartels et al. [109]. Klecka and Gibson [110] demonstrated reversible inhibition of the same enzyme in a different bacterium, and suggested that the inhibition results from chelation of Fe(II), which is essential for activity of the enzyme.

Several groups have isolated *Pseudomonas* species, capable of utilizing 3-chlorobenzoate as the sole source of carbon and energy [97]. The *Pseudomonas* sp., described by Johnston et al. [111], degrades 3-chlorobenzoic acid via 3-hydroxybenzoic acid. Dorn et al. [97], on the other hand, found *Pseudomonas* sp. which converted 3-chlorobenzoic acid into 3- and 4-chlorocatechol as intermediates. They were further catabolized via an *ortho*-cleavage pathway.

Knackmuss and his co-workers [112-116] have studied the effects of various substituents upon enzymes, catalyzing the individual reactions of the chlorobenzoate degradative pathway. In *Pseudomonas* sp. B13, for example, they detected two types of catechol 1,2-dioxygenases. One was mainly active in the ring cleavage of catechol and had low affinity for chlorinated catechols, while the second enzyme could efficiently degrade chlorocatechols [112]. Furthermore, two cycloisomerases, catalyzing cyclization of the chloromuconic acids with cleavage of the chlorine substituent, were also isolated from strain B13. One was found to be specific for unsubstituted muconate, whereas the other enzyme was also capable of transforming chloromuconates [115,116]. These two enzymes, i.e., the dioxygenase and the cycloisomerase, were responsible for the growth of strain B13 on 3-chlorobenzoic acid.

From these studies, Knackmuss and his group, using gene transfer techniques, were able to develop strategies, that allowed *Pseudomonas* sp. B13 to degrade a wide range of halogenated aromatic compounds to the corresponding chlorocatechols. In this way, a gene encoding a broad-specificity benzoate 1,2-dioxygenase from *P. putida* mt2, carried on TOL plasmid, was transferred and expressed in strain B13. Since the TOL plasmid enzymes can also convert 4-chloro-, 3,5-dichlorobenzoic acids into the corresponding chlorocatechols, the hybrid strain B13, containing the plasmid, was able to grow on these compounds as well. Genetic exchange between strain B13 and an aniline-utilizing bacterium, containing plasmid encoded enzymes that convert chloroanilines into chlorocatechols, allowed the construction of hybrid strains, capable of degrading chloroaniline. Interestingly, in all of these genetically constructed hybrid strains, production of plasmid encoded enzymes catalyzing *meta*-ring cleavage was switched off by spontaneous mutations [44,97].

Arthrobacter, Nocardia, and *Pseudomonas* species are capable of utilizing 4-chlorobenzoic acid as the sole source of carbon and energy [116-118]. In all these strains, the first step of the catabolic pathway is the replacement of the chlorine substituent by a hydroxyl group, the 4-hydroxybenzoic acid produced is converted into 3,4-dihydroxybenzoic acid by an oxygenase-catalyzed hydroxylation. Ring cleavage of this intermediate may be either of the *ortho* [118] or of the *meta* [117] type depending on the bacterium. Todate, enzymes catalyzing chlorine elimination from the aromatic ring have been identified in four different bacteria [119,120]. In vitro studies show that the enzymes catalyze the incorporation of ^{18}O-label from $H_2{}^{18}O$, into the product of 4-chlorobenzoic acid dechlorination. They are, therefore, the only known examples of enzymes that have been unequivocally shown to catalyze direct substitution of the halogen on the aromatic ring by a hydrolytic mechanism.

The degradation of 2-chlorobenzoic acid by *P. cepacia* also involves removal of the chlorine substituents in the first step [121,122], but here the aromatic ring undergoes oxidation with elimination of CO_2. Catechol, which is the first intermediate of this metabolic pathway, is further degraded by an *ortho*-cleavage pathway. A metabolic pathway for the degradation of 3-chlorobenzoic acid by *Pseudomonas* sp. B13 is shown in Figure 4. Zaitsev and Baskunov [123] reported a similar scheme for the metabolism of 3-chlorobenzoic acid in *Acinetobacter calcoaceticus.*

Chlorotoluene and 4-chlorophenylacetate

Pierce et al. [124,125] reported that *P. cepacia* and several other *Pseudomonas* species were capable of utilizing mono- and dichlorinated toluenes as the sole source of carbon and energy. These strains contained plasmids of about 45 kilobase, which coded for chlorotoluene degradation. Restriction analysis showed that these plasmids, although of similar size, were not identical.

Pseudomonas sp. strain CBS3 is able to utilize 2-chloroacetate, 4-chlorobenzoic acid and 4-chlorophenylacetate [33,68]. An enzyme, catalyzing oxidative removal of chlorine substituent from 4-chlorophenylacetic acid, has been isolated from strain CBS3 [33]. Expectations that this oxygenase, which converts 4-chlorophenylacetic acid into 3,4-dihydroxyphenylacetic acid, would be a non-specific enzyme, proved incorrect. Only low activities were recorded in assays with unsubstituted phenylacetic acid. The 3,4-dihydroxyphenylacetate has been reported to undergo *meta*-cleavage.

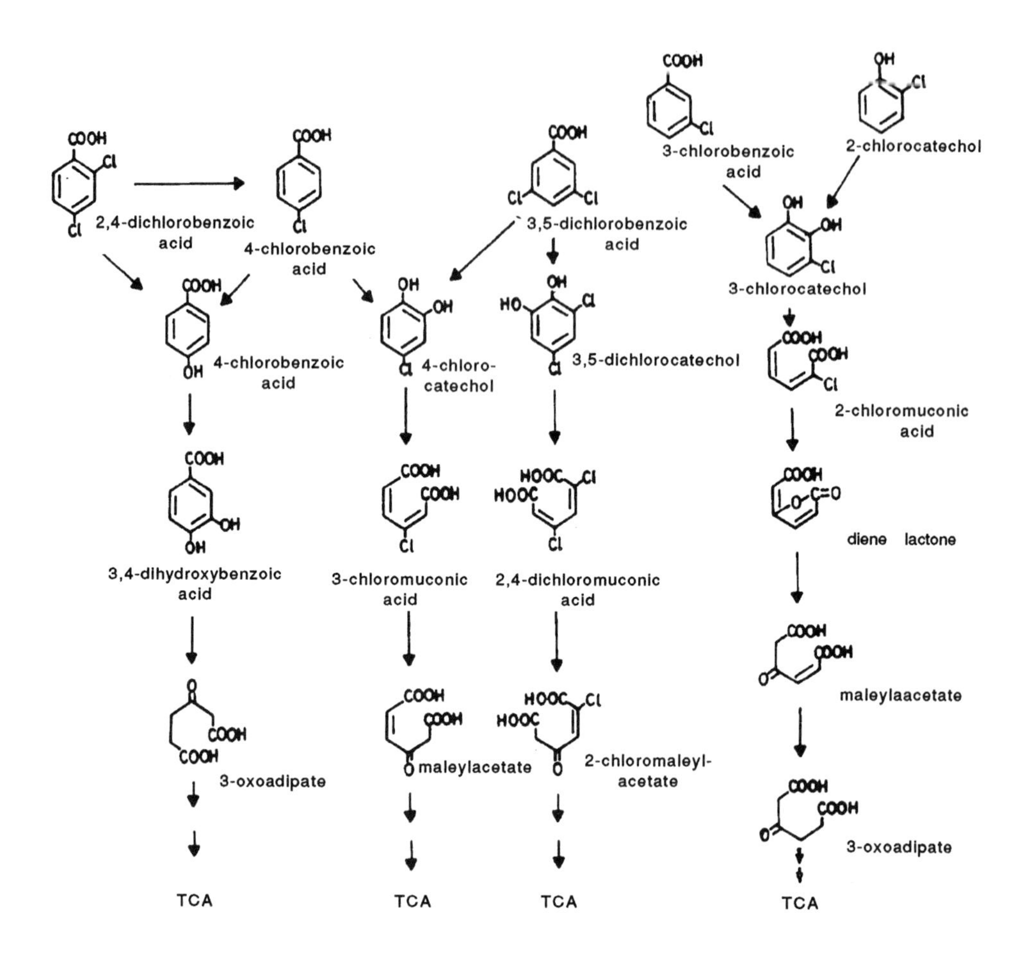

Figure 4. Proposed pathways for the microbial degradation of chlorobenzoic acids and chlorophenols.

Chlorophenoxyalkanoic acids

The chlorophenoxyalkanoic acids constitute yet another economically important group of halogenated aromatic hydrocarbons. They are widely used as herbicides to control dicotyledonous weeds. The most important of them are 2,4-D and its propionic and butyric acid homologs, 4-chloro-2-methylphenoxyacetic acid (MCPA), and 2,4,5-trichlorophenoxyacetic acid (2,4,5-T). In all the cases studied, aerobic biodegradation proceeds by removal of the aliphatic side chain with the formation of the corresponding chlorinated phenol. Thus, formation of 2,4-dichlorophenol from both 2,4-D and 4-(2,4-dichlorophenoxy)-butyric acid, 4-chloro-*o*-cresol from MCPA, and 2,4,5-trichlorophenol from 2,4,5-T has been reported [126]. The subsequent degradative pathways differ in different microorganisms. For example, the biodegradation of 2,4,5-T by two bacterial species may involve the dehalogenase-catalyzed replacement of one chlorine substituent by a hydroxyl group, resulting in the formation of 3,5-dichlorocatechol [127-129,129a]. The enzymes, however, have not been detected. Two pathways have been suggested for biodegradation of 2,4-dichlorophenol: (i) enzyme-catalyzed replacement of one chlorine substituent by a hydroxyl group gives 4-chlorocatechol and (ii) oxidation produces 3,5-dichlorocatechol [130]. Further, degradation of 4-chloro- and 3,5-dichlorocatechol, by one of these bacteria, a *Pseudomonas* sp., involves dioxygenase-catalyzed cleavage of these intermediates between the hydroxyl groups to give the corresponding chloromuconic acids. In contrast, an *Achromobacter* sp., that cleaves chlorocatechols adjacent to the hydroxyl groups, has been isolated [131]. Lactonization of dichloromuconic acid results in the removal of one chlorine substituent to form γ-carboxymethylene-α-chloro-$\Delta^{\alpha\beta}$-butenolide, which is finally converted into 2-chloro-4-oxo-2-hexenedioic acid (Figure 5). Similar catabolic pathways have been suggested for the degradation of chloromuconic acids, derived from 2,4-D and MCPA [129-133].

Among the chlorinated aromatic hydrocarbons, 2,4-D and related chlorophenoxyalkanoic acids have become substantial environmental pollutants because of their world-wide employment as herbicides against broad leafed plants since the early 1940's [83]. The use of herbicides, to minimise crop yield losses due to weeds, has become an integral part of modern agriculture. There is a continuous search for new herbicides, that are highly effective and, at the same time, safe for mammals. A new class of herbicides, that fulfils these criteria, acts by interfering with specific amino acid biosynthetic pathways in plants [134]. 2,4,5-T is another herbicide, that has been used extensively for the last several decades not only for weed control on range lands, pastures, and pathways, but also as a growth regulator to delay colouration of lemons, to increase the size of citrus fruits, and to reduce deciduous fruit drop. Use of Agent Orange during Vietnam war (containing 2,4-D and 2,4,5-T) and as herbicide in United States and other countries for defoliation and control of poison ivy, poison oak and various broad leafed weeds has created toxicological problems [135,136], hence its use has been restricted in the United States and other countries.

2,4-D and related phenoxyalkanoic herbicides are highly effective against broad leafed plants and have found application under a wide variety of situations; from weed control in lawns, gardens, cereal crops, and pastures to defoliants in forestry and warfare. In the early 1980's, production of 2,4-D in USA alone was approximately 13 million pounds per year. The fact that large quantities of 2,4-D and its related herbicides are manufactured and applied each year, necessitates the importance of

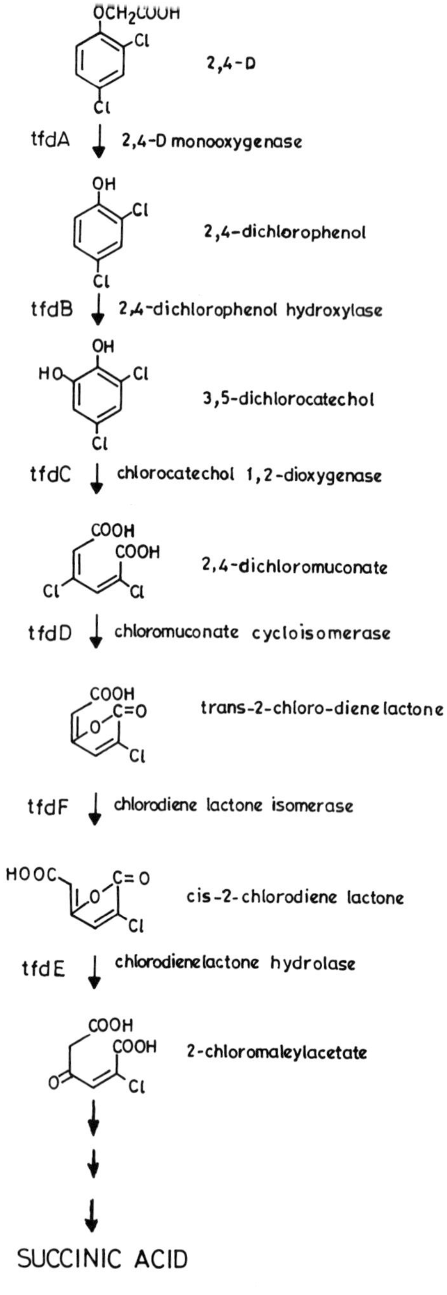

OCH_2COOH — Cl — Cl — 2,4-D

tfdA — 2,4-D monooxygenase

OH — Cl — Cl — 2,4-dichlorophenol

tfdB — 2,4-dichlorophenol hydroxylase

OH — HO — Cl — Cl — 3,5-dichlorocatechol

tfdC — chlorocatechol 1,2-dioxygenase

COOH — COOH — Cl — Cl — 2,4-dichloromuconate

tfdD — chloromuconate cycloisomerase

COOH — C=O — O — Cl — trans-2-chloro-diene lactone

tfdF — chlorodiene lactone isomerase

HOOC — C= O — O — Cl — cis-2-chlorodiene lactone

tfdE — chlorodienelactone hydrolase

COOH — COOH — O — Cl — 2-chloromaleylacetate

SUCCINIC ACID

Figure 5. Proposed pathways for the microbial degradation of 2,4-D.

having effective means of treating production wastes, and a thorough understanding of the fate of these chemicals in the environment.

A large variety of microorganisms have been isolated, which are capable of degrading 2,4-D and related compounds. Those found to degrade any of the phenoxyacetic acids can often degrade other structurally related compounds as well. Results on their ability to degrade MCPA and 2,4,5-T, two compounds of interest, closely related to 2,4-D, are also given, wherever available.

Reports on the biodegradative pathway of 2,4-D have shown that the ether bond in 2,4-D is cleaved oxidatively by a 2,4-D monooxygenase [137]. The resulting 2,4-dichlorophenol is then hydroxylated by a 2,4-dichlorophenol hydroxylase to yield 3,5-dichlorocatechol [137], which is further cleaved by catechol 1,2-dioxygenase to 2,4-dichloromuconic acid [131]. Further steps, including lactonization, delactonization, and chloride release to yield tricarboxylic acid cycle intermediates have been discussed more recently by Pieper et al. [80].

Several workers have established that 2,4-D gradually disappears from the soil [138-140] and that a biological factor is the agent for inactivation [141,142]. Newman and Thomas [143] reported that crude cultures of some bacteria inactivated 2,4-D and some undefined pure cultures exhibited a marginal detoxifying effect. Enrichment techniques employed by Audus [141,144] yielded an organism of the type *Bacterium globiforme* which, in pure culture, decomposed 2,4-D. Jensen and Peterson [142] isolated two organisms, one identified as *Flavobacterium aquatile* and the other similar to that of Audus, which would carry out the decompositions. Neither of the investigators found it possible to effect the decomposition in a synthetic medium with the pure culture.

Intact cells of a soil bacterium, a strain of *Arthrobacter*, were capable of converting 2,4-D and other phenoxyacetates to the corresponding phenols. The 2,4-dichlorophenol generated from 2,4-D and a number of other phenols were, in turn, oxidized by the microbial cells [145]. From this microorganism an enzyme preparation was obtained which cleaves the ether linkage between the phenolic and fatty acid moieties of the phenoxyalkanoate substrate to yield the free phenol. Thus 2,4-D, 4-chloro-2-methyl-, 2- and 4-chlrophenoxyacetates are converted enzymatically to produce compounds with chromatographic characteristics of 2,4-dichlorophenol, 4-chloro-2-methylphenol, and 2- and 4-chlorophenol, respectively [145]. These were the first enzymatic studies of the cleavage of the ether linkage of phenoxy herbicides. Helling et al. [137] observed that resting cells of the same organisms converted ^{18}O ether-labelled phenoxyacetate to phenol, the latter containing all of the original ^{18}O label. The product, formed from the acetate moiety of 2,4-D, has not been identified. However, it was proposed that glyoxylate is the initial product formed enzymatically upon cleavage of the side chain from 2,4-D [131,139].

A chlorophenoxyacetate-degrading *Arthrobacter* sp. contained an enzyme, which converted 2,4-dichlorophenol and other chlorophenols to catechols. Reduced nicotinamide adenine dinucleotide phosphate (NADPH) and O_2 were required for the reaction. The enzyme hydroxylating 2,4-dichlorophenol was partially separated from the catechol-degrading enzyme present in the bacterial extract. The compounds formed from 2,4-dichlorophenol and 4-chlorophenol were identified as 3,5-dichlorocatechol and 4-chlorocatechol, respectively [131]. From the same *Arthrobacter* sp. an enzyme preparation liberated all the chlorine from 4-chlorocatechol but only half the chlorine atoms from 3,5-dichlorocatechol. Two compounds, found in the reaction mixture after incubation of the crude enzyme preparation with 3,5-dichlorocatechol, were characterized by mass spectrometry and infra-red spectroscopy. The analytical data suggested that a product generated from 4-chlorocatechol was γ-carboxymethylene-Δ^{α}-butenolide [131].

Evans et al. [128] reported two *Pseudomonas* strains, which were isolated from soil and metabolized 2,4-D as sole carbon source in mineral salts medium. 2,4-D cultures of *Pseudomonas* I contained 2,4-dichlorophenol, 2-chlorophenol, 3,5-dichlorocatechol, and α-chloromuconate, the last as a major metabolite. Dechlorination at the 4(*p*)-position of the aromatic ring must have, therefore, taken place at some stage before ring fission. *Pseudomonas* NCIB 9340 cultures, metabolizing 2,4-D, contained 2,4-dichloro-6-hydroxyacetate, 2,4-dichlorophenol, 3,5-dichlorocatechol, and an unstable compound, probably α,γ-dichloromuconate. Cell-free extracts of the latter organism, grown in 2,4-D cultures, contained an oxygenase that converted 3,5-dichlorocatechol into α,γ-dichloromuconate, a chlorolactonase that, in the presence of Mn^{2+} ions, converted the dichloromuconate into γ-carboxy-methylene-α-chloro-$\Delta^{\alpha\beta}$-butenolide, and a delactonizing enzyme that gave α-chloromaleylacetate from this lactone.

Sharpet et al. [146] reported an enzyme preparation from 2,4-D-grown *Arthrobacter* sp., which converted 2,4-dichloromuconate to chloromuconate to yield 2-chloro-4-carboxy-methylene-but-2-enolide and separated it from the butenolide-delactonizing enzyme. Tyler and Finn [139] found that the optimal pH range for 2,4-dichlorophenol-degradation (7.1-7.8) is significantly higher than the range for 2,4-D (6.8-6.9).

Several other microorganisms, such as *Acinetobacter*, *Arthrobacter* spp., *Corynebacterium, Flavobacterium,* and other *Pseudomonas* spp., have also been shown to degrade 2,4-D and related phenoxyacetates. Recently, Chaudhry and Huang [69] reported a *Flavobacterium* sp., which degrades 2,4-D and utilizes it as sole source of carbon and energy.

Since biodegradation of 2,4-D and related compounds has generally been observed to be an aerobic process, oxygen supply is expected to be a significant factor in determining biodegradation rates. Williams and Crawford [147] have shown that the biodegradation of 2,4-D generally increases as oxygen supplies are increased. Based on these studies, most experiments on biodegradation of 2,4-D and related compounds have been conducted under aerobic conditions. 2,4-Dichlorophenol has been generally accepted as the first product in the 2,4-D biodegradation pathway.

It is now well established that in microorganisms, these compounds are hydroxylated by a group of enzymes known as *oxygenases*. These enzymes activate molecular oxygen and incorporate oxygen atom(s) into the substrates, making them water soluble and susceptible to further catabolism. By the action of various enzymes, the aromatic compounds are completely mineralized, and the organism derives its energy for growth and multiplication.

As is evident from the literature, considerable information is available on the metabolism and genetics of 2,4-D degradation, but very few reports exist in this literature about the enzymes involved in its degradation. As proposed by Evans et al. [128] and Chaudhry and Huang [69], the ether bond in 2,4-D is cleaved to produce 2,4-dichlorophenol by a 2,4-D monooxygenase [148]. The hydroxylation of 2,4-dichlorophenol to form 3,5-dichlorocatechol has now been well established [149,150]. In the author's laboratory, Radjendirane et al. [151] and Bhat et al. [152] studied the second and third enzymes of the 2,4-D metabolic pathway in *Pseudomonas cepacia* strain CSV90, respectively. Both the enzymes have been obtained in the homogeneous form. The second enzyme, 2,4-dichlorophenol hydroxylase, a flavin-dependent NADPH-requiring monooxygenase, has been studied in detail [153]. The third enzyme of 2,4-D metabolic pathway, i.e., 3,5-dichlorocatechol 1,2-dioxygenase has also been studied in detail [154]. The gene for this dichlorocatechol dioxygenase enzyme has been shown to be present on a high molecular size plasmid pMAB1 in the strain CSV90. It has been cloned, its complete nucleotide sequence has been determined and is now overexpressed in *E. coli* [154].

The degradation of another commonly used herbicide, 2,4,5-T has been less extensively investigated and most of the information, concerning the 2,4,5-T degradation, has been gathered by using reductive (anaerobic) sediments [47,73-77]. As expected, degradation is slow, and the metabolic pathway has not been conclusively established.

Herbicides, 2,4-D and 2,4,5-T are structurally related, the latter having an extra chlorine atom at position 5. Unlike 2,4-D, 2,4,5-T is poorly biodegradable, and persists for long periods, hence constituting a pollution problem [155]. Cometabolism of 2,4,5-T by *Brevibacterium* sp. resulted in the formation of the product, tentatively identified as 3,5-dichlorocatechol. Bacterial cometabolism of 2,4,5-T was also described by Rosenberg and Alexander [156], who proposed a degradation pathway of 2,4,5-T in soil [157]. Reductive dechlorination of 2,4,5-T by anaerobic microorganisms was described by Suflita et al. [130].

Chakrabarty's group reported *P. cepacia* strain AC1100 isolated from a chemostat enrichment, which was found to use 2,4,5-T or 2,4,5-trichlorophenol as the sole source of carbon and energy [73,74,76]. In addition, resting cells of strain AC1100 oxidize or dehalogenate pentachlorophenol, pentafluorophenol, tetra-, tri- and dichlorophenols, 2,4-D and so on [73]. The bromine and fluorine derivatives are dehalogenated much more readily than the iodine derivatives. Several metabolic products were detected by gas chromatography in the culture medium with 2,4,5-T. Gas chromatographic-mass spectrometric (GCMS) analysis of 2,4,5-T metabolites, after methylation, revealed the presence of the following compounds: a product tentatively identified as hydroxy-2,4,5-T; three isomers of dichlorophenoxyacetic acids (one of them tentatively identified as 2,4-dichloro-6-hydroxyphenoxyacetic acid); 2,4,5-trichlorophenol, and three isomers of dichlorocatechol, including 3,5-dichlorocatechol, 4-chlorophenol succinic acid. Dechlorination of 2,4,5-T by an anaerobic methanogenic consortium, grown on 3-chlorobenzoic acid, occurs at the *para*-position to form 2,5-dichlorophenoxyacetic acid as described by Suflita et al. [130].

Three types of dehalogenation mechanisms are employed in the microbial metabolism of 2,4,5-T: Simultaneous hydroxylation and dehalogenation, reductive dehalogenation, and elimination of the halogen as the hydrogen halide after ring fission. Although some common intermediates are formed during the degradation of 2,4-D and 2,4,5-T, AC1100 was unable to grow on 2,4-D, suggesting high specificity of the initial enzyme or enzymes or induction specificity in 2,4,5-T metabolism.

GENETIC STUDIES ON THE BIODEGRADATION OF CHLORINATED AROMATICS

Microorganisms are well-known for their adaptability. An important example of such adaptability is the ability of microorganisms to utilize a variety of xenobiotics, when such compounds are released into the environment in large amounts [158]. Microorganisms, encountering a new organic chemical in their environment, may acquire new catabolic genes, needed for the degradation of that compound, from other microorganisms through conjugation, or transformation events, or they may modify existing genes through mutation processes. These altered metabolic functions can be regarded as a system of *genetic adaptation* to new substrates and, as such, it presents a good model for studying the evolution of enzymes and catabolic pathways.

The present state of research increasingly strengthens the emerging concept that, like the multiple drug resistance markers both in enteric and other pathogenic bacteria, the metabolic versatility of soil bacteria, with few exceptions, is determined totally, or partly, by the resident plasmids [75].

Biphenyl/polychlorinated biphenyl degradative genes

Since the identification of polychlorinated biphenyl (PCB) residues in the environment in 1966, these widespread and persistent pollutants have become a great global concern. They are recognized to be present in great abundance in the ecosystem, similar to 1,1,1-trichloro-2,2'-bis(4-chlorobiphenyl) ethane (DDT) and its metabolites. It has been shown that biphenyl-utilizing bacteria are ubiquitously distributed in the environment and can cometabolize many PCB congeners to chlorobenzoic acid through oxidative routes [19-22,159,160] (Figure 6). Molecular oxygen is introduced at the 2,3 position of the nonchlorinated or less chlorinated ring to produce a dihydrodiol compound (2,3-dihydroxy-4-phenylhexa-4,6-diene) by the action of biphenyl dioxygenase (product of the *bph*A gene). The dihydrodiol is then dehydrogenated to a 2,3-dihydroxybiphenyl by the action of a dihydrodiol dehydrogenase (product of *bph*B). The 2,3-dihydroxybiphenyl is cleaved at the 1,2 position by a 2,3-dihydroxybiphenyl dioxygenase (2,3-DHBPO; product of *bph*C) to produce the *meta*-cleavage compound, 2-hydroxy-6-oxo-6-phenylhexa-2,4-dienoic acid. The *meta*-cleavage compound is hydrolyzed to the corresponding benzoic acid by a hydrolase (product of *bph*D). Thus, at least, four enzymes are involved in the oxidative degradation of PCBs to chlorobenzoic acids. Most of the biphenyl-utilizing strains cannot degrade chlorobenzoic acids any further and, therefore, the corresponding chlorobenzoates accumulate during PCB catabolism. Biphenyl/PCB-degrading bacteria have been isolated from various soils of different places. They are usually Gram-negative soil bacteria, that include species of *Pseudomonas*, *Achromobacter*, *Alcaligenes*, *Acinetobacter*, and *Moraxella*. Gram-positive bacteria, such as *Arthrobacter* spp., have also been isolated [22].

Furukawa and his group cloned a gene cluster, encoding biphenyl/PCB-degrading enzymes from the chromosomal DNA of *P. pseudoalcaligenes* KF707 [161]. The cloned XhoI 7.2 kb DNA fragment contained the genes *bph*A (encoding dioxygenase), *bph*B (encoding dihydrodiol dehydrogenase), and *bph*C (encoding 23DHBPO). *P. aeruginosa* PAO1161, carrying recombinant plasmid, pMFB2 (pKT230 containing the 7.2 kb *bph*ABC operon at the unique XhoI site) converted biphenyl/PCB into *meta*-cleavage yellow compounds [161-164]. The *bph*D (encoding the ring *meta*-cleaving hydrolase) has also been cloned. *E. coli* JM109, carrying recombinant plasmid pFYD177, readily converted the ring *meta*-cleavage yellow compound to benzoic acid. Subcloning, deletion analysis, and transposon mutagenesis of the cloned *bph* genes revealed that the biphenyl/PCB catabolic *bph* operon of *P. pseudoalcaligenes* KF707 is organized as presented in Figure 6. Biphenyl dioxygenase encoded by *bph*A, would most likely be a multicomponent enzyme, that usually consists of several nonidentical subunits. The *bph*B gene is located downstream of *bph*A and extends ca. 1 kb, and the *bph*C gene

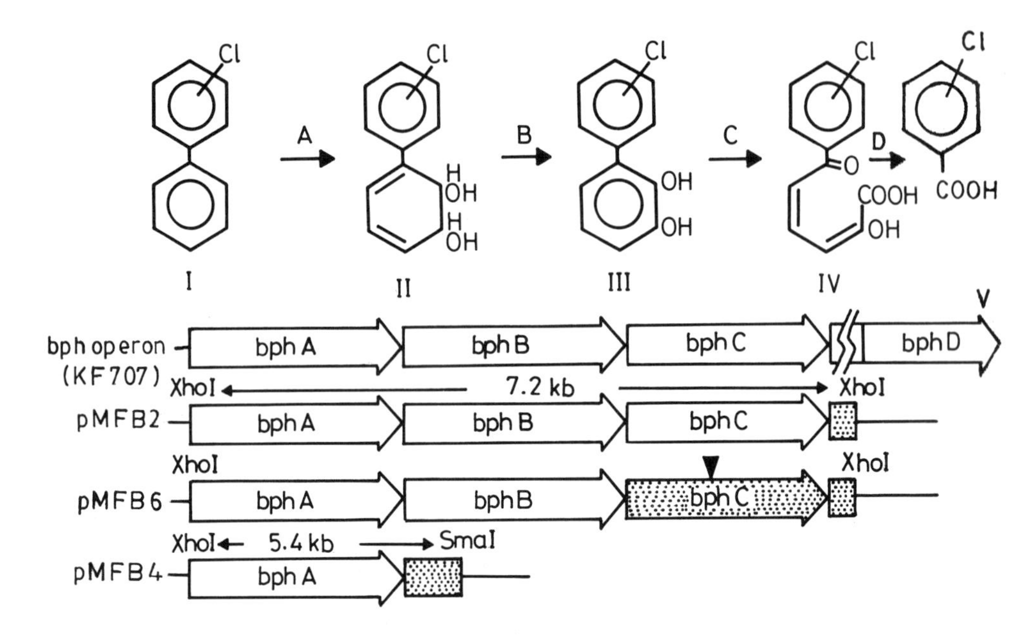

Figure 6. Degradative pathway of biphenyl/PCB by soil bacteria and *bph* genes from *P. pseudoalcaligenes*. Top (compounds): I, Biphenyl/ PCB; II, 2,3-dihydroxy-4-phenylhexa-4,6-diene (dihydrodiol compound); III, 2,3-dihydroxybiphenyl; IV, 2-hydroxy-6-oxo-6-phenylhexa-2,4-dienoic (ring *meta*-cleavage compound); V, benzoic acid. Enzymes: A, biphenyl dioxygenase; B, dihydrodiol dehydrogenase; C, 2,3-dihydroxybiphenyl dioxygenase; D, *meta*-cleavage compound hydrolase. Bottom: Organization of the *bph* operon of *P. pseudoalcaligenes* KF707.

(ca. 1 kb) immediately follows *bph*B. The *bph*D gene is not located just downstream of *bph*C. There is an extra (ca. 3 kb) DNA segment (*bph*X) between *bph*C and *bph*D. The function of putative *bph*X gene has not yet been elucidated. The *bph* operon in *P. pseudoalcaligenes* KF707 is, thus, organized as *bph*ABCXD. The transcriptional start site of the *bph* operon (KF707) has been determined by reverse transcriptase mapping. This transcriptional initiation site is located 104 base pairs (bp) upstream from the start codon of the *bph*A cistron.

2,4-D and 3-chlorobenzoic acid degradative plasmids

Many synthetic chlorinated aromatic and aliphatic hydrocarbons have been released into the environment as herbicides, pesticides, solvents, plasticizers, and industrially useful compounds. While their release has resulted in acute pollution problems in many parts of the

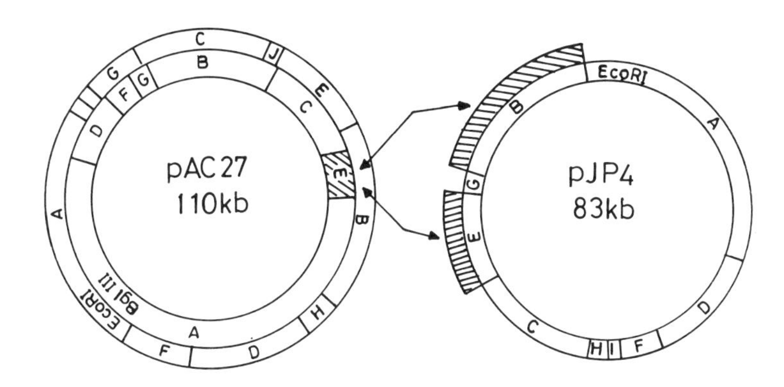

Figure 7. Physical maps of plasmids pAC27 and pJP4, carrying chlorocatechol degradative genes in common. The extent of homology is shown by hatched areas, as indicated by arrows.

world, it has also enabled microorganisms to evolve degradative pathways for simple chlorinated compounds, occasionally through elaboration of degradative plasmids [6,75,163] (Figure 7).

The first plasmid, encoding the degradation of the synthetic compound, 2,4-D, was described by Pemberton and Fisher [79]. It was not isolated from *Pseudomonas* sp. initially. Don and Pemberton [70] described the characterization of several other 2,4-D degradative plasmids, some with the ability to allow slow utilization of 3-chlorobenzoate. Such a typical plasmid is pJP4, originally isolated from *Alcaligenes eutrophus,* which could be transferred to *Pseudomonas* species because of its wide host range (p1 incompatibility group), transmissibility, and ability to confer resistance to mercury ions, which could be conveniently scored. Weightman et al. [165], as well as Pemberton and colleagues [70,71, 166,167], have constructed a physical map of this 84 kb plasmid and located the key genes, specifying chlorocatechol metabolism, involved in both 2,4-D and 3-chlorobenzoate degradation, on a single *Eco*RI fragment (fragment B) by transposon mutagenesis. Weightman et al. [165] also presented evidence that the genes, allowing conversion of 3-chlorobenzoate to chlorocatechol, were presumably chromosomal in *P. putida* and *Alcaligenes eutrophus*.

Weightman et al. [165] cloned the *Eco*RI fragments B, E, F, and I of plasmid pJP4 as part of the broad host range plasmid vector pKT230, and demonstrated that transfer of this plasmid to *P. putida* allowed the host cells to utilize 3-chlorobenzoate, but not 2,4-D. This suggested that, while all the genes, allowing utilization of 3-chlorocatechol in *P. putida,* were present on these fragments (pJP4 *Eco*RI fragments B, E,

F, and D, the other genes needed for complete 2,4-D metabolism were present somewhere else. Since the chlorocatechol degradative pathway, coded by pJP4 is quite similar (Figure 8) to that encoded by a 3-chlorobenzoate-degradative plasmid pWR1 [168], these authors attempted to delineate the homology of the chlorocatechol degradative genes by hybridizing ^{32}P-labelled pJP4 and pKJ11 (vector pKT231, containing the cloned *Hind*III-G fragment of pJP4) with restriction enzyme-digested fragments of plasmid pWR1 under high stringency. No homology was detected by these authors, under such conditions, between pWR1 fragments and pJP4 or the *Hind*III-G fragment of pJP4. Homology (Figure 8) between the pJP4 *Eco*RI-B and E fragments and a *Bgl*II-E of

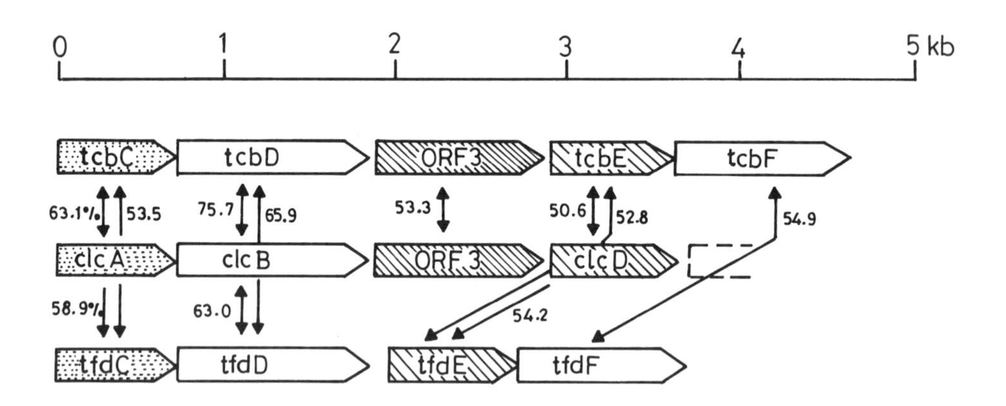

Figure 8. Organization of the gene clusters *tcb*CDEF, *clc*ABD, and *tfd*CDEF, which encode enzymes for chlorocatechol metabolism. The ORFs represent different open reading frames of the gene clusters. Similar patterns indicate DNA and amino acid sequence homology between equivalent genes of the three gene clusters, and percentages of identity between deduced amino acid sequences are shown. Overall DNA sequence identities between gene clusters are found to lie in the region (56.7 to 72.1%).

Symbol:

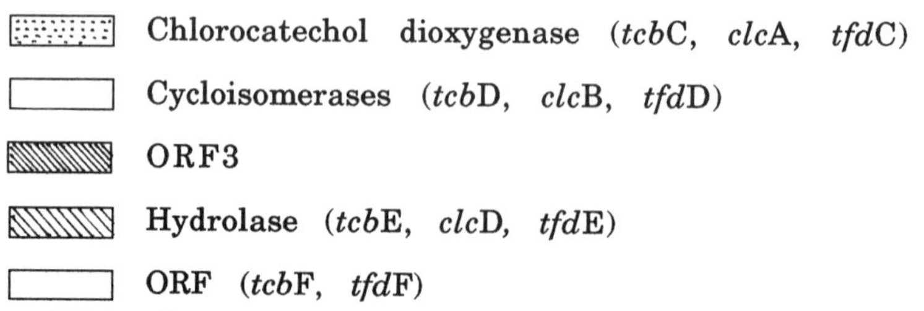

Chlorocatechol dioxygenase (*tcb*C, *clc*A, *tfd*C)

Cycloisomerases (*tcb*D, *clc*B, *tfd*D)

ORF3

Hydrolase (*tcb*E, *clc*D, *tfd*E)

ORF (*tcb*F, *tfd*F)

The ORF *tfd*F encodes a putative *trans-cis* isomerase.

another 3-chlorobenzoate degradative plasmid pAC27 [44], was observed by Ghosal et al. [47]. Chatterjee and Chakrabarty [45] demonstrated that the 117 kb pAC25 plasmid, originally described by Chatterjee et al. [44], and the 111 kb pWR1 are highly homologous, presumably identical, with pWR1, harbouring a deletion of about 6 kb. A 7 kb deletion from pAC25 has resulted in the formation of the 110 kb plasmid pAC27. Ghosal et al. [47] cloned various restriction fragments of plasmid pAC27 and demonstrated that the key genes for 3-chlorocatechol degradation, including the regulatory genes, were present on the *Eco*RI-B fragment of the plasmid pAC27. A 4.3 kb *Bgl*II-E fragment was shown to harbour all the structural genes for chlorocatechol degradation, but allowed only slow growth with 3-chlorobenzoate, presumably because of the absence of a regulatory sequence, that controls efficient expression of the structural genes.

The localization of individual 2,4-D degradative genes (*tfd*), present on the plasmid pJP4, has been reported by Don and Pemberton [70,71] and Don et al. [72]. Transposon mutagenesis of pJP4 by transposons Tn5 and Tn1771 enabled them to localize five genes for enzymes, involved in the 2,4-D catabolic pathway, as proposed by Evans et al. [127,128]. Four of the genes, *tfd*B, *tfd*C, *tfd*D, and *tfd*E encoded 2,4-dichlorophenol hydroxylase, dichlorocatechol 1,2-dioxygenase, dichloromuconate cycloisomerase, and chlorodienelactone hydrolase, respectively. No function had been assigned to *tfd*F which might encode a *trans*-chlorodiene-lactone isomerase. Inactivation of the genes *tfd*C, *tfd*D, and *tfd*E, which encode the enzymes involved in the transformation of dichlorocatechol to chloromaleylacetic acid, prevented the host strain JMP134 from degrading both 3-chlorobenzoate and 2,4-D, thus, indicating that the pathways were converging after the formation of chlorocatechols, and that the conversion of 3-chlorobenzoate to 3-chlorocatechol was specified by chromosomal genes.

Amy et al. [169] reported the characterization of some aquatic bacteria, isolated from rivers, streams, ponds, and activated sewage and found most of these belonging to genus *Alcaligenes*; some of them were found to contain plasmids, specifying 2,4-D degradation. No detailed genetic analysis has been carried out. Frantz and Chakrabarty [170] reported the organization and complete nucleotide sequence of the pAC27 gene cluster, specifying 3-chlorocatechol degradation. Chaudhry and Huang [69] isolated a new plasmid, pRC10 from a *Flavobacterium* sp. This plasmid shows considerable differences in size and restriction patterns from pJP4. It is a 45 kb plasmid, carries genes essential for the degradation of 3-chlorobenzoate, and 2-methyl-4-chlorophenoxy acetate (MCPA), imparts resistance to mercury and encodes utilization of 2,4-D. Comparison with plasmid pJP4 showed strong homology with the regions, containing 2,4-D-degradative genes; the first two genes responsible for the 2,4-D degradation, *tfd*A and *tfd*B have been cloned as a subfragment of the

*Eco*RI A fragment of pRC10. Expression of pRC10 in *P. putida* (Nld^r) and *A. eutrophus* JMP228, showed the cloned fragment, coding for *tfd*A and *tfd*B. In the author's laboratory a new 2,4-D degradative plasmid pMAB1 has been reported from *Pseudomonas cepacia* strain CSV90. The molecular size of this plasmid is about 90 kb. Restriction digestion patterns are apparently different from those of pJP4 and pRC10 [154].

More recent studies have focused on the organization, cloning, and characterization of the *tfd* genes, particularly on the plasmid pJP4 [171-173]. Like TOL and NAH plasmids, the expression of *tfd* genes has also been found to be regulated by *regulatory elements*. It has been shown that *tfd*R regulates the expression of *tfd*A and the *tfd*CDEF operon, but not *tfd*B [174]. The promoters for *tfd*A and the operon seem to be homologous [175].

van der MeeR and his group [176-178] have reported a new plasmid p51 from a *Pseudomonas* sp., encoding genes for the degradation of 1,2-dichloro-, 1,4-dichloro- and 1,2,4-trichlorobenzene. They showed that the genes for trichlorocatechol-degrading enzymes are in the form of an operon *tcb*CDEF [177], and that this operon was regulated by *tcb*R. They also characterized this regulatory gene *tcb*R, and the promoter regions, where TcbR exerts its activity and suggested that TcbR belongs to the LysR family of transcriptional activator proteins [178]. The chlorocatechol degradative genes, present on the plasmid p51, might code for enzymes with more relaxed substrate specificities and could be the best candidate genes for transfer to other genera and, in turn, would help to develop strains, which could detoxify mono-, di- and trichlorocatechols simultaneously. The organization of the different chlorocatechol degradative operons is shown in Figure 8.

CONCLUSIONS

Halogenated aromatic compounds are being extensively used in agriculture as herbicides and pesticides, and also find use in manufacturing industries as solvents and heat insulators. Chlorinated aromatics are of a major public concern about their possible hazardous effects on humans and the general life on this planet. The chlorinated compounds are often recalcitrant or insoluble in aqueous systems, they escape degradation. However, microorganisms exposed to these toxic compounds have evolved the ability to utilize a few of them. Bacteria from different genera have been shown to degrade chlorinated aromatics. Majority of the microorganisms harbour plasmids, which code for degradative enzymes. By understanding the biochemistry and genetics of plasmid-borne degradative functions, and by using the recombinant DNA techniques, it is possible to characterize the appropriate genes and transfer them to construct improved strains, with enhanced degradative ability against several toxic compounds. One can develop the so called "*superbugs*" which can detoxify the hazardous chemicals in the natural environment. These strains can be tested for survival and stability under various natural and laboratory conditions.

Although the risk of releasing of such strains into the atmosphere is not known, the prospects for the construction of catabolic pathways to effect mineralization and detoxification of halogenated compounds are encouraging. The most effective means of restricting pollution lie in reducing the use of recalcitrant compounds and replacing them with nonrecalcitrant ones. The genetically engineered microorganisms could be useful in decontaminating the toxic waste in contained environmental situations, such as biodegradation reactors, waste dump sites, and water treatment systems.

ACKNOWLEDGEMENTS

The work in the author's laboratory was supported by the Indian Institute of Science, Bangalore, India. M.A.B. received the financial support from the Council of Scientific and Industrial Research, New Delhi, India.

REFERENCES

1 Keith LH. In: Identification and Analysis of Organic Pollutants in Water. Ann Arbor, Michigan: Ann Arbor Publishers, 1976.
2 Schneider MJ. In: Persistent Poisons: Chemical Pollutants in the Environment. New York: New York Academy of Sciences, 1979.
3 Dagley S. In: Degradation of Synthetic Organic Molecules in the Biosphere. Washington DC: Printing and Publishing Office, National Academy of Science, 1972.
4 Hegeman GD. In: Degradation of Synthetic Organic Molecules in the Biosphere. Washington DC: Printing and Publishing Office, National Academy of Science, 1972.
5 Reineke W. In: Gibson DT, ed. Microbial Degradation of Organic Compounds. New York: Marcel Dekker Inc, 1984.
6 Ghosal D, You IS, Chatterjee DK, Chakrabarty AM. Science 1985; 228: 135-142.
7 Reineke W, Knackmuss HJ. Ann Rev Microbiol 1988; 42: 263-287.
8 Choudhry GR, Chapalamadugu S. Microbiol Rev 1991; 55: 59-79.
9 Heitkamp MA, Franklin W, Cerniglia, CE. Appl Environ Microbiol 1988; 54: 2549-2555.
10 Pal D, Weber JB, Overcash MR. Resid Rev 1980; 74: 45-98.
11 Clark RR, Chian ESK, Griffin RA. Appl Environ Microbiol 1979; 37: 680-685.

12 Furukawa K, Tomizuka N, Kamibayashi A. Appl Environ Microbiol 1983; 46: 140-145.
13 Kopecky A, Omen GS, Hollaender A. In: Genetic Control of Environmental Pollutants, New York: Plenum, 1984.
14 Kong HL, Sayler GS. Appl Environ Microbiol 1983; 46: 666-672.
15 Sylvestre M. Appl Environ Microbiol 1980; 39: 1223-1224.
16 Sylvestre M, Masse R, Massier F, Fauteux J, Bisaillon JG, Beaudet R. Appl Environ Microbiol 1982; 44: 871-877.
17 Sylvestre M, Masse R, Ayotte C, Messier F, Feuteux J. Appl Microbiol Biotechnol 1985; 21: 192-195.
18 Masse R, Messier F, Peloquin L, Ayotte C, Sylvestre M. Appl Environ Microbiol 1984; 47: 947-951.
19 Furukawa K, Tonomura K, Kamibayashi A. Appl Environ Microbiol 1978; 35: 223-227.
20 Furukawa K, Matsumura F. J Agric Food Chem 1976; 24: 251-256.
21 Ahmed M, Focht DD. Can J Microbiol 1973; 19: 47-52.
22 Furukawa K, Chakrabarty AM. Appl Environ Microbiol 1982; 44: 619-626.
23 Yagi O, Sudo R. J Water Pollution Control Federation 1980; 52: 1035-1043.
24 Takase I, Omori T, Minoda Y. Agric Biol Chem 1986; 50: 681-686.
25 Dmochewitz S, Ballschmiter K. Chemosphere 1988; 17: 111-121.
26 Eaton DC. Enz Microb Technol 1985; 7: 194-196.
27 Adriaens HP, Kohler E, Kohler-Staub D, Focht DD. Appl Environ Microbiol 1989; 55: 887-892.
28 Barton MR, Crawford RL. Appl Environ Microbiol 1988; 54: 594-595.
29 Alexander M. Biotechnol Bioeng 1973; 15: 611-647.
30 Pfander FK, Alexander A. J Agric Food Chem 1973; 21: 397-401.
31 Golovleva LA, Skryabin GK, Leisinger T, Cook AM, Hutter R, Nuesch J. London: Academic Press, 1981.
32 Bumpus JA, Aust SD. Appl Environ Microbiol 1987; 53: 2001-2008.
33 Markus A, Klages U, Krauss S, Lingens F. J Bacteriol 1984; 160: 618-621.
34 Morris CM, Barnsley EA. Can J Microbiol 1982; 24: 499-507.
35 Pierce GE, Robinson JB, Facklam TJ, Rice JM. Dev Ind Microbiol 1982; 23: 407.
36 Pierce GE, Robinson JB, Garrett GE, Sojka SA. Dev Ind Microbiol 1984; 25: 597-602.
37 DeBont JAM, Vorage MJAW, Hartmans S, VanDentweel WJJ. Appl Environ Microbiol 1986; 52: 677-680.
38 Haigler BE, Nishino MF, Spain JC. Appl Environ Microbiol 1988; 54: 294-301.
39 Ultmanns RH, Rast HG, Reineke W. Appl Microbiol Biotechnol 1988; 28: 609-616.

40 Reineke W, Knackmuss HJ. Appl Environ Microbiol 1984; 47: 395-402.
41 Schraa G, Boone BL, Jetten MSM, Neerven ARW, Colberg PJ, Zehnder AJB. Appl Environ Microbiol 1986; 52: 1374-1381.
42 Spain J, Nishino SF. Appl Environ Microbiol 1987; 53: 1010-1019.
43 Chatterjee DK, Chakrabarty AM. FEMS Symposia 1981; 12: 213-219.
44 Chatterjee DK, Chakrabarty AM. Molec Gen Genet 1982; 188: 279-285.
45 Chatterjee DK, Chakrabarty AM. J Bacteriol 1983; 153: 532-534.
46 Chatterjee DK, Kellogs ST, Hamada S, Chakrabarty AM. J Bacteriol 1981; 146: 639-646.
47 Ghosal D, You IS, Chatterjee DK, Chakrabarty AM. Proc Natl Acad Sci USA 1985; 82: 1638-1642.
48 Hartmann J, Reineke W, Knackmuss HJ. Appl Environ Microbiol 1979; 37: 421-428.
49 Horowitz A, Suflita JM, Tiedje JM. Appl Environ Microbiol 1983; 45: 1459-1465.
50 Marks TS, Smith ARW, Quirk AV. Appl Environ Microbiol 1984; 48: 1020-1025.
51 Apajalathi JHA, Salkinja-Salonen MS. Appl Microbiol Biotechnol 1986; 25: 62-67.
52 Boyd SA, Shelton DR. Appl Environ Microbiol 1984; 47: 272-277.
53 Brown EJ, Pignatello JJ, Martinson MM, Crowford RL. Appl Environ Microbiol 1986; 52: 92-97.
54 Bruhn CR, Bayly RC, Knackmuss HJ. Arch Microbiol 1988; 150: 171-177.
55 Crawford RL. Mohn WW. Enz Microb Technol 1985; 7: 617-620.
56 Ide A, Niki Y, Sakamoto F, Wantanabe I, Wantanage H. Agric Biol Chem 1972; 36: 1937-1944.
57 Moos IP, Kirsch EJ, Wukasch RF, Grady Jr CPL. Aerob Water Res 1983; 17: 1575-1584.
58 Murty NBK, Kaufman DD, Fries GF. J Environ Health Sci 1979; 14: 1-14.
59 Pignatello JJ, Martinson MM, Steiert JG, Carlson RE, Crawford RL. Appl Environ Microbiol 1983; 46: 1024-1031.
60 Saber DL, Crawford RL. Appl Environ Microbiol 1985; 50: 1512-1518.
61 Schmidt E, Hellwig M, Knackmuss HJ. Appl Environ Microbiol 1983; 46: 1038-1044.
62 Stanlake GJ, Finn RK. Appl Environ Microbiol 1982; 44: 1421-1427.
63 Steiert JG, Crawford RL. Trends Biotechnol 1985; 3: 300-305.
64 Steiert JG, Crawford RL. Biochem Biophys Res Commun 1986; 141: 825-830.

65 Steiert JG, Pignatello JJ, Crawford RL. Appl Environ Microbiol 1987; 53: 907-910.
66 Valo R, Apajalahti J, Salkinoja-Salonen M. Appl Environ Microbiol 1985; 21: 313-319.
67 Saxena A, Zhang R, Bollag JM. Appl Environ Microbiol 1987; 53: 390-396.
68 Klages U, Markus A, Lingens F. J Bacteriol 1981; 146: 64-48.
69 Chaudhry AR, Huang GH. J Bacteriol 1988; 170: 3897-3902.
70 Don RH, Pemberton JM. J Bacteriol 1981; 145: 681-686.
71 Don RH, Pemberton JM. J Bacteriol 1985; 161: 466-468.
72 Don RH, Weightman AJ, Knackmuss HJ, Timmis KN. J Bacteriol 1985; 161: 85-90.
73 Karns JS, Kilbane JJ, Duttagupta S, Chakrabarty AM. Appl Environ Microbiol 1983; 46: 1176-1181.
74 Karns JS, Duttagupta S, Chakrabarty AM. Appl Environ Microbiol 1983; 46: 1182-1186.
75 Karns JS, Kilbane JJ, Chatterjee DK, Chakrabarty AM. In: Omenn GS, Hollaender A, eds. Genetic Control of Environmental Pollutants. New York: Plenum Press, 1984.
76 Kilbane JJ, Chatterjee DK, Karns JS, Kellogg ST, Chakrabarty AM. Appl Environ Microbiol 1982; 44: 72-78.
77 Kilbane JJ, Chatterjee DK, Chakrabarty AM. Appl Environ Microbiol 1983; 45: 1697-1700.
78 Kilpi S, Backstrom V, Korhala M. FEMS Microbiol Lett 1980; 8: 177-182.
79 Pemberton JM, Fisher PR. Nature 1977; 268: 732-733.
80 Pieper DH, Reineke W, Engesser KH, Knackmuss HJ. Arch Microbiol 1988; 150: 95-102.
81 Pierce GE, Facklum TJ, Rice JM. Dev Ind Microbiol 1981; 22: 401-408.
82 Pierce GE, Robinson JB, Facklam TJ, Rice JM. Dev Ind Microbiol 1992; 23: 407-417.
83 Sinton GL, Fan LT, Erickson LE, Lee SM. Enz Microb Technol 1986; 8: 395-403.
84 Reineke W. J Basic Microbiol 1986; 9: 551-567.
85 Crosby DG. Pure Appl Chem 1981; 53: 1051-1080.
86 Schenk TR, Muller F, Morsberger F, Otto MK, Lingens F. J Bacteriol 1989; 171: 5487-5491.
87 Suzuki T. J Environ Sci Health, Part B 1977; 12: 113.
88 Edgehill RU, Finn RK. Appl Environ Microbiol 1983; 45: 1122-1125.
89 Rippen G, Klopffer W, Frische R, Gunther K. Exotoxicol Environ Safety 1984; 8: 363-377.
90 Schwarzenbach RP, Ginger W, Hoehn E, Schneider JR. Environ Sci Technol 1983; 17: 472-479.

91 Bouwer EJ, McCarty PL. Environ Sci Technol 1982; 16: 836-843.
92 Kuhn EP, Colberg P, Schnoor L, Wanner O, Zehnder AJB, Schwarzenbach RP. Environ Sci Technol 1984; 19: 961-968.
93 de Bont JAM, Vorage MJAW, Hartmans S, van den Tweel WJJ. Appl Environ Microbiol 1986; 52: 677-680.
94 Haigler BE, Nishino NF, Spain JC. Appl Environ Microbiol 1988; 54: 294-301.
95 van der MeeR JR, Roelofsen W, Schraa G, Zehnder AJB. FEMS Microbiol Ecol 1987; 45: 333-341.
96 Gibson DT, Koch JR, Kallio RE. Biochemistry 1968; 7: 2653-2662.
97 Dorn E, Hellwig M, Reineke W, Knackmuss HJ. Arch Microbiol 1974; 99: 61-70.
98 Weisshaar MP, Franklin FCH, Reineke W. J Bacteriol 1987; 169: 394-402.
99 Oltmanns RH, Rast HG, Reineke W. Appl Microbiol Biotechnol 1988; 28: 609-616.
100 Krockel L, Focht DD. Appl Environ Microbiol 1988; 53: 2470-2475.
101 Chapman PJ. In: Omenn GS, ed. Environmental Biotechnology: Reducing the Risks from Environmental Pollution through Biotechnology. New York: Plenum Press, 1988; 81-95.
102 Fathepure BZ, Tiedje JM, Boyd SA. Appl Environ Microbiol 1988; 54: 327-330.
103 McCann J, Simmon V, Streitwieser D, Ames BN. Proc Natl Acad Sci USA 1975; 72: 3190-3193.
104 Digeronimo MJ, Nikaido M, Alexander M. Appl Environ Microbiol 1979; 37: 619-625.
105 Suflita JM, Horowitz A, Shelton DR, Tiedje JM. Science 1982; 218: 1115-1117.
106 Suflita JM, Robinson JA, Tiedje JM. Appl Environ Microbiol 1983; 45: 1466-1473.
107 Spokes JR, Walker N. Arch Microbiol 1974; 96: 125-134.
108 Horwath RS, Alexander M. Appl Microbiol 1970; 20: 254-258.
109 Bartels I, Knackmuss HJ, Reineke W. Appl Environ Microbiol 1984; 47: 500-505.
110 Klecka GM, Gibson GT. Appl Environ Microbiol 1981; 41: 1159-1165.
111 Johnston HW, Briggs GG, Alexander M. Soil Biol Biochem 1972; 4: 187-190.
112 Dorn E, Knackmuss HJ. Biochem J 1978; 174: 73-84.
113 Dorn E, Knackmuss HJ. Biochem J 1978; 174: 85-94.
114 Reineke W, Knackmuss HJ. Biochim Biophys Acta 1978; 542: 412-423.
115 Schmidt E, Knackmuss HJ. Biochem J 1980; 192: 331-337.

116 Reineke W, Jeenes DJ, Williams PA, Knackmuss HJ. J Bacteriol 1982; 150: 195-201.
117 Ruisinger S, Klages U, Lingens F. Arch Microbiol 1976; 110: 253-256.
118 Klages U, Lingens F. FEMS Microbiol Lett 1979; 6: 201-203.
119 Klages U, Lingens F. Zentralbl. Bacteriol Parasitenkd Infektionskr Reihe C 1980; 1: 215-223.
120 Muller R, Thiele J, Klages U, Lingens F. Biochem Biophys Res Commun 1984; 124: 178-182.
121 Zaltsev GM, Karasevich AN. Mikrobiologiya 1984; 53: 75-80.
122 Thiele J, Muller R, Lingens F. FEMS Microbiol Lett 1987; 41: 115-119.
123 Zaltsev GM, Baskunov BP. Mikrobiologiya 1985; 54: 203-208.
124 Pierce GE, Robinson JB, Colaruotolo JR. Deve Ind Microbiol 1983; 24: 499-507.
125 Pierce GE, Robinson JB, Garrett GE, Sojka SA. Dev Ind Microbiol 1984; 25: 597-602.
126 Macrae IC, Alexander M, Rovira AD. J Gen Microbiol 1963; 32: 69-76.
127 Evans WC, Smith BSW, Moss P, Fernly HN. Biochem J 1971; 122: 509-517.
128 Evans C, Smith BSW, Fernley HN, Davis JI. Biochem J 1971; 122: 543-551.
129 Horvath RS. Bull Environ Cont Toxicol 1970; 5: 537-541.
129a Horvath RS. Biochem J 1970; 119: 871-876.
130 Suflita JM, Stout J, Tiedje JM. J Agric Food Chem 1984; 32: 218-221.
131 Bollag JM, Helling CS, Alexander M. J Agric Food Chem 1968; 16: 826-828.
132 Gaunt JK, Evans WC. Biochem J 1971; 122: 519-526.
132a Gaunt JK, Evans WC. Biochem J 1971; 122: 533-542.
133 Gamar Y, Gaunt JK. Biochem J 1971; 122: 527-531.
134 Steenson TI, Walker N. J Gen Microbiol 1957; 16: 146.
135 Grant WF. Mutat Res 1979; 65: 83-119.
136 Hanify JA, Metcalf P, Nobbs CL, Worsley KJ. Science 1981; 212: 349-351.
137 Helling CS, Bollag JM, Dawson JE. J Agric Food Chem 1968; 16: 538-539.
138 Tiedje JM, Duxbury JM, Alexander M, Dawson JE. J Agric Food Chem 1969; 17: 1021-1026.
138a Tiedje JM, Alexander M. Journal of Agric Food Chem 1969; 17: 1080-1084.
139 Tyler JE, Finn RK. Appl Microbiol 1974; 28: 181-184.
140 Nutman PS, Thornton HG, Quastel JH. Nature 1945; 155: 498-500.

141 Audus LJ. Pl Soil 1949; 2: 31-36.
142 Jensen HL, Petersen HI. Acta Agric Scandinavia 1972; 2: 215-231.
143 Newman AS, Thomas JR. Soil Sci Soc Proc (USA) 1949; 14: 160-164.
144 Audus LJ. Pl Soil 1951; 3: 170-192.
145 Loos MA, Roberts RN, Alexander M. Can J Microbiol 1967; 13: 679-690.
146 Sharpet KW, Duxbury JM, Alexander M. Appl Microbiol 1973; 27: 445-447.
147 Williams RT, Crawford RL. Can J Microbiol 1983; 29: 1430-1437.
148 Streber WR, Timmis KN, Zenk MH. J Bacteriol 1987; 169: 2950-2955.
149 Beadle CA, Smith ARW. Euro J Biochem 1982; 123: 323-332.
150 Liu T, Chapman PJ. FEBS Lett 1984; 173: 314-318.
151 Radjendirane V, Bhat MA, Vaidyanathan CS. Arch Biochem Biophys 1991; 288: 169-176 [addendum 296; 354 (1992)].
152 Bhat MA, Ishida T, Horiike K, Vaidyanathan CS, Nozaki M. Arch Biochem Biophys 1993; 300: 738-746.
153 Radjendirane V. Ph.D. Thesis, Indian Institute of Science, Bangalore, India.
154 Bhat MA. Ph.D. Thesis, 1992; Indian Institute of Science, Bangalore, India.
155 Mikesell MD, Boyd SA. J Environ Quality 1985; 14: 337-340.
156 Rosenberg A, Alexander M. J Agric Food Chem 1980; 28: 297-302.
157 Rosenberg A, Alexander M. J Agric Food Chem 1980; 28: 705-709.
158 Alexander M. Science 1981; 211: 132-138.
159 Furukawa K. Tomizuka N, Kamibayashi A. Appl Environ Microbiol 1979; 38: 301-310.
160 Furukawa K. Tonomura K, Kamibayashi A. Agric Biol Chem 1979; 43: 1577-1583.
161 Furukawa K, Miyazaki T. J Bacteriol 1986; 166: 392-398.
162 Furukawa K, Arimura N. J Bacteriol 1987; 169: 924-927.
163 Furukawa K, Arimura N, Miyazaki T. J Bacteriol 1987; 169: 427-429.
164 Taira K, Hayase N, Arimura N, Yamashita S, Miyazaki T, Furukawa K. Biochemistry 1988; 27: 3990-3996.
165 Weightman AJ, Don RH, Lehrbach PR, Timmis KN. In: Omenn GS. Hollaender A, eds. Genetic Control of Environmental Pollutants New York: Plenum, 1984; 47-80.
166 Pemberton JM, Corney B, Don RH. In: Timmis KN. Puhler A. eds. Plasmids of Medical, Environmental and Commercial Importance Amsterdam: Elsevier Biomedical Press, 1979; 287-299.

167 Pemberton JM. Int Rev Cytol 1983; 84: 155-183.
168 Reineke W, Nackmuss HJ. Nature 1979; 277: 385-386.
169 Amy PS, Sahulke JW, Frazier LM, Seidler RJ. Appl Environ Microbiol 1987; 49: 1237-1245.
170 Frantz B, Chakrabarty AM. Proc Natl Acad Sci USA 1987; 84: 4460-4464.
171 Ghosal D, You IS, Molec Gene Genet 1988; 211: 113-120.
172 Ghosal D, You IS. Gene 1989; 83: 225-232.
173 Perkins EJ, Gordon MP, Caceres D, Lurquin PF. J Bacteriol 1990; 172: 2351-2359.
174 Harker AR, Olsen RH, Seidler RJ. J Bacteriol 1989; 171: 314-320.
175 Perkins EJ, Bolton GW, Gordon MP, Lurquin PF. Nucleic Acids Research 1988; 16: 7200.
176 van der MeeR JR, van Neerven ARW, de Vries EJ, de Vos WM, Zehnder AJB. J Bacteriol 1991; 173: 6-15.
177 van der MeeR JR, Eggen RIL, Zehnder AJB, de Vos WM. J Bacteriol 1991; 173: 2425-2434.
178 van der MeeR JR, Frijters ACJ, Leveau JHJ, Eggen RIL, Zehnder AJB, de Vos WM. J Bacteriol 1991; 173: 3700-3708.

Biotransformations: Microbial Degradation of Health Risk Compounds
Ved Pal Singh, editor
© 1995 Elsevier Science B.V. All rights reserved.

Microbial degradation of azo dyes

John A. Bumpus

Center for Bioengineering and Pollution Control and
Department of Chemistry and Biochemistry, University of Notre Dame,
Notre Dame, Indiana 46556, U.S.A.

INTRODUCTION

Azo dyes and pigments are extremely versatile colourants. Approximately one-half of all known dyes, are azo dyes; making them the largest group of synthetic colourants.

The popularity and widespread use of azo dyes is due to several factors [1]. As a group, they are *colour-fast* and encompass the entire visible spectrum, and many are easily synthesized from inexpensive and easily obtained starting materials. Azo dyes are also typically amenable to structural modification, and representative azo dyes can be made to bind most synthetic and natural textile fibers.

Azo dyes are also the most common synthetic colourants released into the environment. Their widespread use, coupled with the fact that azo dyes are not readily degraded in most biological treatment systems makes this class of chemicals a significant environmental problem [2].

Selected aspects of environmental problems, that are associated with the use of azo dyes, are summarized herein, and the recent literature, documenting the *biodegradation* of azo dyes by bacteria and fungi, is reviewed.

ENVIRONMENTAL CONTAMINATION BY AZO DYES

Compared with other classes of chemicals, azo dyes make a relatively small contribution to the total mass of pollutants released into the environment. However, because they are highly coloured, contamination by these chemicals, when it occurs, is readily apparent. Moreover, in locales, where dyes are synthesized and used, their presence in ground water and surface water may represent important pollution problems. For example, the Coosa River Basin and its tributaries in Northwest Georgia contain approximately 50% of all carpet dyeing wastewater in the United States; concentrations of acid azo dyes in Coosa River Basin water samples often being present in the parts per billion to the low parts per million level [3]. Sediment samples from the same source revealed the presence of several acid azo dyes as well as several disperse azo dyes in the parts per billion to parts per million range [3].

Azo dye contamination of water in developing nations is, perhaps, an even greater problem. Due to their low cost, versatility and synthetic accessibility, azo dyes are often the colourants of choice. As a case in point, it has been noted [4] that there are approximately 500 dyeing industries in Tirupur in Southern India and that factories located on the Noyyal River release coloured effluents containing untreated azo dyes directly into the river, whereas those not located on the river, deposit wastewater in open fields or ponds. Quantities released into the environment are so high that photosynthesis in surface water is inhibited and a considerable amount of ground water is visibly coloured [4].

TOXICITY OF AZO DYES

The number of azo dyes, synthesized and/or in commerce, is enormous. Similarly, the literature describing the *toxicity* of these compounds is substantial. In general, dyes are relatively nontoxic as judged by acute oral toxicity studies. Out of 4,461 dyes (all classes) tested, only 44 (~1%) had an LD_{50} less than 250 mg/kg while 314 (7%) had an LD_{50} between 250 and 2000 mg/kg, 434 (9.7%) had an LD_{50} between 2000 and 5000 mg/kg and 3,669 had an LD_{50} greater than 5000 mg/kg [2]. Such assays, however, do not assess long term problems which might arise from exposure to such compounds. Indeed, studies have shown that some azo dyes are known carcinogens [5-9]. Thus, in addition to acute toxicity, studies to determine LD_{50}, several techniques have been used to assess *mutagenicity* and *carcinogenicity* of azo dyes. The *Salmonella typhimurium* plate incorporation assay (*Salmonella*/microsomes) has been used extensively to demonstrate the mutagenicity of these and other compounds [10]. Interestingly, some azo dyes that are known carcinogens are not mutagenic in the *Salmonella*/microsomes assay. Trypan Blue (I) is an example [11,12].

The reduction product, 3,3'-dimethylbenzidine, of Trypan Blue (I) is carcinogenic. Indeed, benzidine derivatives are especially potent and important reduction products of some azo dyes [13,14]. The bicyclic 2-naphthylamine is also an important carcinogenic reduction product of many azo dyes. Thus assessment of the *genotoxicity* of azo dyes require that studies be included in which the dyes are reduced (typically dithionite reduction is used) and the genotoxicity of the reduction products assessed [11]. Another approach to this problem is the use of the flavin mononucleotide (FMN) modified *Salmonella*/microsomes assay in which FMN is included in the assay to reduce azo linkages [12]. Genotoxicity of azo dyes has also been studied using the mouse lymphoma $TK^{+/-}$ assay to test dyes for their ability to induce unscheduled DNA synthesis [15]. Ingestion of azo dyes whose reduction produces include benzidine or benzidine derivatives often cause bladder cancers in rodents

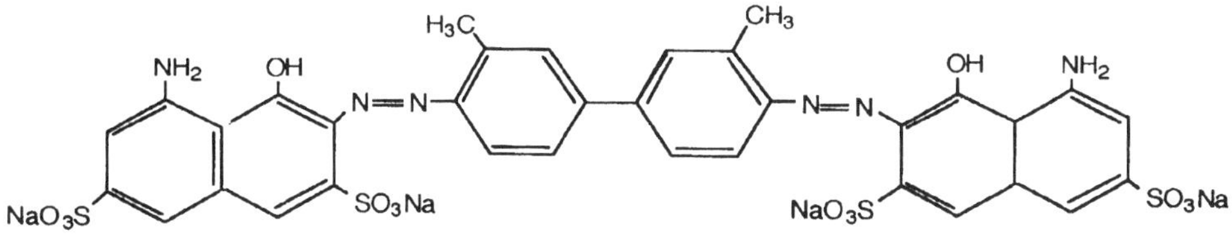

TRYPAN BLUE

I

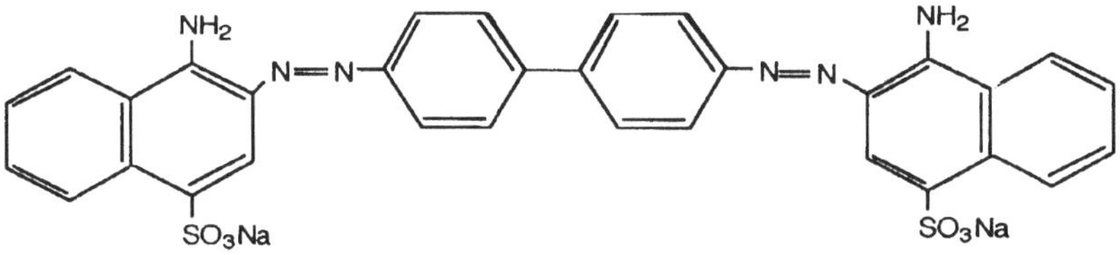

CONGO RED

II

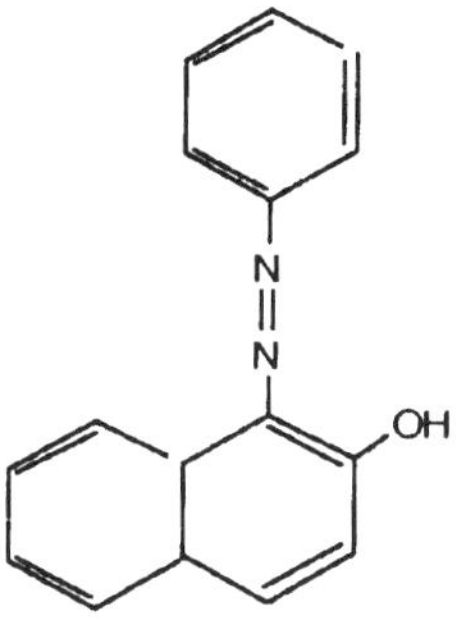

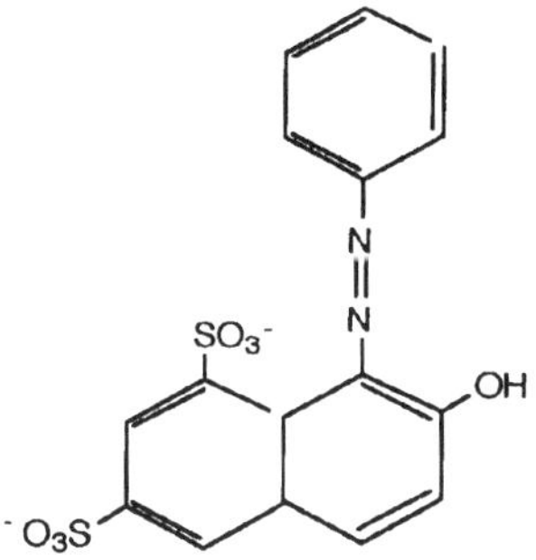

SOLVENT YELLOW 14

III

ACID ORANGE 10

IV

[13,14]. More recent studies have shown that such exposure can also reduce fertility. When female and male mice were exposed to Congo Red (II) *in utero* gonadal development was adversely affected in both sexes [16]. However, only females displayed reduced fertility [16].

It should be stressed that not all azo dyes are toxic. Nor do all azo dyes produce toxic (i.e., carcinogenic or mutagenic) metabolites upon reduction. It should also be noted that strategies exist to produce less toxic azo dyes. For example, in addition to being a carcinogenic reduction product of some azo dyes, 2 naphthylamine was also (not unexpectantly) used in the commercial synthesis of such dyes. In general, 2-napthylamine is no longer used in the large scale commercial synthesis of azo dyes. Instead, modern synthetic routes to many newer azo dyes use the structurally similar sulfonated napthylamines to manufacture sulfonated derivatives of dyes formerly made from 2-naphthylamine. This is based on the empirical observation that sulfonated azo dyes and aromatic amine reduction products are often less toxic than the corresponding unsulfonated compound. For example, Solvent Yellow 14 (III) is known to be carcinogenic, whereas its sulfonated analog, Acid Orange 10 (IV), is not [9].

In addition to sulfonic acid groups, substitution of the parent compound with other electrophilic groups (hydroxyl and carboxyl groups, for example) tend to decrease the toxicity of the substituted derivative relative to the unmodified parent compound [17].

BIODEGRADATION OF AZO DYES BY BACTERIA

Wastewaters, containing azo dyes, are generally resistant to remediation by bacteria. Although many examples of azo dye metabolism have been reported, most occur under anaerobic conditions, resulting primarily in the reduction of the azo linkage, forming the corresponding aromatic amines which are themselves a disposal problem, as many are carcinogenic. The early literature concerning microbial degradation of azo dyes has been reviewed [18]. Among the bacteria, reported to degrade these dyes, are lactic acid bacteria, *Proteus* sp., *Enterococcus* sp., *Streptococcus faecalis*, *Bacillus pyocyaneous*, *Bacillus subtilis*, *Bacillus cereus*, *Pseudomonas* spp., *Aeromonas hydrophilia* var. 24B, *Caprococcus catus*, *Fusobacterium prausnitzii*, *Bacteroides thetaitaomicron*, *Bifidobacterium infantis*, *Eubacterium biforme*, *Peptostreptococcus productus*, *Citrobacter* sp., and *Fusobacterium* sp.

Anaerobic bacterial azo reductases appear to be rather non-specific. This observation, coupled with the fact that many aromatic amines are readily degraded under aerobic conditions, has led to the proposal that coupled systems might be effective in achieving the complete biodegradation of azo dyes. In such systems azo dyes would be reduced anaerobically, followed by subsequent aerobic treatment to further degrade to carbon dioxide, the aromatic amines were formed as intermediary metabolites. This sequential approach has been pursued recently, using a bacterial consortium (6A2NS), adapted to grow on amino and hydroxynaphthalene-2-sulfonates [19,20]. Initially, the bacterial consortium was grown aerobically in the presence of the azo dye Mordant Yellow 3 (V) without biodegradation of the dye.

However, substantial degradation occurred when the culture was made anaerobic; nearly stoichiometric amounts of the corresponding aromatic amines were formed. Further degradation did not occur until the culture was reaerated, whereupon the amines were degraded to Krebs cycle intermediates by strain BN6, a member of the bacterial consortium known to be able to use a variety of aromatic amines as its sole carbon, nitrogen, and energy source. In addition to Mordant Yellow 3 (V), Acid Yellow 21 (VI), Amaranth (VII), Tartrazine (VIII), and 4-Hydroxyazobenzene-4'-Sulfonic Acid (IX) were assayed for their ability to be metabolized in this system.

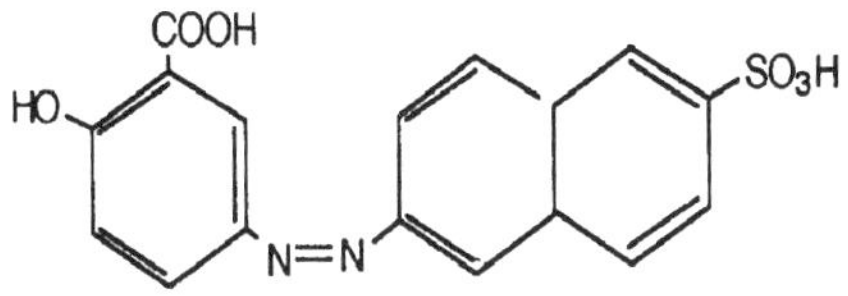

MORDANT YELLOW 3

V

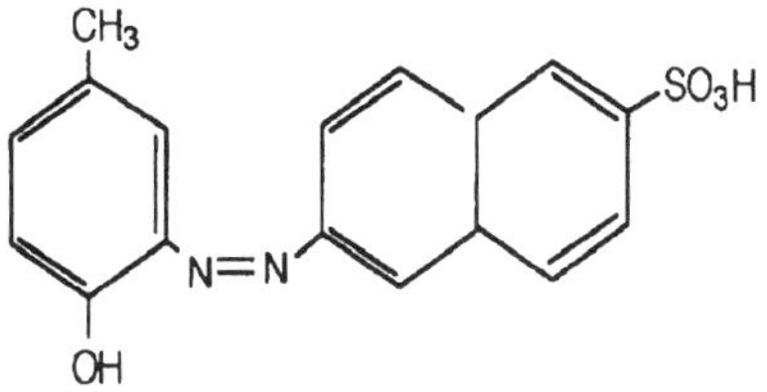

ACID YELLOW 21

VI

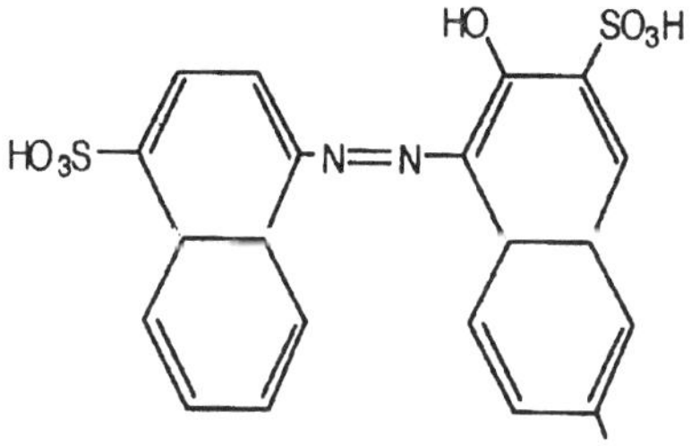

AMARANTH

VII

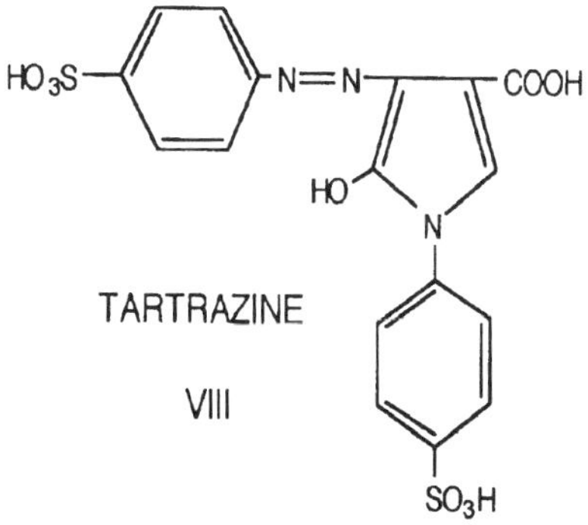

TARTRAZINE

VIII

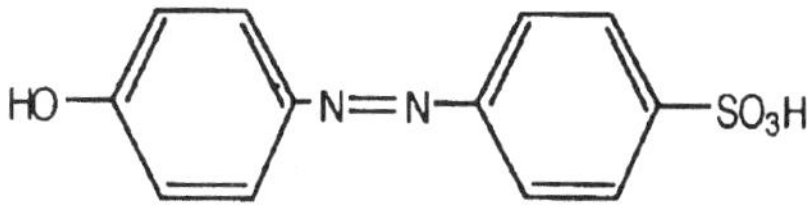

4-HYDOXYAZOBENZENE-4'-
SULFONIC ACID

IX

Under anaerobic conditions, in the presence of 10 mM glucose, all of the dyes tested, except Tartrazine, were decolourized completely. Only 16% of the Tartrazine was decolourized. Evidence was also presented which suggested that the role of glucose was to promote formation of reducing equivalents (e.g., NADH and $FADH_2$) used for azo dye reduction.

A considerable effort to isolate aerobic bacteria capable of complete degradation of azo dyes has, in large part, been unsuccessful. Possibly the most successful effort, in this regard, focused on the use of a soil microorganism (*Pseudomonas* sp.) able to utilize 4,4'-Dicarboxyazobenzene (X) as its sole source of carbon and nitrogen [21]. Attempts were made to adapt continuous cultures of this isolate to utilize the more complicated azo dyes, Orange I (XI) and Orange II (XII).

Carboxylated (instead of sulfonated) Orange I (XI) and Orange II (XII) derivatives were prepared and incubated in cultures, containing 4,4'-Dicarboxyazobenzene (X) for 100 and 400 generations, respectively. Resulting cultures utilized the carboxy derivatives of Orange I (XI) and Orange II (XII), but not the original sulfonated dyes, as sole carbon, nitrogen, and energy sources [21,22].

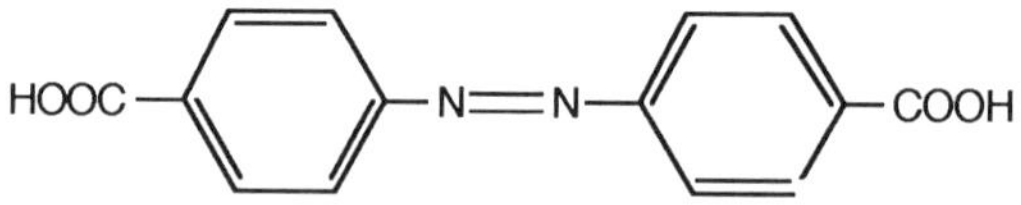

HOOC—C6H4—N=N—C6H4—COOH

4,4'-DICARBOXYAZOBENZENE

X

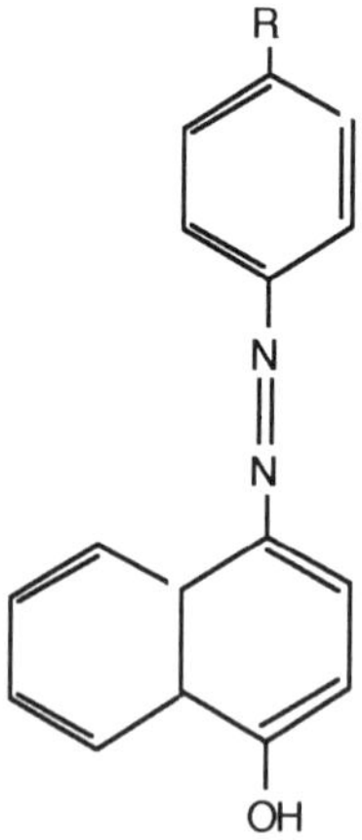

R= COO^- or SO_3^-

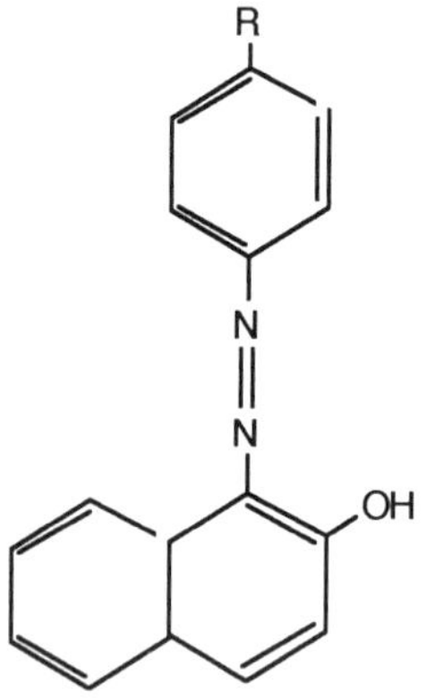

ORANGE I AND
CARBOXYLATED ORANGE I
XI

ORANGE II AND
CARBOXYLATED ORANGE II
XII

The ability of *Pseudomonas cepacia* acclimated to degrade azo dyes has also been studied [23]. This bacterium was able to mediate reduction of *p*-Aminoazobenzene (XIII) to form aniline and *p*-phenylenediamine. Subsequent acetylation of these compounds resulted in formation of acetanilide, *p*-aminoacetanilide and *p*-phenylenediacetamide. Interestingly, *o*-aminophenol, *m*- and *p*-acetamidophenol and 3,4-dihydroxyacetaniline were also identified as metabolites, suggesting that oxidative metabolism may have a major role in the subsequent biodegradation of this dye [24]. Although aromatic ring cleavage and formation of Krebs cycle intermediates have been claimed [24], no published evidence is yet available to support this position. Some success has been noted regarding the ability of *P. cepacia* to decolourize *p*-Aminoazobenzene (XIII), Acid Red 88 (XIV) and Direct Blue 6 (XV) in a multistage rotating biological contactor [24].

Biodegradation of selected azo dyes has also been shown to occur in some lignin-degrading *Streptomyces* spp. [26-28]. This research has been published in conjunction with research documenting the degradation of azo dyes by lignin-degrading fungi. Therefore, as a matter of convenience, degradation of azo dyes by these microorganisms is discussed in the following section, which reviews the recent literature concerning the biodegradation of azo dyes by fungi.

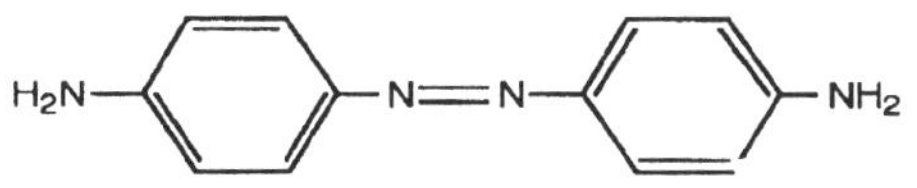

H_2N—N=N—NH_2

P-AMINOAZOBENZENE

XIII

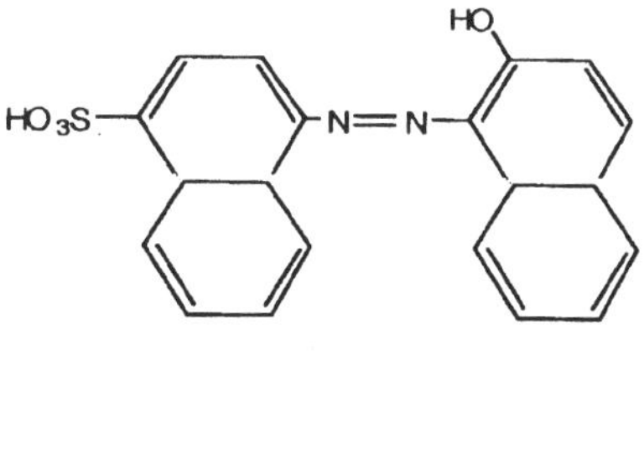

HO
HO_3S N=N

ACID RED 88

XIV

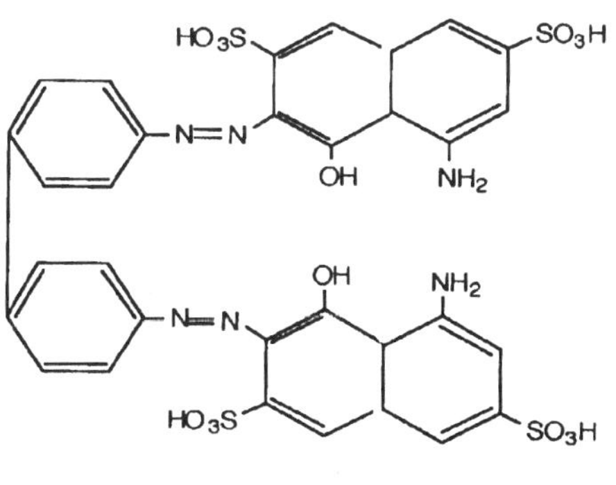

HO_3S SO_3H
N=N
OH NH_2
OH NH_2
N=N
HO_3S SO_3H

DIRECT BLUE 6

XV

BIODEGRADATION OF AZO DYES BY FUNGI

Until recently, the biodegradation of azo dyes by fungi received less attention than biodegradation of these compounds by bacteria. This is unfortunate as some fungi clearly have remarkable biodegradative abilities. The wood-rotting basidiomycete, *Phanerochaete chrysosporium,* is able to degrade a wide variety of environmentally persistent organopollutants to carbon dioxide in aerobic, ligninolytic cultures. Included among the "*difficult-to-degrade*" compounds, degraded by this fungus, are DDT, 2,4,5-T, Benzo[a]pyrene, polychlorinated biphenyls, and Lindane [29]. *P. chrysosporium* is also able to decolourize a variety of highly coloured compounds and mixtures of compounds, such as aqueous waste (Kraft bleach plant effluent) generated by the paper making industry [30]. Other studies have shown that several polymeric dyes [31], and many triphenylmethane dyes, including crystal violet, are degraded by *P. chrysosporium* [32].

Because of its ability to degrade such a variety of structurally diverse organic compounds, including many highly coloured ones, it was logical to determine if azo dyes were also degraded by *P. chrysosporium*. Indeed, it was shown that culture fluid from ligninolytic cultures of *P. chrysosporium,* containing Orange II (XII), Tropaeolin O (XVI) and Congo Red (II), were all decolourized extensively (96-100%) during 5 days of incubation [33]. Initially, decolourization was clearly due, in part, to adsorption of the dyes to the fungal mycelium, as determined by visual inspection. However, it was apparent that, once bound, the dyes were subsequently further degraded such that after 5 days, no Tropaeoline O or Orange II could be detected in methanol extracts of mycelial mats. In contrast to Tropaeolin O and Orange II, Congo Red appeared to be somewhat more persistent in that 6% of the dye, initially present, was extracted from the mycelial mats after 5 days of incubation. Moreover, the culture fluid of ligninolytic cultures of *P. chrysosporium*, containing Congo Red, were not completely decolourized after 12 days of incubation. The ability of nonligninolytic cultures of *P. chrysosporium* to degrade Tropaeoline O, Orange II, and Congo Red was also assessed. Results showed that culture fluid, containing all three dyes, was decolourized extensively (87 - 93%) during 5 days of incubation. Adsorption of dyes to fungal mycelium was also apparent. However, once bound, subsequent decolourization was not as rapid as in ligninolytic cultures. Mycelial mats were visibly coloured after 5 days of incubation, and methanol extracts revealed that 18%, 11%, and 49% of the Tropaeolin O, Orange II, and Congo Red, initially present, was respectively adsorbed to the mycelium.

The inital investigation of the ability of *P. chrysosporium* to degrade azo dyes [33] showed that all three of the dyes assayed were, indeed, degraded. Subsequent studies [34] suggest that all azo dyes are not susceptible to biodegradation by this fungus. Of the 18 dyes studied, only 8 were degraded in aerobic ligninolytic cultures of this fungus. The dyes studied represented a cross section of commercially used azo dyes. Of interest was the finding that degradation appeared to be a function of the class of dye. Reactive, disperse, sulfur, and vat dyes were decolourized whereas naphthol and acid dyes were not decolourized. Also of interest was the finding that large and highly sulfonated dyes, such as Reactofix Golden Yellow (XVII), underwent extensive decolourization.

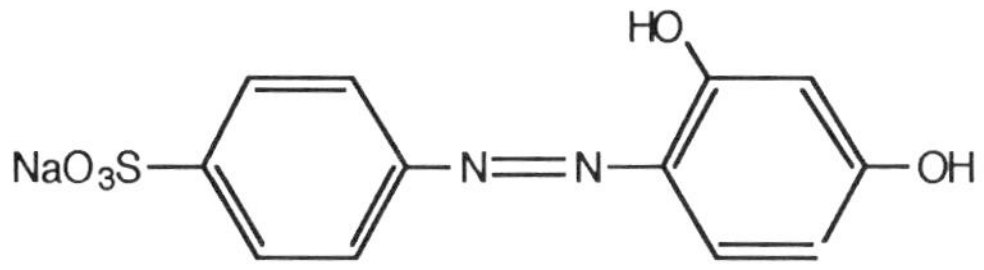

TROPAEOLIN O

XVI

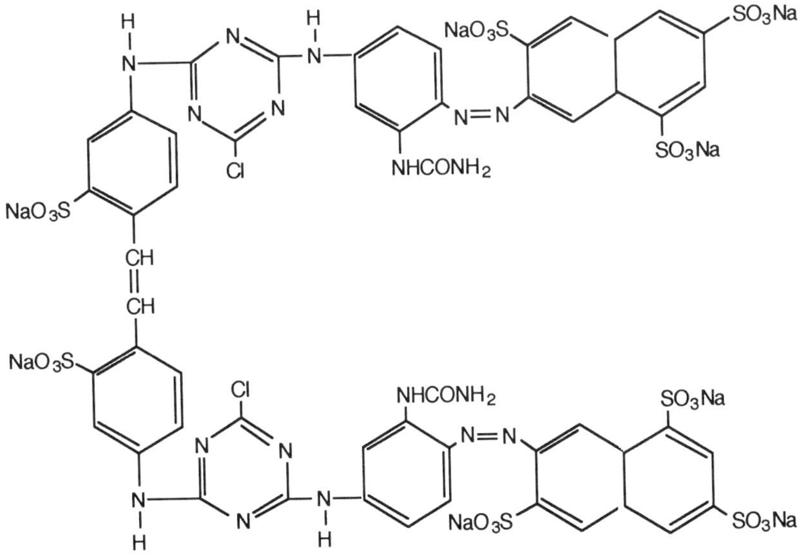

REACTOFIX GOLDEN YELLOW

XVII

Many biodegradation studies have been performed, using ^{14}C-labelled compounds. In such studies, degradation to $^{14}CO_2$ can be assessed by flushing the cultures with air or oxygen and forcing the culture atmosphere through a CO_2 trap containing scintillation cocktail. The amount of ^{14}C-labelled compound, completely degraded to $^{14}CO_2$, can then be determined by liquid scintillation spectrometry. Such studies are important, as they are able to demonstrate whether or not a complete degradation pathway exists for the compound under study. Although initial findings demonstrated that *P. chrysosporium* could decolourize a variety of azo dyes, the question of whether such compounds were degraded to CO_2 was left open. This issue, however, has been addressed by studies using seven (III, XVIII - XXIII) relatively simple ^{14}C-labelled azo dyes [35].

In all cases, substantial degradation (23.1 - 48.1%) to $^{14}CO_2$ was observed in ligninolytic cultures of *P. chrysosporium*. In general, degradation to $^{14}CO_2$ was less extensive in nonligninolytic culture. However, the amount of mineralization observed clearly demonstrated that biodegradation of these compounds was not due exclusively to enzymes of the lignin-degrading system. Although ^{14}C-azo dyes were degraded to $^{14}CO_2$ in nonligninolytic cultures, it should be emphasized that the extent of degradation observed was always greater in ligninolytic cultures. These studies also showed that aromatic rings, containing hydroxyl, amino, acetamido, and nitro substituents, were degraded to a greater extent than were unsubstituted aromatic rings.

In other studies, it was shown that five ^{14}C-labelled azo dyes, Orange I (XI), Orange II (XII), Acid Yellow 9 (XXIV), 4-(3-Methoxy-4-Hydroxyphenylazo) Benzenesulfonic Acid (XXV), and 4-(2-Sulfo-3'-Methoxy-4-Hydroxyazobenzene-4-Azo-Benzenesulfonic Acid Mono Sodium Salt (XXVI), all of which were synthesized from ^{14}C-sulfanilic acid as well as ^{14}C-sulfanilic acid itself, were degraded (17.2 - 34.8%) to $^{14}CO_2$ by *P. chrysosporium* [28].

In contrast, *Streptomyces chromofuscus* was able to degrade only three of the five dyes to $^{14}CO_2$, and the extent of degradation was only 1.1 - 3.6%. Nevertheless, this observation is significant, as it demonstrates that a complete pathway for the biodegradation of some azo dyes appears to exist in *S. chromofuscus*. Unlike *P. chrysosporium*, however, *S. chromofuscus* was unable to degrade ^{14}C-sulfanilic acid to $^{14}CO_2$.

The remarkable biodegradative abilities of *P. chrysosporium* are due, at least in part, to the nonspecific lignin-degrading system of this fungus that is expressed in nutrient limited cultures (usually nitrogen or carbon limitation is used). Briefly, lignin is a large, water insoluble, non-repeating aromatic heteropolymer, formed by the seemingly random polymerization of phenylpropanoid monomers [36]. The resultant lignin polymer contains at least 12 or 13 different C-C and C-O bonds connecting various monomers. Moreover, several chiral carbons atoms are present

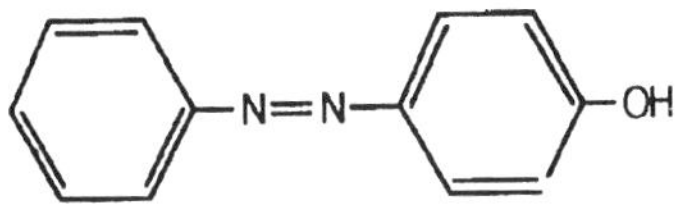

4-PHENYLAZOPHENOL

XVIII

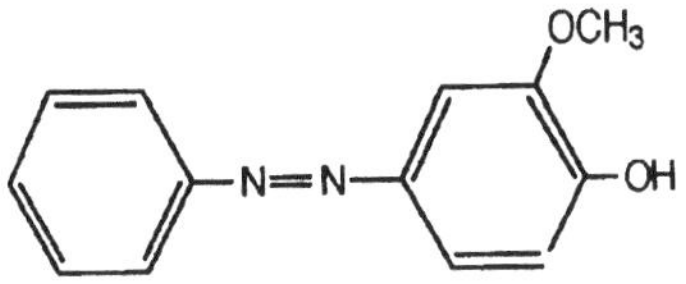

4-PHENYLAZO-2-METHOXYPHENOL

XIX

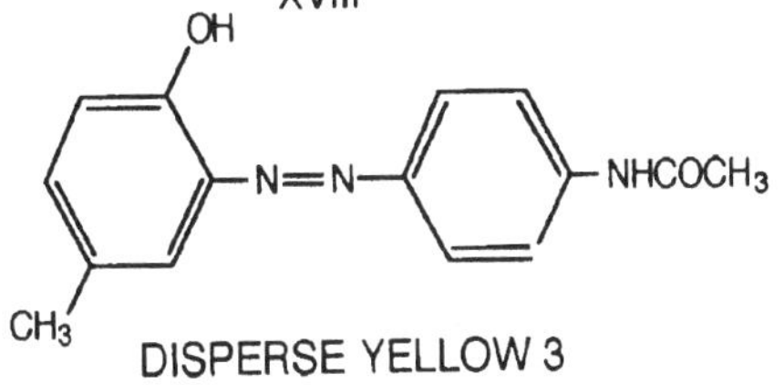

DISPERSE YELLOW 3

XX

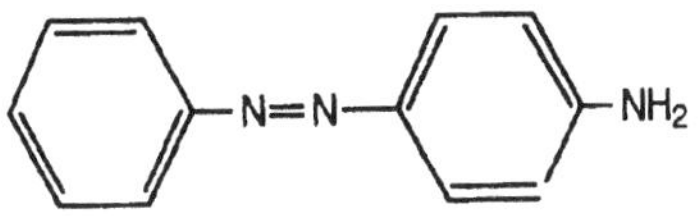

4-PHENYLAZOANILINE

XXI

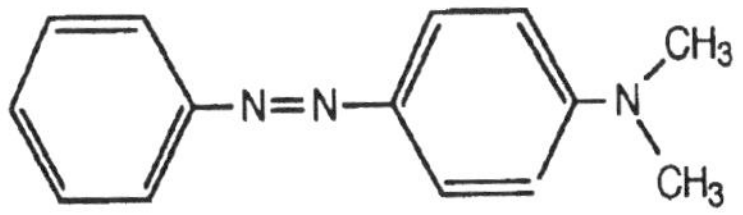

N,N-DIMETHYL-4-PHENYLAZOANILINE

XXII

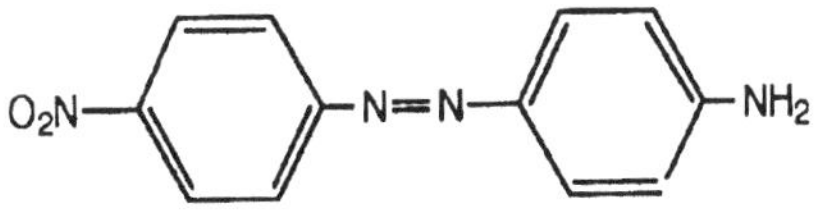

DISPERSE ORANGE 3

XXIII

in lignin. All of these properties combine to make lignin one of the most environmentally persistent naturally occurring compounds known. Only a relatively few groups of organisms are known to be able to cause substantial biodegradation of lignin. Although the details of lignin degradation are still emerging, it is thought that the initial oxidation and depolymerization reactions of *P. chrysosporium* are mediated by lignin peroxidases (ligninases) and Mn peroxidases, the two families of isozymes, that are secreted by the fungus in nutrient limited cultures [37]. Interestingly, lignin peroxidases have been shown to catalyze the initial oxidation of many environmentally persistent organic compounds [32,38-43]. Although they have received less attention in this regard, it is likely that Mn peroxidases of this fungus also have a major role in oxidation of environmentally persistent organic compounds.

It has been reported that lignin peroxidase from *P. chrysosporium* has the ability to partially decolourize Orange II (XII) and Tropaeolin O (XVI) [33], but not Congo Red (II). More recent investigations [44], however, have shown that Congo Red is indeed decolourized by lignin peroxidases. Methyl Orange (XXXII) has also been shown to be decolourized by lignin peroxidases [44]. In other studies, lignin peroxidase from this

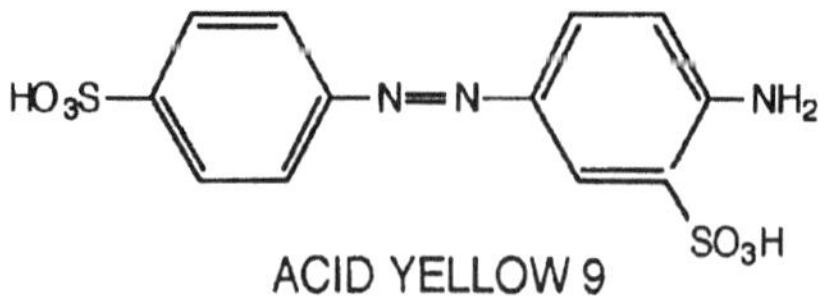

ACID YELLOW 9

XXIV

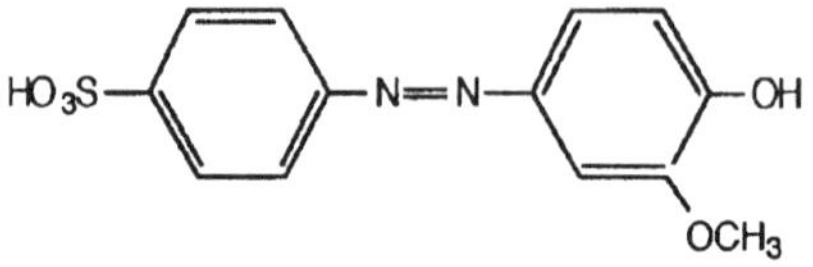

4-(3 -METHOXY-4-HYDROXYPHENYLAZO)BENZENESULFONIC ACID

XXV

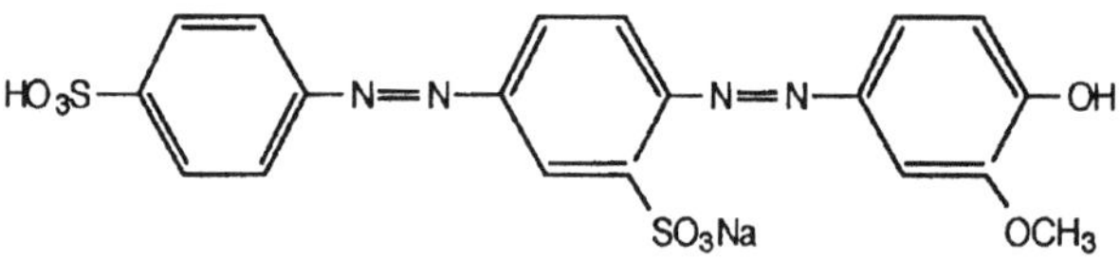

4-(2-SULFO-3'-METHOXY-4-HYDROXYAZOBENZENE-4-AZO-BENZENESULFONIC ACID
MONOSODIUM SALT
XXVI

fungus catalyzed decolourization of Acid Yellow 9 [45]. In contrast, Mn peroxidase from *P. chrysosporium* did not decolourize Acid Yellow 9. Mn peroxidase did, however, decolourize two newly synthesized azo dyes (XXV, XXVI) which contained guaiacol groups. Possibly, the most interesting and important observation, concerning the oxidation of azo dyes by lignin peroxidases, is the fact that veratryl alcohol, a natural product formed by the fungus and a substrate for lignin peroxidases is required for efficient decolourization (oxidation) of azo dyes by these enzymes [45]. During their reaction cycle, lignin peroxidases, like other peroxidase, undergo initially a two-electron oxidation in the presence of hydrogen peroxide, its oxidizing cosubstrate, to form a reactive intermediate known as compound I. Compound I is able to catalyze one-electron oxidations of a variety of organic compounds. In this process, compound I is reduced to compound II which also is able to catalyze a variety of one-electron oxidations. In the case of veratryl alcohol oxidation, the alcohol is thought to undergo two successive one-electron oxidations to form veratrylaldehyde [37]. In the case of azo dyes, it was shown that oxidation of Biebrich Scarlet (XXVII) and Tetrazine (XXVIII) by lignin peroxidase was limited and terminated rapidly.

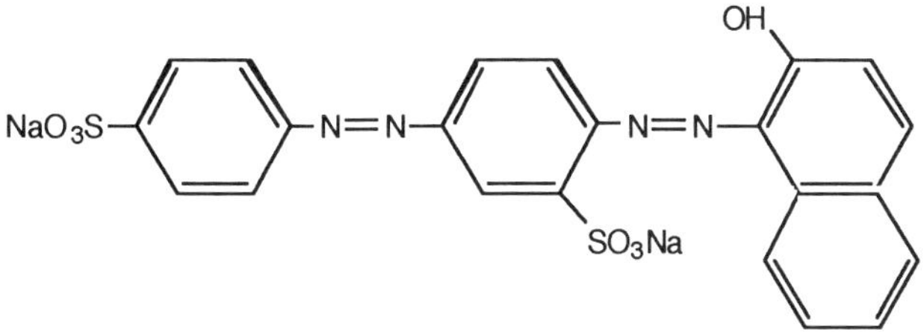

BIEBRICH SCARLET

XXVII

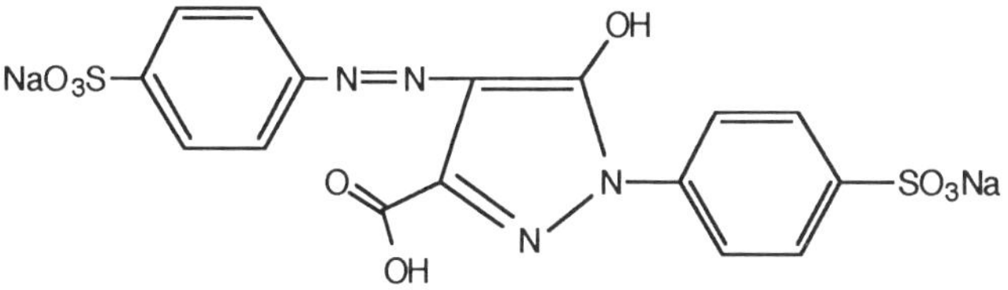

TETRAZINE

XXVIII

Subsequent investigation revealed that lignin peroxidase intermediate compound II accumulated in such reactions, and that compound II, unlike compound I, could not mediate oxidation of the azo dyes under study. When veratryl alcohol was added to these reaction mixtures, oxidation of both the dyes by lignin peroxidase was rapid and extensive [45]. Apparently, veratryl alcohol is oxidized by compound II which, is in turn, reduced to the resting state and may then participate in another round of the catalytic cycle. It is well known that addition of veratryl alcohol accelerates lignin peroxidase-mediated oxidation of several other organic compounds, that are not substrates (or are poor substrates) for this enzyme in the absence of veratryl alcohol [38, 46, 47]. Two explanations have been put forward to explain this phenomenon; one contending that veratryl alcohol radical, formed during oxidation of veratryl alcohol, functions as a secondary oxidant to catalyze one-electron oxidations of compounds, that are not good substrates in the absence of veratryl alcohol [46]. Although controversial, this theory has been given renewed credence by the observation that low steady-state levels of veratryl alcohol radical are indeed formed during oxidation of

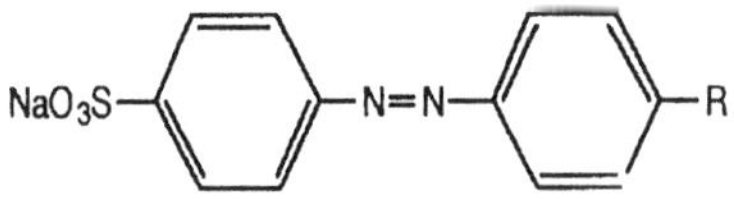

$NaO_3S-C_6H_4-N{=}N-C_6H_4-R$

XXIX $R = -OH$
XXX $R = -OCH_3$
XXXI $R = -N(CH_2CH_3)_2$
XXXII $R = -N(CH_3)_2$
XXXIII $R = -NH_2$

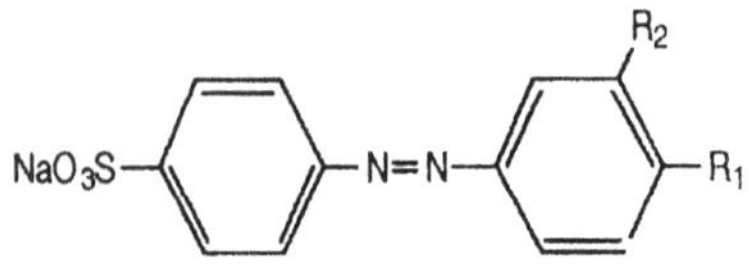

$NaO_3S-C_6H_4-N{=}N-C_6H_3(R_1)(R_2)$

XXXIV $R_1 = -OH$, $R_2 = -CH_3$
XXXV $R_1 = -OH$, $R_2 = -OCH_3$
XXXVI $R_1 = -OCH_3$, $R_2 = -OCH_3$
XXXVII $R_1 = -OH$, $R_2 = -Cl$
XXXVIII $R_1 = -NH_2$, $R_2 = -SO_3Na$
XXXIX $R_1 = -OH$, $R_2 = CH_3\text{-}CH\text{-}CH_2CH_3$

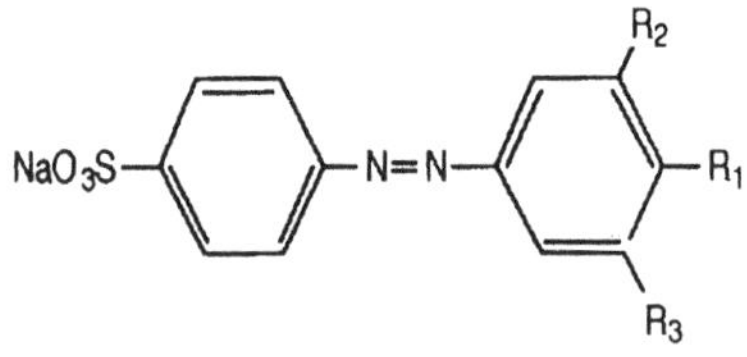

$NaO_3S-C_6H_4-N{=}N-C_6H_2(R_1)(R_2)(R_3)$

XL $R_1 = -OH$, $R_2 = -CH_3$, $R_3 = -CH_3$
XLI $R_1 = -OH$, $R_2 = -OCH_3$, $R_3 = -OCH_3$
XLII $R_1 = -OH$, $R_2 = -F$, $R_3 = -F$

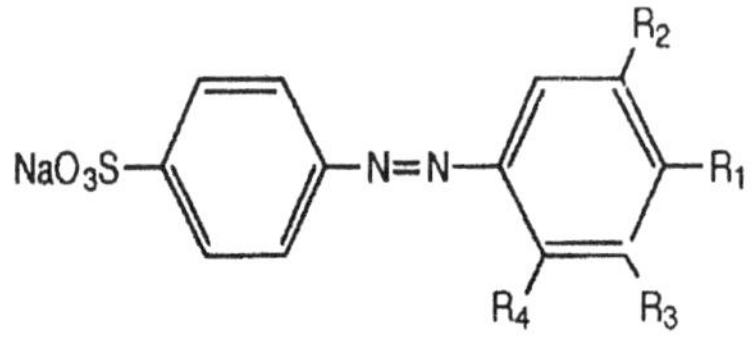

$NaO_3S-C_6H_4-N{=}N-C_6H(R_1)(R_2)(R_3)(R_4)$

XLIII $R_1 = -CH_3$, $R_2 = -CH_3$, $R_3 = -H$, $R_4 = -OH$
XLIV $R_1 = -H$, $R_2 = -CH_3$, $R_3 = -H$, $R_4 = -OH$
XLV $R_1 = -H$, $R_2 = -CH_2CH_3$, $R_3 = -H$, $R_4 = -OH$
XLVI $R_1 = -H$, $R_2 = -CH_3$, $R_3 = -OCH_3$, $R_4 = -OH$

veratryl alcohol by lignin peroxidases [48]. Another explanation suggests that veratryl alcohol functions to protect lignin peroxidases from inactivation by hydrogen peroxide in the presence of a poor reducing cosubstrate [38,49]. Although hydrogen peroxide is a required oxidizing cosubstrate, it may, in the presence of poor reducing cosubstrates, compete with such substrates and react with compounds I and II to form reactive oxygen compounds (hydroxyl radical, superoxide, etc.) which may, in turn, react with critical sites on the enzyme, resulting in inactivation of the enzyme. The observation, that some chemicals, such as certain azo dyes, are oxidized by compound I but not compound II,

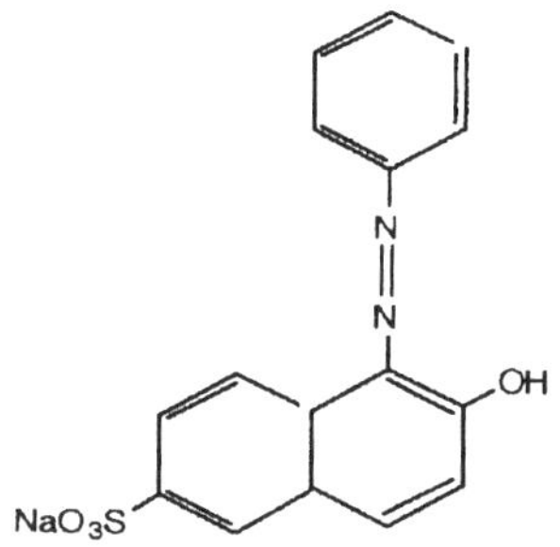

ACID ORANGE 12

XLVII

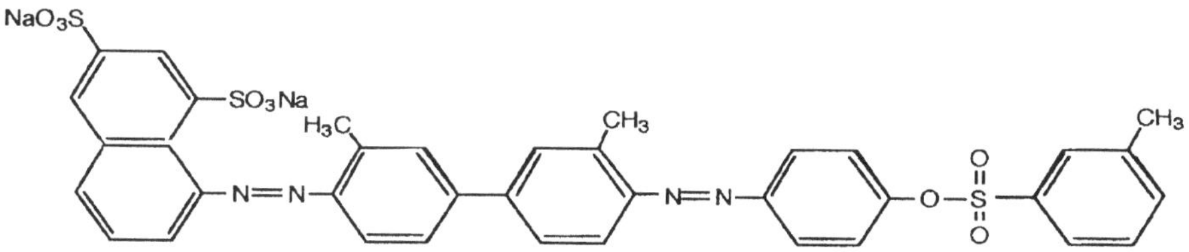

RS(H/C)

XLVIII

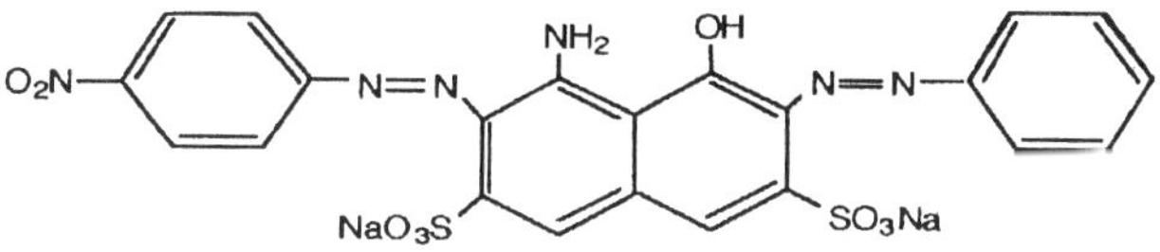

10B(H/C)

XLIX

and that veratryl alcohol prevents accumulation of compound II (which acts as a dead-end form of the enzyme) [45], suggests that this mechanism may be the predominant mechanism by which veratryl alcohol promotes oxidation of chemicals that are not substrates (or are poor substrates) in its absence.

The influence of substituents on the degradability of several relatively simple azo dyes by *P. chrysosporium* and by *Streptomyces* spp. has been investigated [27]. Of the twenty-two dyes studied (XI, XII, XXIV, XXIX -XLVII), all were decolourized by *P. chrysosporium*. In most cases, per cent decolourization ranged from 60 - 99%. In contrast, several *Streptomyces* spp. were shown to be less effective.

None of the *Streptomyces* spp. (*S. rochei* A14, *S. rochei* A15, *S. chromofuscus* A11, *S. diastaticus* A12, and *S. diastaticus* A13) tested were able to decolourize Acid Yellow 9 (XXIV). However, five monosulfonated, two ring, mono azo dyes were decolourized by these bacteria. In all cases, these dyes had a hydroxyl group in the *para*-position, relative to the azo linkage, and at least one methoxy and/or one alkyl group in an *ortho*-position, relative to the hydroxyl group. Of the azo dye derivatives of naphthol studied, *Streptomyces* spp. decolourized Orange I (XI), but not Orange II (XII) or Orange 12 (XLVII).

Interestingly, *P. chrysosporium* decolourized all three. However, Orange II (XII) and Orange 12 (XLVII) were decolourized more effectively than Orange I (XI).

Although, most of the research, concerning the biodegradation of azo dyes by fungi, has focused on the use of *P. chrysosporium*, several *Myrothecium* spp. and several *Ganoderma* spp. were shown to be able to mediate substantial decolourization of Orange II and two relatively complex azo dyes, designated as RS(H/C) (XLVIII) and 10B(H/C) (XLIX) [50]. Of interest is the fact that many *Myrothecium* spp. and *Ganoderma* spp. are like *P. chrysosporium*, white rot fungi.

Very limited information is available, concerning the ability of other classes of fungi to remediate water contaminated with azo dyes. It is interesting to note [51] that the ascomycete, *Neurospora crassa* (strain 74A) was able to decolourize water containing the diazo dye, Vermelho Reanil, which was present in concentrations within the range (16-32 μg ml^{-1}) found in industrial effluents. It was unclear, however, if the fungus was able to metabolize the dye or if water decolourization was due solely to adsorption by fungal mycelium.

In an investigation concerning biodegradation of Reactive Red 22 (L) by bacteria [52], a stable consoritum of four bacterial species (*Pseudomonas aeruginosa, Pseudomonas oryzihabitans, Acinetobacter calcoaceticus,* and *Citrobacter freundii*) was developed from a mixture of soil and sewage microorganisms, that was acclimated to Reactive Red 22 (L) under aerobic conditions. Reactive Red 22 (L) did not appear to serve as a sole carbon source for the bacterial consortium.

However, extensive decolourization occurred when glucose was present in culture. Seven metabolites of this dye were identified. Although no aromatic ring cleavage products were identified, the authors [52] suggested that complete biodegradation to carbon dioxide may occur in this system. In another study [53], three azo dyes, Diamira Brillian Orange RR (LI), Direct Brown M (LII), and Eriochrome Brown R (LIII), were decolourized aerobically by *Pseudomonas* S-42, isolated from activated sludge.

Decolourization also occurred in cell-free extracts by an enzyme purified from this bacterium. The enzyme, an azoreductase, appears to have a broader specificity than the one previously studied by Zimmermann et al. [22].

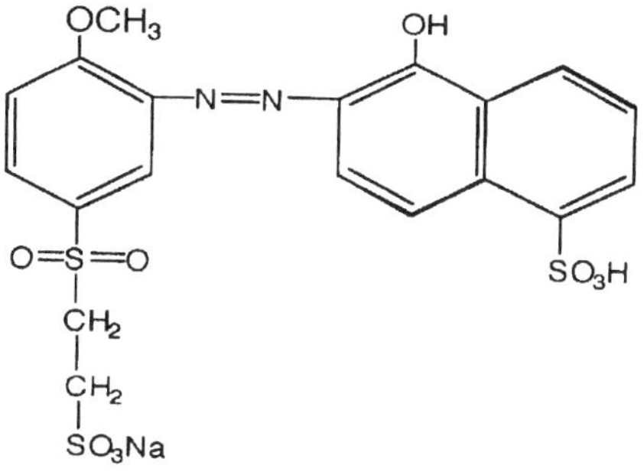

OCH$_3$ OH
N=N
O=S=O SO$_3$H
CH$_2$
CH$_2$
SO$_3$Na

REACTIVE RED 22

L

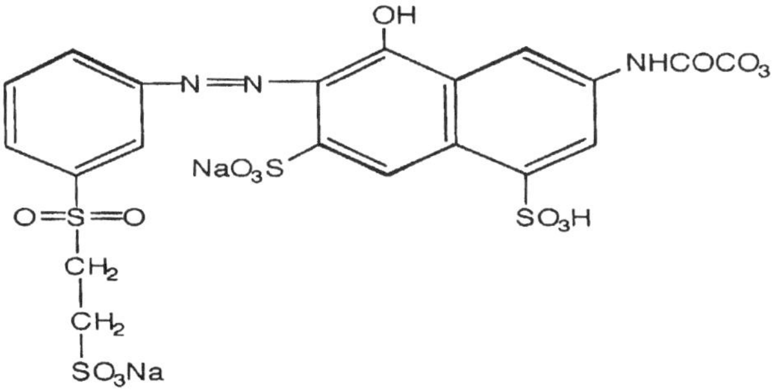

OH
N=N NHCOCO$_3$
NaO$_3$S
O=S=O SO$_3$H
CH$_2$
CH$_2$
SO$_3$Na

DIAMIRA BRILLIANT ORANGE RR

LI

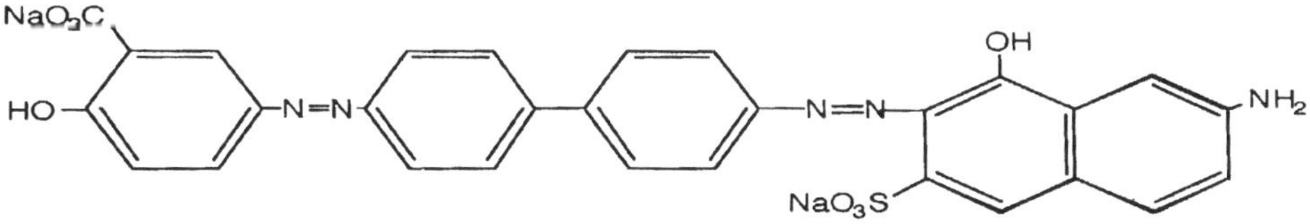

NaO$_2$C OH
HO N=N N=N NH$_2$
NaO$_3$S

DIRECT BROWN M

LII

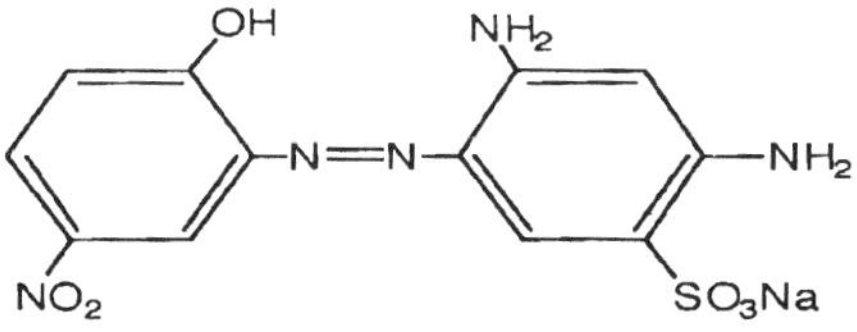

OH NH$_2$
N=N NH$_2$
NO$_2$ SO$_3$Na

ERIOCHROME BROWN R

LIII

CONCLUSIONS

Azo dyes, as a group, are resistant to biodegradation by microorganisms [2,18]. However, research during the past 10 years has clearly shown that a variety of microorganisms and approaches hold promise for the development of effective systems for the treatment of water (and possibly soils, sediments, and sludges), contaminated with these colourants. For bacterial systems, the most promising approach appears to utilize a sequential system in which the azo dye is initially reduced in anaerobic culture, and the aromatic amines, thus generated, are further metabolized under aerobic conditions [19]. Until recently, azo dyes were not thought to be degraded by microorganisms, unless an anaerobic step was included in the process. Although some specialized bacteria have been developed, which are able to degrade some simple azo dyes in aerobic culture [21], this was considered to be an exception. Furthermore, these unique bacteria are very specific and degrade only those azo dyes to which they have become adapted. The finding, that *Pseudomonas* S-42 [53] can decolourize three structurally diverse azo dyes, suggests that it may be possible to develop or discover other strains of bacteria, with an even broader spectrum of biodegradative abilitites. The discovery, that the white rot fungus *P. chrysosporium* degrades a wide variety of azo dyes, provides an entirely new approach to the study of azo dye biodegradation [33]. Possibly, the most interesting finding is that the initial degradation of many azo dyes, in this system, is not a reduction of the azo linkage, but rather an oxidation reaction, mediated by lignin peroxidases or Mn peroxidases that are secreted by the fungus during idiophasic metabolism. The observation, that other white rot fungi (*Myrothecium* spp. and *Ganoderma* spp.) degrade azo dyes, suggests that this ability may be widespread among such fungi [50].

ACKNOWLEDGEMENTS

Research in the author's laboratory is supported by NIEHS grant ESO 4492.

REFERENCES

1 Zollinger H. Color Chemistry: Synthesis, Properties and Applications of Organic Dyes and Pigments. Weinheim: VCH Verlagsgesellschaft, 1987; pp.367.

2 Anliker R. Ecotoxicol Environ Safety 1979; 3: 59-74.

3 Tincher WC, Robertson J.R. Textile Chemist and Colorist 1982; 14(12): 41-47.

4 Navasivayam C, Yamuna RT. Chem Technol Biotechnol 1992; 53: 153-157.
5 Combes RD, Haveland-Smith RB. Mutat Res 1982; 98: 101-248.
6 Longstaff E. Dyes and Pigments 1983; 4: 243-304.
7 Longstaff E, McGregor DB, Harris WJ, Roberston JA, Poole A. Dyes and Pigments. 1984; 5: 65-82.
8 Delclos KB, Tarpley WG, Miller EC, Miller JA. Cancer Research 1984; 44: 2540-2550.
9 Cameron TP, Hughes TJ, Kirby PE, Fung VA, Dunkel VC. Mutat Res 1987; 189: 223-261.
10 Ames BN, McCann J, Yamasaki E. Mutat Res 1975; 31: 347-364.
11 Brown JP, Dietrich PS. Mutat Res 1983; 116: 305-315.
12 Prival MJ, Mitchell VD. Mutat Res 1982; 97: 103-116.
13 Haley TJ. Clinical Toxicol 1975; 1: 13-42.
14 National Institute for Occupational Safety and Health. Center for Disease Control. 1980. Special Hazard Review of Benzidine-Based Dyes. DHHS (NIOSH) Publication 80-109. National Institute for Occupational Safety and Health, Cincinnati, OH.
15 Clive D, Spector JFS. Mutat Res 1975; 31: 17-29.
16 Gray Jr LE, Ostby JS, Kavlocl RJ, Marshall R. Fundament Appl Toxicol 1992; 19: 411-422.
17 Weisburger JH, Weisburger EK. Chemicals as Causes of Cancer. Chem Eng News 1966; February 7: 124-142.
18 Meyer U. In: Leisinger T, Cook AM, Nüesch J, Hütter R, eds. Microbial Degradation of Xenobiotics and Recalcitrant Molecules London: Academic Press, 1981; 371-385.
19 Haug W, Schmidt A, Nörtemann B, Hempel DC, Stolz A, Knackmuss H-J. Appl Environ Microbiol 1991; 57: 3144-3149.
20 Glässer A, Liebelt U, Hempel DC. Design of a Two-Stage Process for Total Degradation of Azo Dyes. DECHEMA Biotechnology Conferences 5-VCH Verlagsgesellschaft 1992; 1085-1088.
21 Kulla HG. In: Leisinger T, Cook AM, Nüesch J, Hütter R, eds. Microbial Degradation of Xenobiotics and Recalcitrant Molecules. London: Academic Press, 1981; 387-389.
22 Zimmermann T, Kulla HG, Leisinger T. Eur Biochem 1982; 129: 197-203
23 Idaka E, Ogawa T, Horitsu H. Bull Environ Contam Toxicol 1987; 39: 100-107.
24 Idaka E, Ogawa T, Horitsu H. Bull Environ Contam Toxicol 1987; 39: 108-113.
25 Ogawa T, Yatome C. Bull Environ Contam Toxicol 1990; 44: 561-566.
26 Paszczynski A, Pasti MB, Goszczynski S, Crawford DL, Crawford RL. Enzyme Microb Technol 1991; 13: 378-384.
27 Past-Grigsby MB, Paszczynski A, Goszczynski S, Crawford DL, Crawford RL. Appl Environ Microbiol 1992; 58: 3605-3613.

28 Paszczynski A, Pasti-Grigsby MB, Goszszynski S, Crawford RL, Crawford DL. Appl Environ Microbiol 1992; 58: 3598-3604.
29 Bumpus JA, Tien M, Wright D, Aust SD. Science 1985; 228: 1434-1436.
30 Sundaman G, Kirk TK, Chang H-M. Fungal Decolorization of Kraft Bleach Plant Effluent: Fate of the Chromophoric Material. TAPPIJ 1981; 64(9): 145-148.
31 Glenn JK, Gold MH. Appl Environ Microbiol 1983; 45: 1741-1747.
32 Bumpus JA, Brock BJ. Appl Environ Microbiol 1983; 54: 1143-1150.
33 Cripps C, Bumpus JA., Aust SD. Appl Environ Microbiol 1990; 56 : 1114-1118.
34 Capalash N, Sharma P. World Microbiol Biotechnol 1992; 8: 309-312.
35 Spadaro JT, Gold MH, Renganathan V. Appl Environ Microbiol 1992; 58: 2397-2401.
36 Crawford DL. Lignin Biodegradation and Transformation. New York: Wiley.
37 Tien M. Critic Rev Microbiol 1987; 15: 141-168.
38 Haemmerli SD, Leisola MSA, Sanglard D, Fiechter A. Biol Chem 1986; 261: 6900-6902.
39 Mileski GJ, Bumpus JA, Jurek MA, Aust SD. Appl Environ Microbiol 1988; 54: 2885-2889.
40 Hammel KE, Kalyanaraman B, Kirk TK. Biol Chem 1986; 261: 16948-16952.
41 Hammel KE, Tardone PJ. Biochemistry 1988; 27: 6563-6568.
42 Valli K, Gold MH. Bacteriol 1991; 173: 345-352.
43 Valli K, Brock BJ, Joshi D, Gold MH. Appl Environ Microbiol 1992; 58: 221-228.
44 Ollikka P, Alhonmäki K, Leppänen V-M, Glumoff T, Raijola T, Suominen I. Appl Environ Microbiol 1993; 59: 4010-4016.
45 Paszczynski A, Crawford RL. Biochem Biophys Res Commun 1991; 178: 1056-1063.
46 Harvey PJ, Schoemaker HE, Palmer JM. FEBS Lett 1986; 195: 242-246
47 Tuisel H, Grover TA, Bumpus JA, Aust SD. Arch Biochem Biophys 1992; 293: 287-291.
48 Gilardi G, Harvey PJ, Cass, AEG, Palmer JM. Biochim Biphys Acta 1990; 1041: 129-132.
49 Valli K, Wariishi H, Gold MH. Biochemistry 1990; 29: 8535-8539.
50 Mou D-G, Lim KK, Shen HP. Biotechnol Adv 1991; 9: 613-622.
51 Corso CR, de Angelis DF, de Oliveira JE, Kiyan C. Eur Appl Biotechnol 1981; 13: 64-66.
52 Rakmi AR, Terahima Y, Ozaki H. Inst Chem Eng Sym Ser 1990; 116: 301-310.
53 Zhipei L, Huifang Y. J Environ Sci China 1991; 3: 89-102.

Biotransformations: Microbial Degradation of Health Risk Compounds
Ved Pal Singh, editor
© 1995 Elsevier Science B.V. All rights reserved.

Microbial degradation of natural rubber

Akio Tsuchii

National Institute of Bioscience and Human-Technology,
Agency of Industrial Science and Technology, Tsukuba city,
Ibaragi 305, Japan

INTRODUCTION

Natural rubber (NR) was once the only source of the elastomeric materials required by a wide range of products. Nowadays, a much greater amount of synthetic rubbers than NR is used in an even larger number of products.

Many synthetic polymers, like plastics and rubbers, are highly resistant to microbial degradation. As a result, such polymers are accumulating in the environment in huge quantities. This has led to a growing interest in the development of degradable plastics with enhanced bio-degradability and photo-degradability in landfills and composts [1].

NR is today obtained mainly from *Hevea* trees on plantations in tropical Asia, from where it is exported throughout the world as an industrial raw material. As noted by Thomas Edison, however, many plants in temperate climates, like the goldenrod and the dandelion, also contain rubber in small amounts [2,3]. Consequently, rubber-degrading microorganisms can be expected to be present widely in the natural environment.

NR is also degraded by solar ultraviolet rays [4]. Thus, NR has been degraded since prehistoric times by the action of both microorganisms and sunlight. The fact that NR is not only a renewable resource but also an enviromentally degradable material would be appreciated more.

RUBBER-DEGRADING MICROORGANISMS

Actinomycetes play a major role in degradation of NR [5], while some strains of fungi and bacteria are also known to attack rubber. Microorganisms, capable of degrading NR, cannot degrade synthetic rubbers other than synthetic isoprene rubber [6,7]. Although there have been a number of reports concerning microbial degradation of various synthetic rubbers, degradation of the hydrocarbon polymer has not been demonstrated yet.

MICROBIAL DEGRADATION OF UNVULCANIZED RUBBER

Microbial attack of raw rubber was first reported early in 1914. Thin films of NR, floating on an aqueous medium, were shown to be disintegrated by some actinomycete strains to a certain extent, that could not be ascribed to the disappearance of impurities from the rubber [8]. Spence and van Niel [9] reported that NR, in the latex state, was degraded by some actinomycete strains, and that a rubber weight loss of up to 70% was observed after a 28-day cultivation period [9]. Thin films of NR on agar plates were also found to be degraded by strains of *Streptomyces* and *Nocardia*, and the weight loss reached 52% after the cultivation period of a month and a half [10]. It was reported that thin strips of NR and synthetic isoprene rubber, with a diameter of 0.5 mm, were decomposed completely by a strain of *Nocardia* in 56 days [11].

Blake and Kitchin [12] found that colonies of Gram-positive micrococci developed on thin films of NR, and that the surface was extremely pitted by action of bacteria as well as actinomycetes [12]. A 20 per cent weight loss of rubber films by the action of a bacterial strain has also been reported [10].

Unlike rubber-degrading actinomycete colonies, fungal colonies on latex-agar plates are not surrounded by transparent zones [5]. Superficial growth of fungi on NR films has been observed with a negligible consumption of the rubber [10]. Growth of the molds on rubber may have proceeded at the expense of the non-rubber constituents, like proteins in these cases.

The latex of NR was found to be attacked by some *Aspergillus* and *Penicillium* strains, with a 32% weight reduction after a month cultivation [13]. Williams [14] reported that a 2 mm thick smoked sheet of NR was degraded by a strain of *Penicillium* and lost approximately 13% of its dry weight over 56 days' cultivation [14]. A fungal growth of 4.1 mg protein/ mm^2 was obtained at the same time. It was observed that ground particles of NR (between 0.8 and 2 mm in diameter) were attacked by a strain of *Cladosporium,* and that 6 successive treatments of 20 days each caused a decrease in molecular weight, estimated by gel permeation chromatography (GPC), from 2×10^6 to 1×10^5 [15].

Production of 7.6 mg of protein from 5 g of nitrile rubber by a mixed culture of bacteria was reported [16]. A strip of styrene-butadiene rubber (1.5 x 2.5 x 0.2 cm) was buried in soil, and a significant growth of fungi (2.8 μg protein/mm^2) was detected after a 9-month incubation [14]. From these observations, however, it was very difficult to estimate the degradation of polymer itself.

Antoine et al. [17] reported that a terpolymer of acrylonitrile, methylacrylate, and butadiene was bioconverted by a strain of either *Nocardia* or *Penicillium* [17]. They found that, after 6 months of incubation, the terpolymer was transformed into both a lower molecular weight form and a second form insoluble in dimethylsulphoxide. It was observed that a strain of *Moraxella* grew on a 1,4-type polybutadiene with an average molecular weight of 2,500 and degraded 44% of the oligomer in 5 days [18]. However, a polymer of butadiene with average molecular weight of 17,000 was not attacked by the bacterial strain.

MECHANISMS OF NR DEGRADATION

Infrared spectroscopy analysis of microbiologically deteriorated vulcanized NR by Cundell and Mulcock [19] indicated the following structural and chemical changes: the presence of the hydroxyl and carbonyl structures at 3,500 cm^{-1} and 1,720 cm^{-1}, respectively [19]; decrease in unsaturation at 890 cm^{-1}, and appearance of a broad peak in the 1,000 cm^{-1} region, which might represent ether, epoxide, or peroxide group. Biochemical oxidation of NR tire tread, ground to about 100 mesh sieve size, was also reported [20]. After *bioexposure* of the ground particles in a perfusion reactor for 30 and 60 days, the infrared curve showed a great increase of structures related to C=O in acid, aldehyde and ketone.

Spence and van Niel [9] noted that, when a sterile latex of NR was degradad by soil microorganisms, the relative viscosity of the dilute rubber solution became lower [9] : it dropped from 20 sec (control) to 3 sec, showing that the residual rubber was thoroughly deteriorated, soft and "dead."

It was observed that isoprene oligomers, with molecular weight from 10^3 to 10^4, accumulated during microbial growth on a latex glove. From the chemical structure of the oligomers, it was supposed that the strain of *Nocardia* cleaved NR at the double bond shown by a wavy line in the formula in Figure 1 [11]. The net-work of the vulcanizate was considered to have been attacked directly by the biological action during microbial degradation of the glove, with the oligomers produced by the scission of polymeric chains being used by the organism as growth substrate.

$$-CH_2-\overset{CH_3}{\overset{|}{C}}=CH-CH_2-CH_2-\overset{CH_3}{\overset{|}{C}} \wr CH-CH_2-CH_2-\overset{CH_3}{\overset{|}{C}}=CH-CH_2-$$

$$\downarrow O_2$$

$$-CH_2-\overset{CH_3}{\overset{|}{C}}=CH_2-CH_2-\overset{CH_3}{\overset{|}{C}}=O + O=CH-CH_2-CH_2-\overset{CH_3}{\overset{|}{C}}=CH-CH_2-$$

Figure 1. Schematic representation of natural rubber-cleaving reaction.

The formation of clear zones surrounding actinomycete colonies on a latex-agar plate can be regarded as an indication of extracellular enzymatic decomposition [9].

Quite recently, Tsuchii and Takeda [21] have reported that rubber-degrading enzyme was secreted in the extracellular culture medium by a strain of *Xanthomonas* [21]. The latices of natural and synthetic isoprene rubber are degraded by the crude enzyme, but no reaction is observed on the latices of other kinds of synthetic rubber. Isoprene oligomers, with average molecular weight of 10^4 (acetonyl polyprenyl acetoaldehyde, $A_LP_nA_t$), are produced by random scissions of NR in endowise form, and $A_LP_nA_t$ is further degraded to form mainly 12-oxo-4,8-dimethyltrideca-4,8-diene-1-al (acetonyl diprenyl acetoaldehyde, $A_LP_2A_t$).

MICROBIAL DEGRADATION OF VULCANIZED RUBBER

Rubber products, made of NR, are known to be rather susceptible to biological attack, and synthetic rubbers are preferred in certain types of rubber goods to be used in moist air, the domestic water supply system, and soils [22].

Microbial degradation of synthetic rubbers will be a subject of further study. A rubber product is made from a number of complex ingredients, and smaller molecules in a synthetic polymer (e.g., stearate, process oils, and waxes in vulcanized synthetic rubber) may be decomposed by microorganisms. A clear distinction must be made between the superficial growth of microorganisms on non-rubber constituents in a synthetic polymers and the *biodegradation* of the rubber hydrocarbon [23].

Many components in vulcanizates, such as accelerators, fillers, oils, and antioxidants, are known to affect microbial activities, and protection of rubber goods against microorganisms, by the addition of chemicals with microbiocidal activities, has been an important research area. However, the use of microbiocides is beyond the scope of this review, and extensive references have been given by Zyska [24,25]. The present review mainly covers the degradation of the polymer itself.

DETERIORATION OF RUBBER PRODUCTS

As far as it is known, the microbial attack of vulcanized rubber was first investigated in 1942, when oxygen consumption and carbon dioxide production were observed during microbial growth [26,27]. When buried in soil, the rubber insulation of electric cables loses its insulating properties [12,28]. With NR compounds, loss of electrical resistance of the insulation is accompanied by visible pitting caused by actinomycetes, and fungal hyphae penetrating the insulation.

Microbial *corrosion* of vulcanized NR sealing rings was observed in the underground pipelines for water supply in areas of Holland, Australia, New Zealand, and America. Since then, a number of investigations have been made to protect the rubber gaskets from the attack of microorganisms. Pure cultures of certain strains of *Streptomyces* apparently attack thin strips of vulcanized NR, leading to a marked decrease in tensile strength after 12 months [5]. The presence of large population of *Streptomyces* spp. was found in the deteriorated rubber rings [6].

In 1968, Leeflang tested different rubber compounds for their resistance to biological attack [29]. The strips were immersed in a basin, through which a slow and constant flow of potable water was maintained, and a piece of deteriorated ring was placed on the bottom of the basin as a source of *Streptomyces*. It appeared that all NR compounds tested were susceptible to corrosion in the long run, but with the exception of synthetic polyisoprene, synthetic rubber compounds were resistant. On the other hand, the addition of 5% of casein to nitrile rubber did not make it susceptible to attack by *Streptomyces* [29]. The method, first used by Leeflang, was an excellent way to estimate the resistance of many compounds in vivo under laboratory conditions, and was used by many investigators as a standard [30]. In areas of Holland, where corrosion was most pronounced, high phosphate level in the dune water and the absence of deliberate chlorination were thought to contribute to the ability of actinomycetes to proliferate [31].

In 1975, Hutchinson et al. reported that the population of actinomycetes, isolated from deteriorated rubber rings, was $5x10^5$-$4x10^6$/g, while that of the organism, isolated from undeteriorated rings was $3x10^3$-$4x10^4$/g [32]. Thiobacilli were also isolated from the rubber rings in the pipelines of municipal sewage, and found to have a population of $22x10^4$/g [32].

Up to 40% loss in weight of a strip of vulcanized NR (0.07 mm thick) after 91 days of soil burial test was reported by Kwiatkowska et al. [33]. In 1986, Williams buried vulcanized NR sheet (15x15x0.2 cm) in soil for 6 months and observed a 4.5% loss in weight, accompanied by a 66% loss in tensile strength [34]. According to a report by Simpson, the weight of a 2 mm thick strip of NR vulcanizate decreased by 10.7% after an immersion period of 2 years, under accelerated test conditions of Leeflang's test basin [35]. In 1988, Kwiatkowska and Zyska found the weight loss of NR vulcanizate sheet to be 8.5% after 28 days of exposure to an *Aspergillus* strain [36], while in 1991, Kajikawa et al. found that thin film of a commercial glove (0.2 mm thick) was completely degraded by a strain of *Nocardia* after a 20-day cultivation period [37].

Of the 31 references published between 1942 and 1972, 94% gave evidence of susceptibility of NR to microbial attack, while 24 to 50% of them reported the resistance of various synthetic rubbers [38].

Blake et al. reported that, when synthetic rubber insulation made of styrene-butadiene rubber was buried in soil, invisible micropores and the presence of fungi inside the rubber walls were detected [28]. On the other hand, neoprene compounds appeared to be inherently resistant to soil exposure [28]. Nickerson and Faber [39] grew a strain of fungus on passenger tire mesh, containing predominantly a styrene-butadiene rubber, and concluded that the organism may have consumed the oils in the tire mesh as their carbon source [39].

EFFECT OF COMPOUNDING INGREDIENTS

The greater resistance of American rubber rings, in comparison with those in the Netherlands, was suggested by Leeflang in 1968, to be due partly to differences in the recipes and curing methods, especially with regard to the use of a greater variety of curatives in larger amounts and, sometimes, in large numbers [29]. The addition of an organic substance that did not take an active part in the curing process, seemed to have little effect in the long run, because of the leaching-out of the inhibitory substances by the water flow in the test basin.

Many investigators have reported that sulphur content can greatly influence microbial attack : low levels allow rapid growth, but with increasing loadings, the deterioration decreases [34,40,41].

A report by Cundell and Mulcock points out that tetramethyl thiuram disulphide (TMTD) sulphur-less cured NR is more resistant to microbial attack than dicumyl peroxide (DCP) cured NR, because of the *biocidal* activity of the curing residue, zinc dimethyl dithiocarbamate [41]. After 18 months of incubation, 6.2 per cent loss in weight of TMTD cured NR was observed in a pure culture experiment, and extraction of the curing residues with acetone from the vulcanizate led to an increased loss of 22% in weight. In contrast, the concentration of the curing agent, without microbiocidal activity, did not influence the rate of deterioration, and the cross-link density of the vulcanized NR was found to have little or no effect [41]. In a soil burial test, Kwiatkowska et al. observed a decrease in network chain density due to activity of microorganisms on vulcanized NR [33]. Increasing concentrations of various kinds of accelerators, such as cyclohexyl benzothiazyl sulphenamide (CBS) and TMTD, gave greater protection to rubber, as most of these were toxic to soil microorganisms [14].

The possible influence of cross-link density on the microbial deterioration of vulcanized NR has been obscured by the microbiocidal activity of the curing agent. In the curing system of CBS-sulphur, however, it was observed that the resistance of the vulcanizates is in good correlation with the cross-link density, regardless of the content of sulphur or CBS (Figure 2) [42].

Using scanning electron microscopy, Reszka et al. showed that high concentration of carbon black gives good protection against microbial

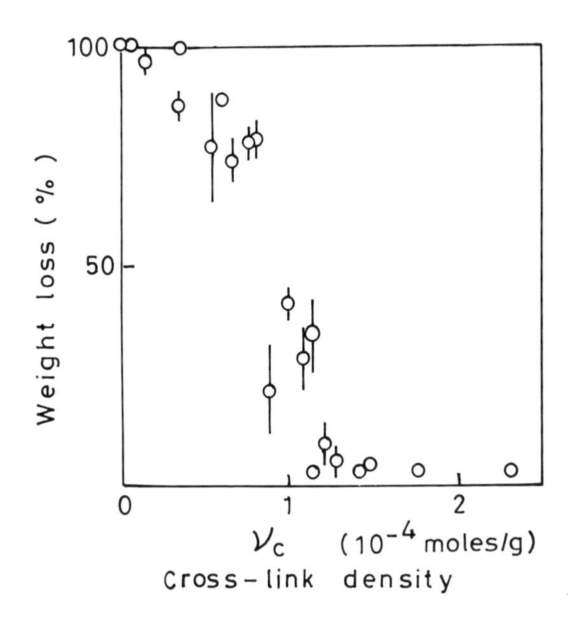

Figure 2. The effect of cross-link density on the microbial degradation of NR vulcanizates.

attack [43]. In 1988, Kwiatkowska and Zyska showed that weight losses of NR vulcanized sheet in soil burial test decreases significantly with carbon black loading of 45 phr [36]. On the other hand, Tsuchii et al. found that specimens with HAF grade carbon black loading were more resistant than those with SRF grade [42].

INDUSTRIAL APPLICATION OF RUBBER-DEGRADING ORGANISMS

In 1975, Nickerson and Faber reported that scrap tyre mesh after *fermentation* by fungal strains was physically and chemically different from the starting material [39] : the particle size was smaller, the material was hydrophilic, and a marked increase in oxygen content was observed. They speculated that the transformations, during fermentation of passenger tire mesh, made the product useful as a *soil conditioner,* since it had ion exchange capacity.

In 1976, Cardarelli used vulcanized NR as binding agent for slow-release pesticide formulations, and detected nematocidal activity over a 3-month test period [44].

Microbial degradation of a relatively large amount of a surgery glove was demonstrated. 2.5 g of the rubber films was added into 3 l of mineral salt medium in a laboratory fermentor every 5 days, and 375 ml of the medium was replaced with fresh medium at the same time. The culture was continued over 150 days, without sterilization of rubber and medium, and the rubber films were completely degraded within 20 days (Figure 3) [37]. Under the same conditions, up to 10 g of rubber added every 5 days was completely degraded.

Soft type NR products, like rubber bands and gloves, are degraded rather rapidly by the organism, so that the microbial process has the potential for early application in practical treatment of waste products. Although hard type products, like automobile tyres, were degraded only at an extremely slow rate and the biological treatment of such a bulky solid waste, like tyres, will be found to entail many difficulties, it is very important, and worth trying to develop a microbial process for waste tyre disposal, because a huge amount of tyres are annually used and discarded.

Figure 3. Microbial degradation of latex gloves. Ten pieces of rubber (0.25 g each), cut from a commercial surgery glove, were added into 3 l of culture medium every 5 days, and all pieces of degraded rubber films were recovered at day-131. I, incubated for 5 days; II, incubated for 10 days; III, incubated for 15 days. Any piece added 20 days before was not found. Small pieces of rubber (2 mm in diameter), at the upper-left in the figure, were edges of the glove.

Tsuchii and Takeda produced isoprene oligomers (acetonyl polyprenyl acetoaldehyde, $A_LP_nA_t$) by enzymatic digestion of NR latex with a very high product specificity for the formation of $A_LP_2A_t$ [21]. Calvin suggested that it is possible to have a part of the hydrocarbon sources obtained from petroleum at present, which will be replaced by those from agricultural plantations of latex-producing green plants in future [45]. If this happens, a rubber refinery for the enzymatic digestion of NR latex may become a useful element of the *green factories*. The isoprene oligomers produced would be used in the chemical and rubber industries as telechilic oligomers and liquid rubber.

CONCLUSIONS

Ecological study of microbial degradation of rubber is still at a primitive stage. The NR vulcanizate has been more or less degraded by microorganisms with both in vivo and in vitro conditions. However, the rate and the extent of microbial degradation may be greatly influenced by the rubber formulations and by environmental circumstances.

Rubber balloons and plastic bags, when disposed off in the natural environment, are now considered to be a danger to wild animals. As tyres wear out in normal use, the tread rubber abrades to particles with diameters between 5 and 100 μm [20]. The amount of tread rubber in atmospheric dust near an expressway has been found to vary from 1.5 to 9.2%, with an overall average of 4.3% [46].

Up till now, the purpose of most studies has been to protect rubber products from microbial *deteriorations*. Waste disposal of used rubber and estimation of the degradation rate in the natural environment will become more important subjects from now onward.

Biochemical or physiological study of rubber-degrading organisms has only just started. Two entirely different organisms, one an actinomycete, the other a bacterium, have been found to degrade NR in exactly the same way via chain scissions [11]. This suggests that the cleaving reaction of the rubber double bonds is common among a variety of microorganisms.

The rubber-degrading reaction has been found to be, at least partly, oxygenase catalyzed [21]. Most of the degrading enzymes for common natural polymers, such as amylases, proteases, and cellulases, catalyze hydrolytic reactions. However, an extracellular lignin-degrading enzyme from white rot fungus was recently reported to be a H_2O_2-dependent oxygenase [47]. Further study of rubber-degrading enzymes may possibly open the way to obtaining organisms with stronger rubber-degrading ability.

Actinomycetes play a major role in degradation of NR [5], while some strains of fungi and bacteria are also known to attack rubber. A strain of *Nocardia* is a very strong decomposer of solid rubber, but the organism dose not produce extracellular enzyme. On the other hand, a *Xanthomonas* sp. secretes an extracellular enzyme, capable of degrading NR in the latex state, and the ability of this bacterial strain to decompose solid rubber is very weak [48]. There are various kinds of rubber-degrading microorganisms with different characteristics. Better understanding of the diversity of microorganisms will lead us to making better use of their versatility.

REFERENCES

1 Tokiwa Y, Ando T, Suzuki T, Takeda K. ACS Symposium Series 1990; 433: 136-148.
2 Hall HM, Long FL. Rubber Content of North American Plants, Carnegie Institution of Washington, Washington DC, 1921.
3 Vanderbilt BM. Thomas Edison, Chemist, American Chemical Society, Washington DC, 1971.
4 Golub MA, Rosenberg ML, Gemmer RV. Rubber Chem Technol 1977; 50: 704.
5 Rook JJ. Appl Microbiol 1955; 3: 302-309.
6 Leeflang KWH. J Am Water Works Assoc 1963; 55: 1523-1535.
7 Cain RB. Soc Gen Microbiol Symp Cambridge: Cambridge University Press, 1992; 48: 293-338.
8 Sohngen NL, Fol JG. Centr Bakt Parasitenk 1914; 40: 87.
9 Spence D, van Niel CB. Ind Eng Chem 1936; 28: 847-850.
10 Nette IT, Pomortseva NV, Koslova EI. Mikrobiologiya 1959; 28: 881-886.
11 Tsuchii A, Suzuki T, Takeda K. Appl Environ Microbiol 1985; 50: 965-970.
12 Blake JT, Kitchin DW. Ind Eng Chem 1949; 41: 1633-1641.
13 Kalinenko BO. Mikrobiologiya 1938; 7: 119-129.
14 Williams GR. Int Biodetn Bull 1982; 18: 31-36.
15 Borel M, Kergomard A, Renard MF. Agric Biol Chem 1982; 46: 877-881.
16 Voegeli HE, Cousminer JJ. Int Biodetn Bull 1978; 14: 119-122.
17 Antoine AD, Dean AV, Gilbert SG. Appl Environ Microbiol 1980; 39: 777-781.
18 Tsuchii A, Suzuki T, Fukuoka S. Agric Biol Chem 1984; 48: 621-625.
19 Cundell AM, Mulcock AP. Dev Ind Microbiol 1975; 16: 88-96.
20 Dannis ML. Rubb Chem Technol 1975; 48: 1011.
21 Tsuchii A, Takeda K. Appl Environ Microbiol 1990; 56: 269-274.
22 Japan Industrial Standards. 1982; JIS K 6353.
23 Alexander M. Biotechnol Bioeng 1973; 15: 611-647.

24 Zyska BJ. In: Rose AH, ed. Economic Microbiology New York: Academic Press Inc, 1981; 6: 323-385.
25 Zyska BJ. In: Biodeterioration. Elsevier Applied Science 1988; 7: 535-552.
26 Zobell CE, Grant CW. Science 1942; 96: 379.
27 Zobell CE. In: Marine Microbiology. Waltham Mass Chronica Botanica Co, 1946; 20.
28 Blake JT, Kitchin DW, Pratt OS. Appl Microbiol 1955; 3: 35-39.
29 Leeflang KWH. J Am Water Works Assoc 1968; 60: 1070-1076.
30 British Standard 1986; BS 2494.
31 Dickenson PB. J Rubb Res Inst Malaya 1969; 22: 165-175.
32 Hutchinson M, Ridgway JW, Cross T. In: Microbial Aspects of the Deterioration of Materials, Technical Series No.9, Academic Press, 1975; 187.
33 Kwiatkowska D, Zyska BJ, Zankowicz LP. In: Biodeterioration. London : Pitman, 1980; 135-141.
34 Williams GR. Int Biodetn Bull 1986; 22: 307-311.
35 Simpson KE. Int Biodetn Bull 1988; 24: 307-312.
36 Kwiatkowska D, Zyska BJ. In: Biodeterioration. Elsevier Applied Science 1988; 7: 575-579.
37 Kajikawa S, Tsuchii A, Takeda K. Nippon Nogeikagaku Kaishi 1991; 65: 981-986.
38 Backer H, Gross H. Material and Organismen 1974; 9: 81.
39 Nickerson WJ, Faber MD. Dev Ind Microbiol 1975; 16: 111-118.
40 Stanescue C, Cirlan V. Commun Stiint Simp Biodeterior Clim 1976; 6(1): 159-165.
41 Cundell AM, Mulcock AP. Int Biodetn Bull 1973; 9: 91-94.
42 Tsuchii A, Hayashi K, Hironiwa T, Matsunaka H, Takeda K. J Appl Polym Sci 1990; 41: 1181-1187.
43 Reszka J, Zyska BJ, Fudalej PS, Reszka KR. Int Biodetn Bull 1975; 11: 71-77.
44 Cardarelli NF. Controlled Release Pesticides Formulations. Florida: CRC Press Inc, 1976.
45 Calvin M. Chem Eng News 1978; March 20: 30-36.
46 Cardina JA. Rubber Chem Technol 1974; 47: 1005-1010.
47 Tien M, Kirk TK. Proc Natl Acad Sci USA 1984; 81: 2280-2284.
48 Tsuchii A. Unpublished data.

Biotransformations: Microbial Degradation of Health Risk Compounds
Ved Pal Singh, editor
© 1995 Elsevier Science B.V. All rights reserved.

Microbial degradation of polyesters

Katsuyuki Mukai and Yoshiharu Doi

Polymer Chemistry Laboratory, The Institute of Physical and Chemical Research (RIKEN), Hirosawa, Wako-Shi, Saitama 351-01, Japan

INTRODUCTION

Plastics have been applied to the wide range of packaging, household, agricultural, marine, architectural, and many other materials. Plastics were developed as strong, light-weight, durable and bioinert materials, and they have replaced natural resources, such as metals, woods, and stones. However, its properties of durability and bioinertness have caused them to accumulate in the environment. The accumulation of abandoned plastics has caused a global environmental problem. Nature is not willing to accept the waste plastics, since majority of plastics are not degraded by microorganisms.

In 1990, about 12 million tons of plastics were produced in Japan, and the production was almost double of the value in 1970. At present, about one hundred million tons of plastics are produced in the world. With increase in production, the amount of plastic wastes has been raised enormously. It is reported that these waste plastics kill many kinds of animals, scar the beautiful scenery, contaminate the coast and disturb the cruise of ships in various environments. In marine environment, especially, one million marine animals are killed every year either by choking on floating plastic items or by becoming entangled in plastic debris. The quantities of the marine debris reaches several hundred thousand tons every year.

In recent years, biodegradable plastics have attracted as environmentally friendly materials to solve the problem of the waste plastics. This paper reports the microbial degradation of polyesters in the marine environments and the properties of extracellular depolymerases from some polyester-degrading microorganisms.

MICROBIAL POLY(HYDROXYALKANOATES)

Organisms produce a number of biological polymers, such as polynucleotides, polypeptides, polysaccharides, polyphosphates, and polyesters. Many scientists have made efforts to elucidate their physiological role and function. The microbial poly(hydroxyalkanoates) (PHAs), a family of polyesters, are synthesized, and accumulated within the cells of a wide variety of microorganisms [1,2]. Poly(3-hydroxybutyrate)

[P(3HB)] is the best studied material of PHA and probably most abundant in the environment (Figure 1a). P(3HB) was firstly discovered in *Bacillus megaterium* by Lemoigne in 1925 [3], and physiological studies of these reserve materials have been performed. When bacteria face the limitation of nutrients, they survive to use polyesters as carbon and energy sources.

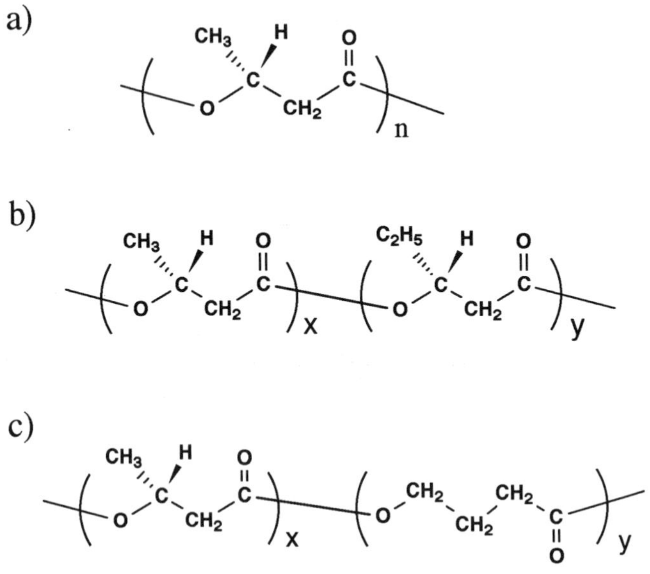

Figure 1. The structures of microbial polyesters used in this study. a) Poly(3-hydroxybutyrate). b) Poly(3-hydroxybutyrate-*co*-3-hydroxyvarelate). c) Poly(3-hydroxybutyrate-*co*-4-hydroxybutyrate).

P(3HB) has a right-handed 2_1 helix with a fiber repeat of 0.596 nm [4]. The comparison of P(3HB) and polypropylene shows the similar structure and physical properties [2]. However, P(3HB) is a stiffer and more brittle material, which shows a very low value (about 5%) of extension to break. *Alcaligenes eutrophus* produces a random copolymer of 3-hydroxybutyrate (3HB) and 3-hydroxyvalerate (3HV) (Figure 1b), when propionic [5,6] or pentanoic acid [7] was fed as the sole carbon source. This copolyester P(3HB-*co*-3HV) is tougher and more flexible than P(3HB) homopolymer, and more suitable for commercial use. Imperial Chemical Industries (ICI) in the United Kingdom has produced P(3HB-*co*-3HV) commercially with a large fermentation process [5,8]. Its ICI's tradename is BIOPOL[R], and German Wella Company firstly put the biodegradable shampoo bottles on the market in 1990. PHA was recognized as oil-price independent plastics with biodegradability, and was also applied to the carrier for drug delivery system (DDS) by another properties of biocompatibility.

Recently, a random copolymer of 3HB and 4-hydroxybutyrate (4HB) (Figure 1c) is produced by *A. eutrophus* from 4-hydroxybutyric acid [9,10], or γ-butyrolactone [11,12]. This copolyester can be made in a wide variety of materials, from hard crystalline plastic to very elastic rubber, depending on the copolymer composition. More recently, the polyester with a high 4HB fraction of 99 mol% is synthesized by *Comamonas acidovorans* from 1,4-butanediol [13].

A remarkable property of the microbial polyesters is their biodegradability in the environments [2]. Table 1 shows the degradation of P(3HB) films in various environments, such as soil, sludge, and seawater. Table 1 indicates that no simple hydrolysis of P(3HB) takes place in water without microorganisms. As PHA is a solid polymer of high molecular weight and incapable of transporting through the cell wall, the microorganisms excrete extracellular PHA depolymerases to degrade environmental PHA into the oligomers and/or monomers, and the resulting products are adsorbed and utilized as nutrients [14]. The extracellular PHA depolymerases have been isolated from some bacteria, such as *Pseudomonas* strain P1 [15], *Pseudomonas lemoignei* [16,17], *Alcaligenes faecalis* T1 [18-20], and *Pseudomonas pickettii* [21]. Recently, PHA depolymerase has been prepared from the eukaryote, *Penicillium pinophilum* [22]. The enzyme from *P. pinophilum* exhibited the similar properties to those of the enzyme from bacteria.

Table 1
Changes in the thickness of microbial poly(3-hydroxybutyrate) films in various environments

Environment	Change in the thickness (μm/week)
Sea water (22°C)	5
Aerobic sewage (25°C)	7
Soil (25°C)	5
Sterile sea water[a]	0

[a]Autoclaved at 121°C for 15 min.

DEGRADATION OF POLYESTERS UNDER MARINE ENVIRONMENTS

Biodegradation experiments under marine environments were carried out at the exposure facility at Kanagawa Prefectural Fishery Experiment Station at Jogashima, Japan [23]. All experimental samples (5 cm x 10 cm in size), placed in nylon nets within a stainless steel cage, were positioned at a water depth of 1.5 m in an outdoor tank (10 x 10 m x 3 m in depth), with fresh sea water continuously flowing through the tank from the Pacific Ocean. The lowest mean temperature was 13°C in February, and the highest was 26.1°C in August, 1990.

The microbial polyester samples were produced by a controlled fermentation. P(3HB) homopolymer was produced from butyric acid by *Alcaligenes eutrophus*. P(3HB-*co*-3HV) and P(3HB-*co*-4HB) copolyesters were produced by *A. eutrophus* from mixed carbon substrates of butyric and pentanoic acid and of fructose and γ-butyrolactone, respectively.

Figure 2 shows the surface erosions of solvent-cast films of five polyester samples for the three-week period 20 June-12 July (22±3°C)

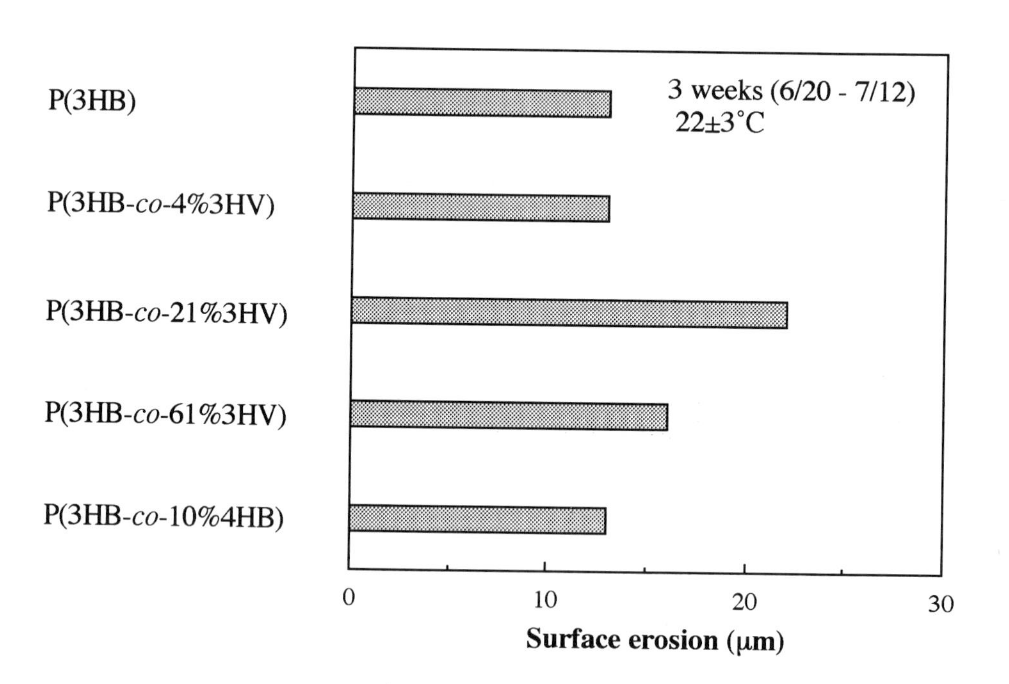

Figure 2. Surface erosion of solvent-cast films (initial thickness: 50-100 μm) of five polyester samples for three weeks (20 June - 12 July) in sea water.

in sea water. The weight and thickness of the films decreased, and the surface of the films was apparently blemished. All samples of microbial polyesters, exposed in sea water, were degraded via surface dissolution. Scanning electron micrographs of the films showed that no appreciable change took place inside the film. After three weeks exposure, all films with various compositions had lost 13-22 mm of film thickness. The rate of surface erosion was almost independent of copolyester compositions, except for P(3HB-*co*-21%3HV).

The films of P(3HB-*co*-4HB) samples were exposed for eight weeks in sea water during two seasons (18 January - 15 March and 21 August - 19 October). Figure 3 shows the surface erosions of the films after eight weeks in sea water. The film erosions of P(3HB-*co*-6%4HB) and P(3HB-*co*-10%4HB), after eight weeks from 18 January to 15 March (14±2°C) were 31 and 33 μm, respectively. In contrast, the films thickness lost, respectively, 55 and 60 μm during eight weeks from 21 August to 19 October (24±3°C). The rate of surface erosion was almost independent on the copolyester composition of P(3HB-*co*-4HB) films, but strongly dependent on the temperature of the sea water.

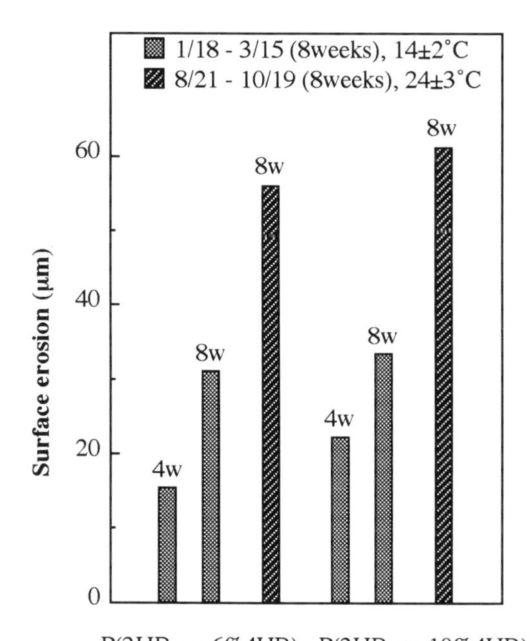

Figure 3. Surface erosion of solvent-cast films (initial thickness: 100-150 μm) of P(3HB-*co*-4HV) samples in sea water for eight weeks during two seasons (18 January - 15 March and 21 August - 19 October).

Figure 4 shows the surface erosion of melt-extruded plates of P(3HB-*co*-3HV) in sea water for 17 weeks from 21 August to 18 December (21±6°C). The P(3HB-*co*-3HV) plates lost 100 - 140 μm of plate thickness in 17 weeks. The rate of surface erosion of the melt-extruded plates was also independent on the compositions of P(3HB-*co*-3HV).

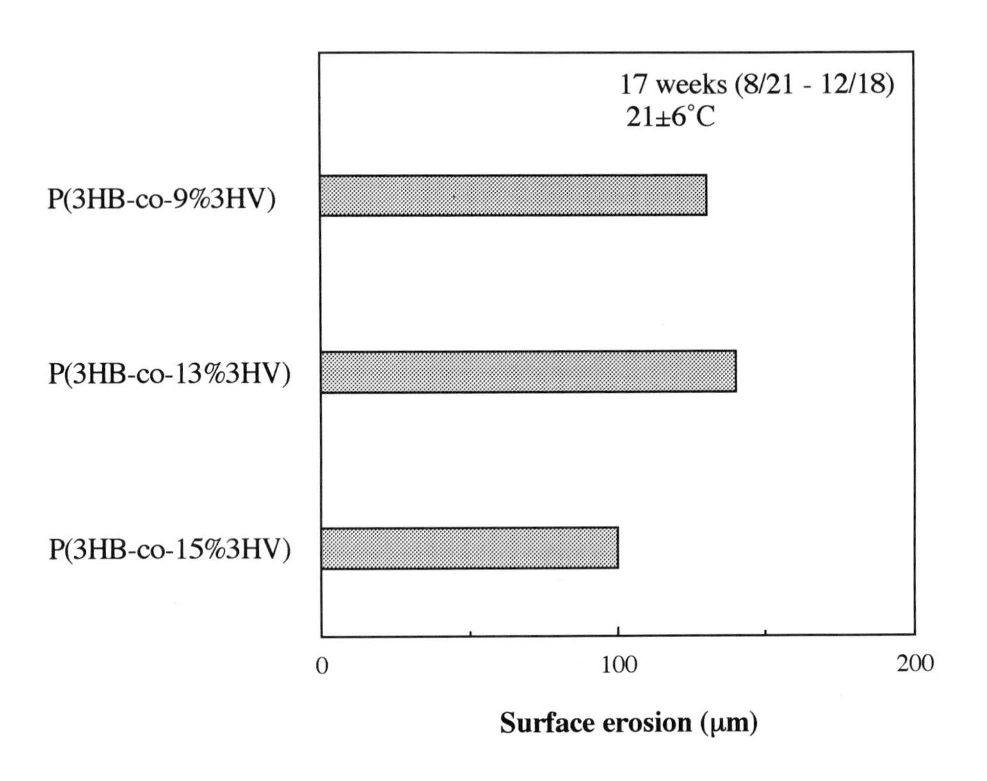

Figure 4. Surface erosions of melt-extruded plates (initial thickness: 2050-2100 μm) of P(3HB-*co*-3HV) samples for 17 weeks (21 August - 18 December) in sea water.

CHANGES IN PHYSICAL PROPERTIES OF POLYESTERS AFTER THE EXPOSURE IN SEA WATER

The changes in molecular weights and mechanical properties of polyesters were studied before and after the exposure in sea water. Table 2 shows the changes of P(3HB-*co*-6%4HB) and P(3HB-*co*-10%4HB) films during the eight-week period, 18 January - 15 March. The number average molecular weight (Mn) of the residual films decreased slightly

Table 2
Changes in thickness, molecular weight and mechanical properties of P(3HB-*co*-4HB) films in sea water during the period 18 January-15 March

Samples[a]	Time (weeks)	Thickness (μm)	Molecular weight		Strain at yield (%)	Stress at yield (MPa)	Strain at break (%)	Stress at break (MPa)
			$Mn \cdot 10^{-3}$	Mw/Mn				
I	0	100	494	2.1	8	25	51	26
I	4	85	398	1.8	5	20	16	n.d.[b]
I	8	69	360	1.9	3	16	25	n.d.[b]
II	0	150	197	2.6	13	19	523	24
II	4	128	186	2.6	6	17	48	n.d.[b]
II	8	117	186	2.6	5	19	50	n.d.[b]

[a]Samples I and II were P(3HB-*co*-6%4HB) and P(3HB-*co*-10%4HB), and initial thickness of films 100 and 150 μm, respectively.
[b]It was difficult to determine accurate values of stress at break.

as the surface erosion proceeded, while the polydispersities (Mw/Mn) remained almost unchanged. The strain at break of the films decreased during the exposure in sea water, and the material gradually turned brittle.

The degradation of melt-spun fiber (260 μm diameter) of P(3HB-*co*-14%3HV) was carried out in sea water for eight weeks from 15 January to 15 March (14±2°C). Table 3 summarizes the weight loss, molecular weight, and mechanical properties of the fiber. The weight of the fiber decreased to only 35% of the initial weight after eight weeks in sea water. The strain and stress at break of the fiber rapidly decreased in sea water, and both values reached zero after eight weeks.

Table 3
Changes in weight, molecular weight, and mechanical properties of the fiber of P(3HB-*co*-14%3HV) in sea water during the period 18 January-15 March

Time (weeks)	Retention of weight (%)	Molecular weight		Strain at break (%)	Stress at break (MPa)	Young modulus (MPa)
		Mnx10^{-3}	Mw/Mn			
0	100	138	2.2	67	110	1100
4	75	136	1.9	13	75	1400
8	35	134	1.9	0	0	-

Note: The initial diameter of monofilament fiber was 260 μm.

ISOLATION OF P(3HB)-DEGRADING MICROORGANISMS FROM SEA WATER

The degradation of microbial polyesters in sea water may involve a simple hydrolytic degradation process in addition to a microbial (enzymatic) degradation. The degradation of P(3HB) and P(3HB-*co*-10%4HB) films was studied at 37°C for four weeks in sea water, that had been autoclaved for 15 min at 121°C. No weight loss of films was observed in the pretreated sea water, indicating that a simple hydrolytic degradation does not contribute to the degradation of microbial polyesters in the marine environment.

An attempt has been made to isolate P(3HB)-degrading microorganisms from the marine environment (sea water), where we carried out the biodegradation tests of polyesters. The sea water sample was filtered through a 0.45 μm Millipore filter, and the filter-adhered microorganisms were incubated aerobically at 30°C in an aqueous medium with P(3HB) granules as a sole carbon source. After 3 days, samples from the liquid culture medium were plated on the mineral agar plates, containing P(3HB) granules as a sole carbon source. Several fungi and bacteria, capable of forming clear zones surrounding the colonies in P(3HB)-mineral agar plates, were isolated. Four bacterial colonies of these P(3HB)-degrading microorganisms showed the best growth and activities to degrade P(3HB) granules in the supernatant of the culture medium. The three of four colonies were same in respect of the bacterial properties. As a result, two types of P(3HB)-degrading bacteria were isolated and labelled YM1004 and YM1006.

Two different isolated bacterial strains, YM1004 and YM1006, were Gram-negative and aerobic rods, which were catalase and oxidase positive. The isolates were identified, using API 20 NE test kit, and further tests, such as the carbon source utilization. On the basis of these test results, strains YM1004 and YM1006 were identified as *Comamonas testosteroni* and *Pseudomonas stutzeri*, respectively.

PURIFICATION AND PROPERTIES OF PHA DEPOLYMERASES

A pure culture of *C. testosteroni* (strain YM1004) grew on P(3HB) as a sole carbon source, and was cultured under aerobic conditions at 30°C on a reciprocal shaker. The growth curve of *C. testosteroni* in an aqueous medium, containing 0.15% P(3HB) granules as a sole carbon source, is shown in Figure 5.

The activity of PHA depolymerase was maximum at the end of logarithmic growth phase. The purification of an extracellular PHA depolymerase from *C. testosteroni* was performed by an ammonium sulphate fractionation and a hydrophobic interaction (Butyl-Toyopearl) column chromatography. The purification steps of the enzyme are summarized in Table 4. After a single column chromatography, the purified enzyme was homogeneous, as judged by polyacrylamide gel electrophoresis, in the presence of sodium dodecyl sulphate (SDS-PAGE).

Table 4
Preparation of PHA depolymerase from *C. testosteroni*

Purification step	Total activity (units)	Total protein (mg)	Specific activity (units/mg)	Yield (%)
1 Culture medium	150	21.6	7	100
2 $(NH_4)_2SO_4$ (0-50%)	61	0.70	86	40
3 Butyl-Toyopearl	31	0.16	191	21

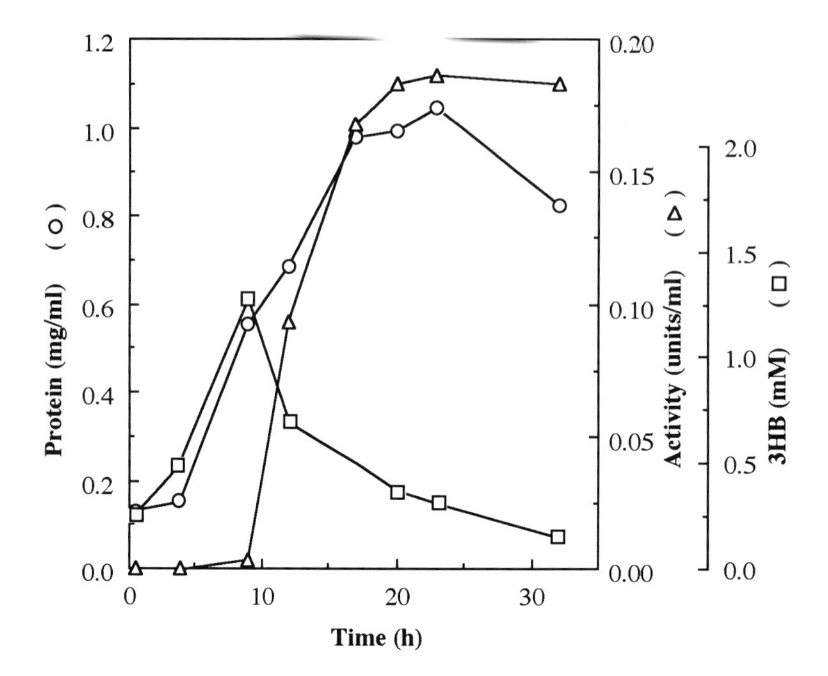

Figure 5. Growth curve of *C. testosteroni* and the excretion of PHA depolymerase. The isolated was grown at 30°C in a liquid culture medium containing 0.15% P(3HB) granules as a sole carbon source. PHA depolymerase activity (Δ), the concentration of 3-hydroxybutyric acid (3HB) (□) in the supernatant, and the protein content (○) in the pellet were determined.

The molecular weight of purified PHA depolymerase was about 50,000, as estimated by SDS-PAGE. This enzyme had an isoelectric point of about 8.4, as determined by isoelectric focusing. Figure 6 shows the pH profile of the activity of P(3HB) degradation by purified enzyme, as determined by the turbidimetric method with P(3HB) granules. The optimum pH for the hydrolysis of P(3HB) granules was in the alkaline region between 9.5 and 10.0 in the glycine-NaOH buffer. The enzyme was drastically inactivated at the values of pH above 10.0. The PHA depolymerase from *C. testosteroni* as well as those from *A. faecalis* T1, *P. lemoignei* and *P. pickettii* was inhibited by the treatment of diisopropylfluorophosphate (DFP), phenylmethylsulphonyl fluoride (PMSF), and dithiothreitol (DTT). This result suggests the presence of serine residues and disulphide bonds at the active site, for the activity.

In addition, a PHA depolymerase from *P. stutzeri* (strain YM1006) also prepared by a hydrophobic interaction column chromatography. The enzyme from *P. stutzeri* showed a single band on the gel of SDS-

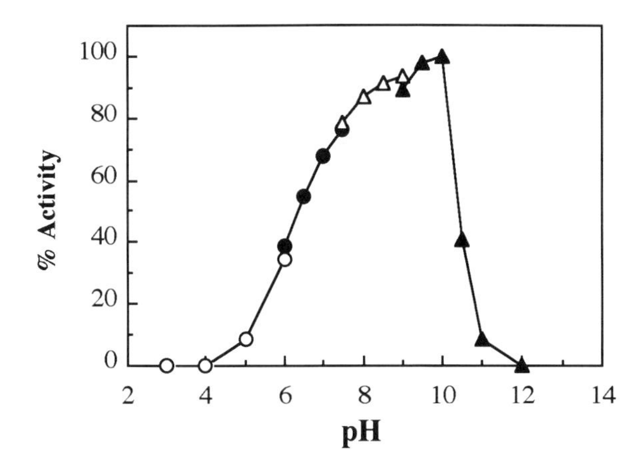

Figure 6. pH profile of PHA depolymerase activity. The activity was determined in 0.1M sodium acetate buffer (○), sodium phosphate buffer (●), Tris-HCl buffer (△) and glycine-NaOH buffer (▲).

PAGE, and the molecular weight was about 60,000. As PHA depolymerases have a high hydrophobicity on the enzyme surface, the utilization of the hydrophobic interaction column chromatography for the preparation of the enzymes is simple and useful.

ENZYMATIC DEGRADATION OF POLYESTER FILMS

Table 5 shows the compositions, molecular weights and crystallinities of polyester samples used in this study, and the weight loss data of microbial polyester films at 37°C by five kinds of PHA depolymerases, which were prepared from *C. testosteroni*, *A. faecalis* T1, *P. lemoignei* and *P. pickettii*. The enzymatic degradation occurred at the surface of polyester films, and the thickness of the film decreased with time. The rate of enzymatic degradation of P(3HB-*co*-3HV) films by all the depolymerases were lower than that of P(3HB) film. The P(3HB-*co*-3HV) films, with high 3HV compositions over 64 mol% 3HV, were hardly eroded by the enzymes from *P. pickettii* and *P. lemoignei*. However, the rates of enzymatic degradation of all P(3HB-*co*-3HV) films by the enzyme from *C. testosteroni* were almost identical, and about one-third higher than that of P(3HB) film.

On the other hand, the rates of enzymatic degradation of P(3HB-*co*-3HB) films by the enzyme from *C. testosteroni* were also identical, and

Table 5
Compositions and properties of P(3HB-*co*-3HV) and P(3HB-*co*-4HB) films, and enzymatic erosion of films by different PHA deploymerases at 37°C

Sample	Composition (mol%)[a]			Molecular weight[b]		Crysta-llinity (%)[c]	Weight loss by depolymerases (mg)[d]				
	3HB	3HV	4HB	Mn-10^{-3}	Mw/Mn		*P. picketti*	*A. faecalis*	*P. lemoignei*A	*P. lemoignei*B	*C. testosteroni*
1	100	0	0	171	3.1	60±5	2.0	2.3	3.7	2.9	2.7
2	96	4	0	186	2.6	57±5	0.5	0.7	0.6	0.4	0.6
3	88	12	0	109	2.5	50±5	1.0	1.4	1.3	1.2	1.1
4	85	15	0	153	2.4	53±5	0.8	1.3	2.4	2.3	0.6
5	79	21	0	238	3.0	59±5	1.2	1.6	1.6	1.1	1.1
6	36	64	0	984	1.8	57±5	0	1.6	0	0	0.5
7	20	80	0	257	3.8	52±5	0	0.4	0	0	0.5
8	100	0	0	171	3.1	60±5	0.3	0.2	0.3	0.3	0.3
9	95	0	5	257	2.3	50±5	1.0	1.5	1.7	1.1	0.6
10	86	0	14	189	2.6	41±5	5.1	6.1	5.7	2.4	0.7
11	83	0	17	366	2.9	47±5	5.2	5.1	3.8	2.6	0.5
12	79	0	21	62	3.7	37±5	2.0	3.4	2.8	1.6	0.5
13	73	0	27	91	4.3	38±5	1.7	2.0	2.0	0.7	0.6
14	3	0	97	92	3.8	37±5	0.1	0.5	0.1	0	0.3

[a]Determined by ^{1}H NMR.
[b]Determined by GPC with polystyrene standards.
[c]Determined by X-ray diffraction.
[d]Enzymatic degradations of samples 1-7 and 8-14 were carried out for 19 and 2h at 37°C, respectively.

twice higher than that of P(3HB) film. Depolymerases from *P. lemoignei* and *P. pickettii* were not able to hydrolyze P(3HB-*co*-97%4HB) films.

PHA depolymerase from *C. testosteroni* had a broad specificity against the hydrolysis of copolyesters with various compositions. The rate of enzymatic degradation of copolyester films depended on the types of copolyesters. The crystallinities of polyester films in Table 5 were determined by analysis of the X-ray diffraction patterns. All films of P(3HB-*co*-3HV) samples had high crystallinities over 50%, through a wide range of compositions from 4 to 80 mol% 3HV. In contrast, the films of P(3HB-*co*-3HB) samples with compositions of 5-97 mol% 4HB had low crystallinities below 50%. In a previous paper [24], we reported that the rate of enzymatic degradation of P(3HB) film by PHA depolymerase from *A. faecalis* T1 decreased with an increase in the crystallinity. The difference in the rate of enzymatic degradation of P(3HB-*co*-3HV) and P(3HB-*co*-4HB) films may be accounted by the difference in the crystallinity, which is at least one of the factors.

HYDROPHOBICITY OF PHA DEPOLYMERASES

The PHA depolymerase from *C. testosteroni* had a very high hydrophobicity, and it was eluted from the hydrophobic interaction column chromatography at more than 40% ethanol. The hydrophobicity of the enzyme was compared with three kinds of PHA depolymerases from *A. faecalis* T1, *P. lemoignei,* and *P. pickettii*. *A. faecalis* T1, *P. lemoignei* and *P. picketti* had been isolated as P(3HB)-degrading bacteria from activated sludge, soil, and atmosphere, respectively. The mixture of purified PHA depolymerases from *A. faecalis* T1, *P. lemoignei, P. pickettii,* and *C. testosteroni* was applied to a Butyl-Toyopearl column and eluted by the linear gradient of 0-50% ethanol. Figure 7 shows the elution profile of four enzymes from the column. The result indicates that the hydrophobicity of enzymes decreases in the following order: *C. testosteroni* > *A. faecalis* T1 > *P. lemoignei* A = *P. pickettii*. It may be concluded that a very high hydrophobicity of the extracellular PHA depolymerase from *C. testosteroni* is essential to strictly adhere the surface of polyester samples in marine environments. This conclusion may be supported by the fact that *A. faecalis* T1, isolated from the activated sludge, secretes a PHA depolymerase with higher hydrophobicity than those of *P. lemoignei* from soil and of *P. pickettii* from atmosphere.

Figure 8 shows the dependence of enzyme concentration on the weight losses of P(3HB) and P(3HB-*co*-97%4HB) films. In this experiment, the hydrophobic depolymerase from *C. testosteroni* was used. The rates of film erosion by the enzyme were strongly dependent on the enzyme concentration. Maximal rates of the enzymatic degradation were observed

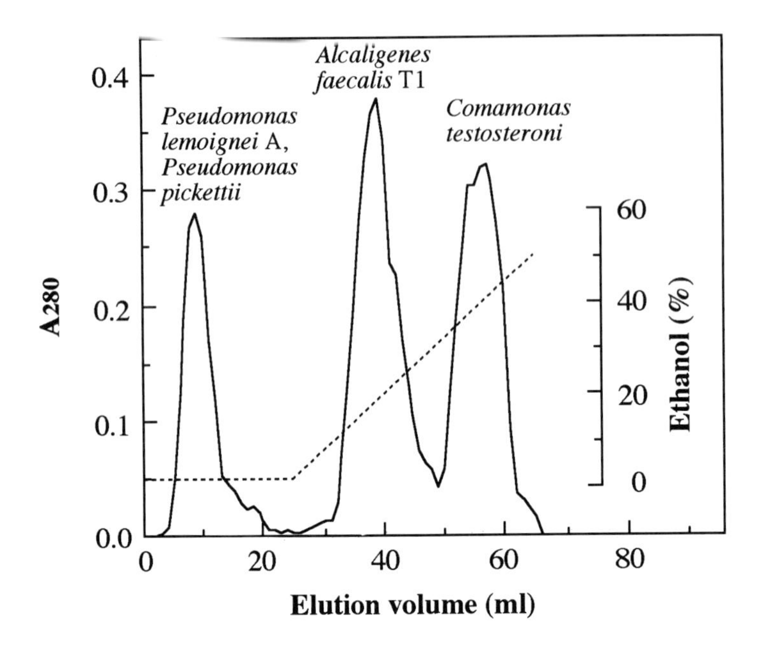

Figure 7. A hydrophobic (Butyl-Toyopearl) column chromatogram of PHA depolymerase. A mixture of purified PHA depolymerases from *P. lemoignei, A. faecalis* T1, *P. pickettii,* and *C. testosteroni* was loaded to the column and eluted with linear gradient of 0-50% ethanol.

at around 0.5 μg protein/ml of enzyme concentration. The enzymatic degradation of both films almost stopped at high concentrations of enzyme over 4.0 μg/ml. This incredible phenomenon for the enzyme reaction was also observed in the enzymatic degradation of P(3HB) granules for the enzyme assays. It has been reported that the PHA depolymerase from *A. faecalis* T1 has a hydrophobic domain to adhere hydrophobic substrates, in addition to a catalytic domain [19,20,25]. The binding domains, separated from the catalytic domains, were also found in cellulase and xylanase [26,27]. In the enzyme degradation of P(3HB) film with a PHA depolymerase from *A. faecalis* T1, a maximal activity of the enzyme was also observed at 4.0 μg/ml. The enzyme molecules with a strong hydrophobicity, may entirely cover the surface of polyesters at high concentrations over 4.0 μg/ml, and an active site of the catalytic domain in the enzyme may be incapable of hydrolyzing polyester molecules on the surface by the entire covering of enzymes, adhereing at a hydrophobic site as binding domain. As a result, excess amounts of enzyme may function as an inhibitor for the enzymatic degradation of PHA samples.

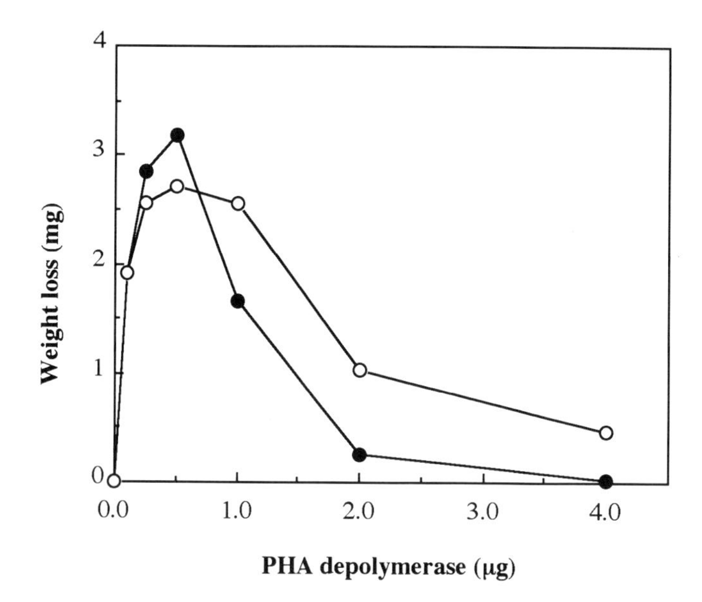

Figure 8. Weight loss of P(3HB) and P(4HB) films for 19 h in the aqueous solution containing different amounts of PHA depolymerase from *C. testosteroni* at 37°C and pH 7.4. P(3HB) films (○) and P(3HB-*co*-97%4HB) films (●).

REFERENCES

1 Anderson AJ, Dawes EA. Microbiol Rev 1990; 54: 450-472.
2 Doi Y. In: Microbial Polyesters. New York: VCH Publishers, 1990.
3 Lemoigne M. Annales of de l'Institut Pasteur 1925; 39: 144-173.
4 Cornibert J, Marchessault RH. J Mol Biol 1972; 71: 735-756.
5 Bryom D. Trends Biotechnol 1987; 5: 246-250.
6 Doi Y, Kunioka M, Nakamura Y, Soga K. Macromolecules 1987; 20: 2988-2991.
7 Doi Y, Tamaki A, Kunioka M, Soga K. Appl Microbiol Biotechnol 1988; 28: 330-334.
8 Holms PA. Phys Technol 1985; 16: 32-37.
9 Doi Y, Kunioka M, Nakamura Y, Soga K. Macromolecules 1988; 21: 2722-2727.
10 Kunioka M, Nakamura Y, Doi Y, Polym Commun 1988; 29: 174-176.
11 Kunioka M, Kawaguchi Y, Doi Y. Appl Microbiol Biotechnol 1989; 30: 569-573.
12 Doi Y, Segawa A, Kunioka M. Macromolecules 1990; 12: 106-111.

13 Kimura H, Yoshida Y, Doi Y. Biotechnol Lett 1992; 14: 445-450.
14 Delafield FP, Doudoroff M, Palleroni NJ, Lusty CJ, Contooulos R.J Bacteriol 1965; 90: 1455-1466.
15 Choudhury AA. Arch Microbiol 1963; 47: 167-200.
16 Lusty CJ, Doudoroff M. Proc Natl Acad Sci USA 1966; 56: 960-965.
17 Mukai K, Yamada K, Doi Y. Int J Biol Macromol 1992; 14: 235-239.
18 Tanio T, Fukui T, Shirakura Y, Saito T, Tomita K, Kaiho T, Masamune S. Eur J Biochem 1982; 124: 71-77.
19 Shirakura Y, Fukui T, Saito T, Okamoto Y, Narikawa T, Koide K, Tomita K, Takemasa T, Masamune S. Biochim Biophys Acta 1986; 880: 46-53.
20 Fukui T, Narikawa T, Miwa K, Shirakura Y, Saito T, Tomita K. Biochim Biophys Acta 1988; 952: 164-171.
21 Yamada K, Mukai K, Doi Y. Int J Biol Macromol 1993; 15: 215-220.
22 Brucato CL, Wong SS. Arch Biochem Biophys 1991; 290: 497-502.
23 Doi Y, Kanesawa Y, Tanahashi N, Kumagai Y. Polym Deg Stab 1992; 36: 173-177.
24 Kumagai Y, Kanesawa Y, Doi Y. Makromol Chem 1992; 193: 53-57.
25 Saito T, Suzuki K, Yamamoto J, Fukui T, Miwa K, Tomita K, Nakanishi S, Odani S, Suzuki J, Ishikawa K. J Bacteriol 1989; 171: 184-189.
26 Gilkes NR, Henrissat B, Kilburn DG, Miller RC, Warren RAJ. Microbiol Rev 1991; 55: 303-315.
27 Wong KKY, Tan LUL, Saddler JN. Microbiol Rev 1988; 52: 305-317.

Biotransformations: Microbial Degradation of Health Risk Compounds
Ved Pal Singh, editor
© 1995 Elsevier Science B.V. All rights reserved.

Degradation of hazardous organic compounds by rhizosphere microbial communities

Todd A. Anderson[a], David C. White[b] and Barbara T. Walton[c]

[a]Pesticide Toxicology Laboratory, Department of Entomology, Iowa State University, Ames, IA, U.S.A.

[b]Center for Environmental Biotechnology, The University of Tennessee, Knoxville, TN, U.S.A.

[c]Environmental Sciences Division, Oak Ridge National Laboratory, Oak Ridge, TN, U.S.A.

INTRODUCTION

The rhizosphere, or root zone of plants, contains a diverse microbial community that contributes to plant health and to soil homeostasis. In addition, recent studies indicate that microorganisms in the rhizosphere can degrade toxicants of concern to human health and the environment. The application of these findings to remediation of chemically contaminated soils will be facilitated by a better understanding of the variables that control biotransformation in the rhizosphere. For example, the increased density and diversity of microorganisms in the rhizosphere, as compared to nonvegetated soils, may be a critical factor for microbial degradation of xenobiotic compounds. In addition, the secretion of readily degradable substrates by roots may facilitate cometabolic transformation of hazardous organic compounds [1]. Other mechanisms can also be invoked to explain how *microbial transformations* occur in the rhizosphere. Collectively, these factors have important implications for the successful use of vegetation to increase the participation of microorganisms in biotransformation of toxicants at hazardous waste sites.

The existing evidence for microbial degradation of xenobiotics in the rhizosphere is examined with special attention to whether a community or single species of microorganism is responsible for biotransformations in the root zone. In addition, previously unpublished data resulting from biochemical analyses of trichloroethylene-degrading microorganisms are presented and discussed with respect to microbial biomass, metabolic activity, nutritional status, and community structure in the rhizosphere.

RHIZOSPHERE MICROBIOLOGY

Plant roots affect the soil in which they grow in a multitude of ways, making the soil more conducive to microbial growth and activity. For example, roots affect soil carbon dioxide and oxygen concentrations, osmotic and redox potentials, pH, and moisture content. The rhizosphere is a zone of increased microbial activity and biomass at the root-soil interface [2]. The large microbial populations in the rhizosphere are sustained by exudation of substances such as carbohydrates and amino acids from the root as well as sloughing of root epidermis. All roots are protected from abrasion by root cap cells which are sloughed off during root growth; sometimes as many as 10,000 cells per plant per day [3]. As the root grows downward, cells in the root cap produce a gel that helps lubricate the root, allowing the root to force its way through the soil. This mucigel, along with other root excretions is classified as root exudate [4]. Both root cap cells and root exudates, are useful sources of nutrients for microorganisms in the root zone or rhizosphere. In addition, microorganisms can stimulate exudation, whereas the absence of bacteria and fungi can lead to less exudate production by the plant [2]. These rhizosphere characteristics typically result in microbial populations an order of magnitude or more above microbial populations in nonvegetated soil. This *rhizosphere effect* is often expressed as the ratio of microorganisms in rhizosphere soil to the number of microorganisms in non-rhizosphere soil, the R/S ratio [5]. R/S ratios commonly range from 5 to 20, but occasionally are as high as 100 and above [6].

The actual composition of the microbial community in the root zone is dependent on root type, plant species, plant age, and soil type [3] as well as other selection pressures, such as foreign chemicals [7-10]. Typically, the rhizosphere is colonized by a predominantly Gram-negative microbial community [6]. In addition, leguminous plants exhibit a greater rhizosphere effect than non-leguminous plants, presumably because of increased N levels in soil where legumes grow. The ability of the plants to select for different rhizosphere microbial communities in both composition as well as size is intriguing from the standpoint of exploring whether this selection translates into differences in the rates of microbial degradation of organic compounds in the rhizosphere.

The rhizosphere, which was first described by Hiltner [11], has been the focus of agricultural research for several years, primarily because of its influence on crop productivity. Several excellent comprehensive reviews on the rhizosphere are available [2,3], and due to their extensive nature, the current review will be limited to discussing findings on rhizosphere microbiology within the context of degradation of organic compounds by rhizosphere microbial communities.

The interaction between plants and microbial communities in the rhizosphere is a complex relationship, that has evolved to the mutual benefit of both groups. In addition to the accepted relationships described thoroughly in the aforementioned reviews, other connections between plants and the microorganisms in their root zones undoubtedly exist. One possible additional relationship is the rhizosphere microbial community's role in protecting the plant from chemical injury. Previous research has shown that plants increase root exudation in the presence of xenobiotic chemicals [12,13]. In hydroponic cultures of corn, the presence of simazine (2-chloro-4,6-bis ethylamino-S-triazine), a preemergence herbicide used for controlling weeds in corn, caused a two-fold increase in exudation of organic acids [14]. In addition, simazine increased the length and weight of roots, but only if microorganisms were also present in the medium. It is not clear whether the increase in exudation is an evolved response by the plant to attract more microorganisms (and possibly degrade the chemical faster) or simply the physiological effect of the chemical on the plant. The tolerance of corn to the herbicidal effects of simazine may be the result of rapid metabolism of the compound by the plant. The tolerance of corn to the herbicidal effects of simazine may be the result of rapid metabolism of the compound by the plant. Most herbicides used to control weeds are readily metabolized by non-target plants [15]. Nonetheless, rhizosphere microbial communities may also play a role in protecting the plant from chemical injury. This idea is further supported by the work of Herring and Bering [16]. They found that the toxic effect of phthalate esters on spinach and pea seedlings could be abated or reversed by the presence of microorganisms in the soil.

MICROBIAL DEGRADATION IN THE ROOT ZONE

Studies of microbial degradation of toxicants in the root zone of plants have included a variety of plant types from diverse taxonomic families (Table 1). These studies have been generated from examination of agrochemicals, aquatic systems, and industrial chemicals and are summarized below. Recently, two reviews on microbial degradation of organic compounds in the rhizosphere and the beneficial effects of vegetation at contaminated sites have been published [17,18].

Agrochemicals

Research on microbial transformations in the rhizosphere has been concerned mainly with agricultural chemicals, such as pesticides and fertilizers. Several researchers [1,7,19-21] have described an increased capacity for mineralization of various pesticides by rhizosphere microbial communities as compared with microbial communities in nonvegetated soil. Occasionally, this increased mineralization capacity is correlated with increased numbers of pesticide-degrading microorganisms. Sandmann and Loos [7] found an increase in the number of 2,4-D (2,4-dichlorophenoxyacetate)-degrading bacteria in the rhizospheres of previously untreated African clover and sugarcane. Similarly, work by Abdel-Nasser and coworkers [8] showed higher microbial counts in the

Table 1
Studies relevant to organic chemical degradation in the rhizosphere

Plant	Family	Chemical	Comments	Refs
Wheat	Gramineae	Mecoprop[A] 2,4-D[B] MCPA[C]	Mixed culture capable of using compounds as a carbon source.	[21]
Sugarcane African clover	Gramineae Fabaceae	2,4-D[B]	Higher population of 2,4-D-degrading microorganisms in the rhizosphere of sugarcane, a plant nonsensitive to 2,4-D, compared with African clover, a plant sensitive to the herbicidal effects of 2,4-D.	[7]
Bush bean	Fabaceae	Diazinon[D] Parathion[E]	Increased mineralization of both compounds in the rhizosphere.	[1]
Rice	Gramineae	Parathion[E]	Increased mineralization in the rhizosphere especially under flooded conditions.	[20]
Tobacco	Solanaceae	MH[F]	MH caused enhanced nitrification and mineralization of organic substances in the rhizosphere.	[50]
Rice	Gramineae	Benthiocarb[G]	Eight-fold increase in heterotrophic bacteria in the rhizosphere of treated rice plants.	[22]
Corn Bean Cotton	Gramineae Fabaceae Malvaceae	Temik[H]	Higher counts of microorganisms in treated vs untreated rhizosphere.	[8]

Wheat Corn	Gramineae Gramineae	Diazinon[D]	Rhizosphere microbial counts increased by 2 orders of magnitude.	[10]
Flax	Linaceae	2,4-D[B]	Ammonifying, nitrifying and cellulose-decomposing bacteria in the rhizosphere increased by 1 to 2 orders of magnitude.	[9]
Corn	Gramineae	Atrazine[I]	Increase in production of atrazine degradation metabolites by rhizosphere microorganisms in the presence of decomposing roots.	[19]
Legumes	Fabaceae	Petroleum	Describes the importance of leguminous plants in reclamating petroleum-contaminated sites.	[51]
Rice	Gramineae	Oil residues	*Bacillus* sp. isolated from rice rhizosphere was capable of growth on oil residues, but only in the presence of root exudates.	[28]
Prairie grasses	Gramineae	PAHs[J]	Increased disappearance of PAHs in vegetated vs. nonvegetated soil columns.	[27]
Lespedeza Loblolly pine Bahia grass Goldenrod Soybean	Fabaceae Pinaceae Gramineae Compositae Fabaceae	TCE[K]	Increased degradation of TCE in rhizosphere soil and increased mineralization of ^{14}C-TCE in soils containing lespedeza, loblolly pine, and soybean.	[29,30]
Reeds	Gramineae	VOCs[L]	Vegetated microbial filters increased removal of both aromatics and aliphatics.	[26]

Table 1 continued

Plant	Family	Chemical	Comments	Refs
Corn Soybean	Gramineae Fabaceae	Surfactants[M]	Rhizosphere treatments significantly increased initial rates of mineralization by a factor of 1.1-1.9.	[52]
Cattails	Typhaceae	Surfactants[M]	Mineralization of surfactants was more rapid in the rhizosphere than in root-free sediments.	[24]
--	--	Organo-chlorines[N]	A rhizosphere-competent fungus was able to degrade a variety of organochlorine compounds.	[53]
--	--	PCBs[O]	Compounds produced by photosynthetic plants were shown to support the growth of PCB-degrading bacteria. The organisms retained their ability to metabolize PCBs.	[54]
Wheatgrass	Gramineae	PCP[P]	Mineralization of pentachlorophenol was enhanced in soils containing wheatgrass.	[55]

[A] 2-(2-Methyl-4-chlorophenoxy)propionic acid.
[B] 2,4-Dichlorophenoxyacetic acid.
[C] 2-Methyl-4-chlorophenoxyacetic acid.
[D] O,O-diethyl-O-(2-isopropyl-6-methyl-4-pyrimidinyl) phosphorothioate.
[E] O,O-diethyl-O-*p*-nitrophenyl phosphorothioate.
[F] Maleic hydrazide (1,2-dihydro-3,6-pyridazinedione).

[G]S-*p*-chlorobenzyl diethylthiocarbamate.
[H]2-Methyl-2-)methylthio) propionaldehyde O-(methylcarbamoyl)oxime.
[I]2-Chloro-4-ethylamino-6-isopropylamino-S-triazine.
[J]Polycyclic aromatic hydrocarbons (benz[a]anthracene, chrysene, benzo[a]pyrene, dibenz[a,h]anthracene.
[K]1,1,2-Trichloroethylene.
[L]Volatile organic compounds (benzene, biphenyl, chlorobenzene, dimethylphthalate, ethylbenzene, naphthalene, *p*-nitrotoluene, toluene, *p*-xylene, bromoform, chloroform, 1,2-dichloroethane, tetrachloroethylene, 1,1,1-trichloroethane).
[M]Dodecyl linear alkylbenzene sulphonate, dodecyl linear alcohol ethoxylate, dodecyltrimethylammonium chloride.
[N]Pentachlorophenol, endosulphan, DDT.
[O]Polychlorinated biphenyls.
[P]Pentachlorophenol.

rhizospheres of corn, beans, and cotton, treated with temik [2-methyl-2(methylthio) propionaldehyde O-(methylcarbamoyl) oxime]. More recently, Sato [22] found an 8-fold increase in heterotrophic bacteria in rice rhizosphere after benthiocarb (S-*p*-chlorobenzyl diethylthiocarbamate) addition as compared with plate counts before addition.

Seibert et al. [19] observed an overall increase in atrazine (2-chloro-4-ethylamino-6-isopropylamino-S-triazine) degradation by rhizosphere microorganisms in the presence of decomposing roots. Also, the concentration of unchanged atrazine was lower in the rhizosphere soil, and the concentration of hydroxyatrazine and two other hydroxylated metabolites were 3-fold higher than concentrations outside the rhizosphere. Studies on $^{14}CO_2$ evolution from ^{14}C-parathion (O,O-diethyl-O-*p*-nitrophenyl phosphorothioate) in rice rhizospheres indicated similar results [20]. Only 5.5% of the ^{14}C-parathion was evolved as $^{14}CO_2$ in unplanted soils while 9.2% was evolved from rice rhizospheres under non-flooded conditions. The rice variety used in this experiment grew better in flooded soil, thus when flooded conditions prevailed, 22.6% of the radiocarbon was evolved as $^{14}CO_2$. Reddy and Sethunathan [20} argued that the close proximity of the aerobic-anaerobic interface in rice rhizosphere under flooded conditions favoured the ring cleavage of parathion.

Parathion and diazinon [O,O-diethyl O-(2-isopropyl-4-methyl-6-pyrimidinyl) phosphorothioate] appear to be degraded in soil initially by cometabolic attack [23], a process that requires the presence of a growth substrate other than the compound being degraded. As indicated earlier, root exudates provide microorganisms with a wide range of organic substrates for use in growth and reproduction, and as energy sources. These factors lead Hsu and Bartha [1] to hypothesize that the rhizosphere would be especially favourable for cometabolic transformations of pesticides. Using radiolabelled diazinon and parathion, they were able to show accelerated mineralization of these compounds in bean rhizospheres. Beans were chosen because of their reported inability to metabolize diazinon [23]. Approximately 18% of the parathion and 13% of the diazinon were mineralized in the bean rhizospheres compared with 7.8% and 5.0% in the root-free soil for parathion and diazinon, respectively. Similar results with diazinon were previously found by Gunner and coworkers [23], although they did not observe an increase in microbial biomass in the rhizosphere after diazinon application. Rather, the diazinon (and probably the root exudates) exerted a selective effect, which resulted in the enrichment of a particular isolate capable of diazinon metabolism.

Lappin et al. [21] found that a microbial community isolated from wheat roots was capable of growth on the herbicide, mecoprop [2-(2-methyl 4-chlorophenoxy) propionic acid] as the sole carbon source. The authors isolated five species, none of which was capable of growth on mecoprop individually. This microbial community was also shown to degrade 2,4-D (2,4-dichlorophenoxyacetic acid) and MCPA (2-methyl-4-chlorophenoxyacetic acid).

Aquatic Systems

The increased degradative capability of rhizosphere microbial communities is not limited to terrestrial plants. Federle and Schwab [24] and Federle and Ventullo [25] have made similar observations of the increased microbial degradation of surfactants in the rhizospheres of aquatic plants. Mineralization of linear alkylbenzene sulphonate (LAS) and linear alcohol ethoxylate (LAE) was more rapid in the rhizosphere of cattails (*Typha latifolia*) than in root-free sediments [24]. Surprisingly, the source of the cattails (plants were obtained from a pristine pond and a pond receiving laundromat wastewater) had no significant influence on the rates of LAS and LAE degradation. Additionally, microbial communities associated with duckweed (*Lemna minor*) readily mineralized LAE, but not LAS. Similar results on microbial degradation of LAS and LAE by the microbiota of submerged plant detritus were obtained by Federle and Ventullo [25].

To assess the possible additional benefits of microbial filters (*biofilms*), containing aquatic vegetation in biotransformation of hazardous organic compounds, Wolverton and McDonald-McCaleb [26] compared removal of a variety of EPA priority pollutants in nonvegetated filters and filters planted with the common reed, *Phragmites communis*. In 24 hours, the nonvegetated microbial filter removed 61-99% and 39-81% of the aromatics (benzene, biphenyl, chlorobenzene, dimethylphthalate, ethylbenzene, naphthalene, *p*-nitrotoluene, toluene, *p*-xylene) and aliphatics (bromoform, chloroform, 1,2-dichloroethane, tetrachloroethylene, 1,1,1-trichloroethane) tested, respectively. The vegetated filter system increased the removal of both the aromatics (81->99%) and aliphatics (49-93%). Although sterile controls for elucidating volatilization rates as well as possible abiotic degradation and adsorption mechanisms were not performed, losses due to volatilization appeared to be minor in these systems.

Industrial Chemicals

Although most of the studies described previously have dealt with agricultural chemicals, they provided evidence for the accelerated microbial degradation of organic compounds in the rhizosphere and also gave an incentive for exploring the possibility of similar results with other hazardous organic compounds. Two recent studies have detailed the accelerated disappearance of nonagricultural chemicals in the root zone; a series of polycyclic aromatic hydrocarbons (PAHs) in prairie grass rhizospheres [27], and the increased degradation of oil residues by microorganisms isolated from oil-polluted rice rhizospheres [28].

Rasolomanana and Balandreau [28] appear to be the first to show enhanced microbial degradation of nonagricultural chemicals by rhizosphere microorganisms. This serendipitous discovery came during studies of improved growth of rice in soil to which oil residues had been applied. The authors hypothesized that the increased growth was brought about by the initial removal of the oil residues from the rhizosphere by microorganisms, utilizing the oil, and isolated a *Bacillus* sp. with the ability to grow on the oil residues, but only in the presence of rice root exudates.

The use of eight prairie grasses for stimulating microbial degradation of four PAHs, benz[a]anthracene, chrysene, benzo[a]pyrene, and dibenz[a,h]anthracene, in soil columns was evaluated by Aprill and Sims [27]. Based on residue analysis of the soil columns, PAH disappearance was consistently greater in the vegetated columns compared with nonvegetated controls. Although sterile soil controls were not included in the experiments, the authors speculated that microbial degradation may account for the increased disappearance of the PAHs in the vegetated columns. However, the rhizosphere effect may have been obfuscated by addition of manure to all soil columns during PAH addition. Root uptake of the PAHs may have also obscured the contribution of microorganisms to the disappearance of PAH from the soil columns. Nonetheless, this research does provide evidence for the accelerated disappearance of hazardous organic compounds in the rhizosphere, and also presents a germane discussion on plant and root biology in relation to stimulating soil microbial activity and enhancing microbial degradation of organic compounds in the root zone.

In order to determine the potential role of rhizosphere microorganisms in biodegradation of trichloroethylene (TCE), we tested rhizosphere soils and nonvegetated soils from a former solvent disposal site [29]. Initial experiments with soil slurries monitored disappearance of TCE from the headspace, utilizing gas chromatography techniques [30]. These initial experiments provided the incentive for more rigorous tests with soil samples and soil-plant systems using ^{14}C-TCE. Mineralization of ^{14}C-TCE to $^{14}CO_2$ was monitored in specially designed Erlenmeyer flasks incubated under vegetated, nonvegetated, and sterile conditions (Anderson and Walton, in preparation). Vegetation tested included four plant species from the contaminated site (*Lespedeza cuneata*, *Solidago* sp., *Paspalum notatum* and *Pinus taeda*), as well as soybean, *Glycine max* germinated from commercially available seeds. In soils containing *L. cuneata, P. taeda*, and *G. max*, the levels of $^{14}CO_2$ produced were significantly greater ($p \leq 0.05$, t-test) than $^{14}CO_2$ production in both nonvegetated and sterile control soils. Radiolabelled CO_2 production in soil containing *Solidago* sp. and *P. notatum* was elevated, however, there was no statistically significant difference ($p \leq 0.05$, t-test) from $^{14}CO_2$ produced in the respective nonvegetated soils.

PHOSPHOLIPID FATTY ACID ANALYSIS

The observed variations in the TCE biodegradation activity of the different rhizosphere microbial communities provided the impetus for further exploration of their composition, activity, and nutritional status. Specific biochemical methods have been developed to assay for indicators of microorganisms in soil and sediment samples. Membrane phospholipids are present in all cells, have a rapid turnover, and are easily extracted from environmental samples and quantified, making them ideal for determining viable microbial biomass [31]. Essentially identical estimates of microbial biomass were found by the membrane phospholipid and direct count methods [32]. The phospholipid fatty acid (PLFA) assay has been used to describe microbial communities from such environmental samples as plant rhizospheres [33] and creosote-contaminated soils and sediments [34]. Detection of specific phospholipid fatty acids can indicate the presence and abundance of certain groups of microorganisms. For example, Nichols and coworkers [35] found relatively high proportions of 18-carbon fatty acids relative to 16-carbon fatty acids in a natural gas-exposed soil column capable of TCE degradation. This indicated the presence of a large population of type II methanotrophic bacteria. Marker fatty acids have also been discovered for, among others, sulphate-reducing bacteria, aerobes, anaerobes, and actinomycetes [31].

Phospholipid fatty acid assays can also indicate metabolic changes in a microbial community. During nutrient deprivation and other environmental stresses, fatty acid ratios can shift. Guckert et al. [36] found increases in the *trans:cis* ratio of monoenoic 16-carbon fatty acids in *Vibrio cholerae* during nutrient starvation. The protocol used to extract the lipids from environmental samples simultaneously extracts other biochemical indicators of nutritional status in microbial communities, for example, poly-β-hydroxyalkanoates [31].

Incorporation of ^{14}C-acetate into microbial lipids is a simple, yet useful technique for measuring heterotrophic microbial activity. The rate of acetate incorporation into microbial phospholipids has been shown to be an accurate and sensitive measure of growth in sediment samples [37]. The technique has been used to measure activity in sewage sludge [38], marine sediments [39], antarctic rock microbiota [40], and soils [41]. Acetate incorporation has also been used to assess the effects of toxicants, both inorganic and organic, on microbial metabolic activity [42].

Methods

The extraction procedure, a modification of Bligh and Dyer [43], was performed on lyophilized soil samples from the rhizosphere as well as from nonvegetated areas. Extracts were fractionated on silicic acid columns into neutral lipids, glycolipids, and phospholipids. Phospholipid fatty acids were derivatized and quantified as described previously [29,44]. Total picomoles of phospholipid was converted to active microbial biomass. A cluster analysis of the phospholipid fatty acid profiles of different rhizosphere and nonvegetated soil samples was performed to help in analyzing the different sample types. The glycolipid fraction from the total lipid extraction, described above, was used to determine the endogenous lipid storage products poly-β-hydroxyalkanoates (PHAs) using the techniques of Findlay and White [45].

In order to determine heterotrophic microbial activity in rhizosphere and nonvegetated soils, incorporation of ^{14}C-labelled acetate into total lipids was measured. Soil samples (2 g) from rhizosphere and nonvegetated areas at the MCB were placed in 15-ml polypropylene tubes together with 0.5 ml of sterile, distilled, deionized water and incubated in the dark. Samples were extracted and fractionated as described above. Liquid scintillation spectrometry was used to quantify the radioactivity in total lipids as well as the radioactivity in the 3 lipid fractions after separation.

Results

Microbial biomass estimates, calculated from PLFA analysis, were fairly consistent with the estimates based on CO_2 efflux [30], and also illustrated the increased microbial biomass associated with the rhizosphere soils. Microbial biomass estimates (mean ± standard deviation) of rhizosphere soils of *Lespedeza cuneata, Paspalum notatum, Pinus taeda* and *Solidago* sp. were 1449 ± 423 μg/g, 3624 ± 522 μg/g, 1252 ± 175 μg/g, and 5825 ± 1176 μg/g, respectively. Microbial biomass estimates in the nonvegetated soil were 680 ± 423 μg/g. Although biomass estimates were useful for describing the microbiological properties of the study soils, they were not good predictors of TCE degradation. Rhizosphere soils from *P. notatum* and *Solidago* sp. had comparatively high levels of microbial biomass, yet both rhizosphere types did not degrade TCE as well as other samples with less microbial biomass.

In addition to microbial biomass, PLFA analysis was also used to determine the microbial community structure of the soil samples from the contaminated site. Although taxonomic characterization of the microorganisms from the soil samples was not undertaken, it was possible to compare qualitative differences between the groups of microorganisms present in the different samples. A dendogram of the cluster analysis results from the PLFA analyses (Figure 1) illustrates the primary clustering of nonvegetated soil samples with *Lespedeza cuneata* rhizosphere soil samples, and *Solidago* sp. rhizosphere soil with *Pinus taeda* rhizosphere soil and *Paspalum notatum* rhizosphere soil samples. Secondary clustering occurred between *Pinus taeda* soil samples and *Paspalum notatum* soil samples. A principal components analysis indicated that the fatty acid which most determined the primary

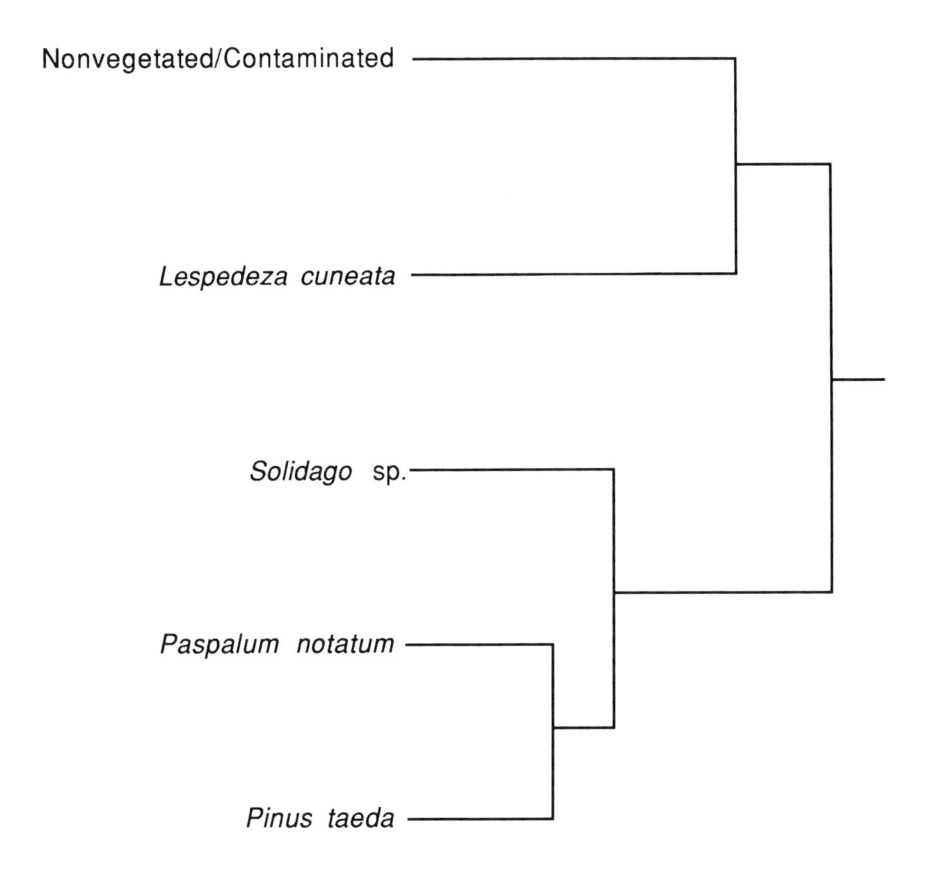

Figure 1. Dendogram of cluster analysis results from phospholipid fatty acid profiles of rhizosphere and nonvegetated soils from the contaminated site. Comparisons of qualitative differences between the groups of microorganisms present in the different samples illustrated the primary clustering of nonvegetated soil samples with *Lespedeza cuneata* rhizosphere soil samples, and *Solidago* sp. rhizosphere soil with *Pinus taeda* rhizosphere soil and *Paspalum notatum* rhizosphere soil samples. Secondary clustering occurred between *Pinus taeda* soil samples and *Paspalum notatum* soil samples.

clustering (separation of the nonvegetated soil and *L. cuneata* rhizosphere soil from the others) was cy19:0. The secondary clustering, which separates *P. notatum* and *P. taeda* from *Solidago* sp., was explained by the greater presence of 10Me16:0 and 10Me18:0 in samples of *P. taeda* and *P. notatum* rhizosphere soil. These fatty acids (10Me16:0 and 10Me18:0) are characteristic of actinomycetes [31].

Overall, the PLFA results were similar in both rhizosphere and nonvegetated soils (Table 2). The most abundant PLFA in both rhizosphere and nonvegetated soils was palmitate (16:0), a common saturated fatty acid. The strong presence of monounsaturated (16:1ω7c and 18:1ω7c), and branched phospholipid fatty acids (i15:0 and i16:0 for example) indicated that the communities of both sample types (rhizosphere and nonvegetated) were composed of Gram-negative and Gram-positive microorganisms in approximately equal abundance. Ratios of *cis* fatty acids to *trans* fatty acids (specifically 16:1ω7c to 16:1ω7t and 18:1ω7c to 18:1ω7t), which can be indicative of stress or unbalanced growth [31], were not significantly different among the rhizosphere and nonvegetated soils. Cyclopropyl fatty acids such as cy17:0 and cy19:0, which can be indicative of unbalanced growth and/or anaerobic microorganisms [46], were present in all sample types. These results are probably best explained by the soil conditions at the contaminated site. Namely that the soil is low in organic carbon, and fluctuates between aerobic and anaerobic conditions. Recent research has also shown that solvent-degrading microorganisms convert *cis* unsaturated fatty acids in their membranes to *trans* fatty acids to avoid substrate toxicity [47]. Previous analyses of a TCE-degradading soil column, enriched with natural gas [48], revealed the strong presence of Type II methanotrophs, as indicated by the fatty acid 18:1ω8c [35]. Because methanotrophs are likely to be found in zones that fluctuate between aerobic and anaerobic conditions, such as surface soils that periodically flood and drain and subsurface soils at the capillary fringe, these bacteria are likely to be present in soils from the site. However, the characteristic fatty acid for Type II methanotrophs (18:1ω8c) was not detected in these soils. It is possible that because enrichments were not performed, the concentration of methanotrophs was below the detection limits for PLFA analysis. Nonetheless, these microorganisms are undoubtedly present, based on their ubiquity in nature and the favorable conditions for their proliferation at the study site.

The microbial metabolic activity of soil samples was determined by measuring the incorporation of ^{14}C-acetate into cellular lipids. The metabolic activity (mean incorporation rate ± standard deviation) of the four rhizosphere samples was significantly greater ($p \leq 0.05$, t-test) than metabolic activity in the nonvegetated soil (1407 ± 10, 905 ± 35, 599 ± 33, 1727 ± 180 pMoles/g soil/h for *L. cuneata*, *P. taeda, P. notatum,* and *Solidago* sp., respectively vs. 445 ± 86 pMoles/g soil/h for the nonvegetated soil). In addition, there were significant differences in ^{14}C-acetate incorporation rates among the four rhizosphere samples ($p \leq 0.05$, t-test) from each other, with *Solidago* sp. rhizosphere soil having the highest rate. The greater metabolic activity of rhizosphere samples is consistent with the biomass estimates reported earlier in this study and also consistent with the findings of others that microbial

Table 2

Phospholipid fatty acid (PLFA) profiles of soil samples collected from the Miscellaneous Chemicals Basin at the Savannah River Site

PLFA[A]	Mole%±SD[B]				
	Nonvegetated	*L. cuneata*	*Solidago* sp.	*P. notatum*	*P. taeda*
i14:0	0.7±0.6	0.7±0.1	0.9±0.1	0.9±0.1	1.0±0.2
14:0	1.4±0.9	1.4±0.2	1.4±0.1	1.3±0.1	1.8±0.3
i15:0	7.1±2.1	8.8±0.4	10.9±0.4	8.6±0.4	10.3±0.2
a15:0	2.7±0.9	3.8±0.3	4.5±0.1	3.8±0.1	5.3±0.4
15:0	1.0±0.7	0.9±0.2	0.9±0.1	1.1±0.0	1.2±0.1
i16:0	4.1±1.0	2.2±0.1	2.4±0.1	2.7±0.2	2.7±0.1
16:1ω9c	1.1±0.7	1.3±0.1	1.7±0.1	1.9±0.2	2.2±0.2
16:1ω7c	5.3±0.5	6.2±0.7	7.6±0.6	5.6±0.1	6.2±0.7
16:1ω7t	0.2±0.3	0.3±0.1	0.3±0.0	0.3±0.0	0.1±0.1
16:1ω5c	2.8±0.8	3.8±0.3	4.9±0.2	4.1±0.1	4.1±0.2
16:0	13.0±2.3	14.9±1.2	14.9±0.4	14.1±0.4	14.8±0.4
i17:1	2.1±0.9	2.3±0.1	2.7±0.1	2.6±0.2	2.7±0.6
10Me16:0	3.8±1.0	4.6±0.8	3.5±0.2	5.6±0.4	5.4±0.4
i17:0	2.4±0.6	1.6±0.1	1.3±0.0	1.6±0.1	1.7±0.1
a17:0	2.4±0.9	1.6±0.1	1.6±0.1	1.7±0.1	1.6±0.3
cy17:0	2.8±1.1	3.8±0.3	3.2±0.2	3.5±0.2	3.5±0.2
18:2ω6	4.5±2.4	3.9±1.1	5.4±1.1	6.6±1.3	5.3±1.0
18:3ω3	2.4±1.5	0.0±0.0	0.9±0.1	1.0±0.2	0.6±0.4
18:1ω9c	5.9±1.8	4.0±0.5	3.9±0.3	3.1±0.1	3.2±0.0
18:1ω7c	6.7±1.2	3.5±0.8	8.4±0.6	7.9±0.1	6.4±0.4
18:1ω5c	0.7±0.6	1.0±0.2	0.7±0.1	0.9±0.0	0.7±0.1
18:0	2.7±0.7	2.9±0.2	2.1±0.6	1.7±0.0	1.6±0.1

Table 2 continued

PLFA[A]	Mole%±SD[B]				
	Nonvegetated	*L. cuneata*	*Solidago* sp.	*P. notatum*	*P. taeda*
br19:1	0.4±0.4	1.3±0.5	0.8±0.0	0.7±0.0	0.7±0.1
10Me18:0	1.8±0.6	0.8±0.1	0.7±0.0	1.1±0.0	1.2±0.4
cy19:0	8.0±2.2	9.7±1.2	4.2±0.2	4.6±0.2	5.2±0.5
20:0	2.5±1.6	2.0±0.2	0.7±0.1	0.5±0.4	0.7±0.5

[A]The shorthand nomenclature used to describe fatty acids is as follows: The number before the colon indicates the number of carbon atoms in the fatty acids. The number after the colon indicates the number of double bonds in the fatty acid chain. The position of the initial double bond is indicated by the number of carbon atoms from the methyl, or ω, end of the molecule. The configuration of the double bond is shown by the 'c' for *cis* and 't' for *trans*. Because almost all unsaturations are *cis*, 'c' is often omitted. Methyl branching can be indicated as iso ('i'; the second carbon from the methyl end), anteiso ('a'; the third carbon) or 'br' (branch) if the position is unknown. When a branch is known but not in the 'i' or 'a' position, it is indicated by the position from the carboxyl end followed by 'Me' before the carbon chain length. Cyclopropane fatty acids are indicated as 'cy' (Adopted from [31]).

[B]For the following fatty acids, mole % values were < 1 for all sample types: a13:0, i15:1, a15:1, 15:1, 16:3, 16:1, 16:1ω7t, 16:1ω13t, 17:1ω6, 17:1, 17:0, 18:3ω6, 18:1ω7t, 18:1, 20:4ω6, 20:5ω3, 20:3ω6, 20:2ω3, 20:1ω9c, 20:0, 23:0, 24:0.

biomass and activity is greater in the rhizosphere [6]. In addition, acetate incorporation appeared to be a good predictor for the ability to degrade TCE in whole-plant experiments. Rhizosphere samples from *L. cuneata* and *P. taeda* had relatively high rates of acetate incorporation and also had the highest TCE degradation rates, while samples of *P. notatum* rhizosphere and nonvegetated soils had lower acetate incorporation rates and lower TCE degradation rates.

Differences in the incorporation of ^{14}C-acetate into the three lipid classes between rhizosphere and nonvegetated soil samples may indicate unbalanced growth in the nonvegetated samples. Soil from nonvegetated areas incorporated ^{14}C-acetate predominantly into neutral and glycolipids (storage lipids) during the first three hours of incubation, possibly indicating unbalanced growth of the microbial community in this soil. Only 20% of the radioactivity was found in the phospholipid (membranes) fraction after 3 hours. In contrast, soils from *L. cuneata, P. taeda* and *Solidago* sp. incorporated the acetate into membrane lipids (phospholipids). Radioactivity in the phospholipid fraction for *P. taeda* and *Solidago* sp. rhizosphere soil was greater than 50% of the total radioactivity in all lipid fractions after 1 hour, and greater than 80% for *L. cuneata* rhizosphere soil after 3 hours. Soils collected from *P. notatum* rhizosphere soil incorporated most of the ^{14}C-acetate into the glycolipid fraction after 1 hour, similar to the results of the nonvegetated soil.

Consistently greater amounts of PHA were detected in rhizosphere soils compared with the nonvegetated soils, although the amounts were significantly greater ($p \leq 0.05$, t-test) only for soils from *Paspalum notatum* rhizosphere. All samples tested contained significant amounts of PHA, however, PHA levels varied among sample types. Mass spectroscopy revealed that almost all of the constituent beta-hydroxy acids from the PHAs was betahydroxybutyrate (PHB) although very small amounts of betahydroxy hexanoate were also detected. Samples of *L. cuneata* and *P. taeda* rhizosphere soils contained elevated levels of PHA compared to PLFA, and were also very capable of degrading TCE in whole-plant experiments. Correlations between TCE degradation rate and PHA production have been observed previously in bioreactors (Ringleberg and Phelps, Center for Environmental Biotechnology, The University of Tennessee, personal communication). Soil samples from *Solidago* sp. rhizosphere and nonvegetated areas at the site had the lowest levels of PHA and also degraded TCE more slowly in whole-plant experiments than rhizosphere soils of *L. cuneata* and *P. taeda*. Only soils from the rhizosphere of *P. notatum* failed to follow the correlation between degradation of TCE and production of PHA. Soils from *P. notatum* rhizosphere had comparatively high levels of PHA, but did not appear to degrade TCE as well as soils from the rhizosphere of *L. cuneata* and *P. taeda*.

APPLICATIONS

The highly versatile metabolic capabilities of fungi and bacteria can be applied to reclaim polluted ecosystems and minimize the potential adverse effects of hazardous chemicals released to the environment. However, a sufficient consortia of microorganisms, capable of degrading the contaminant(s), must be present, and environmental conditions conducive to degradation must be maintained. Environmental conditions onsite may significantly hinder microbial degradation of toxicants. In such cases, microbial degradation may be enhanced by altering conditions through nutrient additions, irrigation, or other interventions. The addition of external carbon sources may be especially important in such cases where the contaminant is degraded cometabolically.

The use of vegetation to enhance the microbial degradation of surface and near-surface soils contaminated with hazardous wastes, such as chlorinated solvents, polycyclic aromatic hydrocarbons (PAHs), or pesticide wastes under field conditions has not been demonstrated. Vegetation may prove to be an important variable, affecting microbial degradation of unwanted chemicals and provide a *cost-effective* remediation strategy for soils containing these compounds [49]. There is growing evidence that rhizosphere treatment systems may be used successfully in the field. Establishing or selectively cultivating vegetation on a contaminated site is a relatively simple site management technique, which could have substantial ramifications for reducing the adverse impact of contaminants on ecosystems. Continued exploration of critical environmental variables, affecting the soil-plant-microbe-chemical relationship, will help to identify situations, in which bioremediation using vegetation may be appropriate. The reviewed literature on rhizosphere microbiology, accelerated microbial degradation of agricultural chemicals in the root zone, and recent research on similar observations with hazardous organic compounds provides additional incentives to explore bioremediation of contaminated soils using vegetation.

CONCLUSIONS

The variety of plants and chemicals studied for evidence of microbial degradation in the rhizosphere strongly suggests that a diverse and synergistic microbial community, rather than a single species, is responsible for biotransformation of toxicants in the rhizosphere. Participation of a microbial community is implicated by (i) the extreme diversity and complexity of toxicants degraded, and (ii) the knowledge that many of these compounds are completely degraded only in the presence of interacting microbial populations (consortia). Moreover, data presented, herein, from phospholipid fatty acid analysis, ^{14}C-

acetate incorporation, and PHA analysis of microorganisms from the rhizosphere are consistent with participation of a microbial community in TCE degradation.

Moreover, the information presented, herein, illustrates the potential for rhizosphere microbial communities to remediate soil systems through biotransformation of hazardous organic compounds in the root zone. Future research in this area should include investigations of the possible role of mycorrhizae in degradation of toxicants and characterization of the microorganisms associated with different plant species and different histories of toxicant exposure. Closer examination of root exudation as a response to chemical challenge in the rhizosphere will also shed light on the complex relationship between plants, microorganisms, soil, and hazardous chemicals in the root zone.

ACKNOWLEDGMENTS

The authors thank A. M. Hoylman and David Ringleberg, The University of Tennessee, Knoxville, TN; A. V. Palumbo, T. J. Phelps, and N. T. Edwards, Oak Ridge National Laboratory, Oak Ridge, TN; T. C. Hazen of the Savannah River Site, Aiken, SC; and E. L. Kruger, Iowa State University, Ames, IA for helpful contributions to this work.

This research was sponsored by the Office of Technology Development, U.S. Department of Energy. Oak Ridge National Laboratory is managed by Martin Marietta Energy Systems, Inc., under contract DE-AC05-84OR21400 with the U.S. Department of Energy. Environmental Sciences Division Publication No. 4035.

REFERENCES

1 Hsu TS, Bartha R. Appl Environ Microbiol 1979; 37: 36-41.
2 Curl EA, Truelove B. The Rhizosphere. Berlin, Heidelberg: Springer-Verlag, 1986; 288 pp.
3 Campbell R. Plant Microbiology. Baltimore: Edward Arnold, 1985; 191 pp.
4 Rovira AD, Foster RC, Martin JK. In: Harley JL, Russell RS, eds. The Soil-Root Interface. New York: Academic Press, 1979; 1-4.
5 Katznelson H. Soil Sci 1946; 62: 343-354.
6 Atlas RM, Bartha R. Microbial Ecology: Fundamentals and Applications. Menlo Park, CA: Benjamin/Cummings, 1993; pp.563.
7 Sandmann ERIC, Loos MA. Chemosphere 1984; 13: 1073-1084.
8 Abdel-Nasser M, Makawi AA, Abdel-Moneim AA. Egypt J Microbiol 1979; 14: 37-44.
9 Abueva AA, Bagaev VB. Izvestiya Timiiryazevsk Skh Akad 1975; 2: 127-130.

10 Gavrilova EA, Kruglov YV, Garankina NG. Tr Vses Nauchno-Issled Instit Skh Mikrobiologii 1983; 52: 67-70.
11 Hiltner L. Arb Dtsch Landwirt Ges 1904; 98: 59-78. In: Curl EA, Truelove B. The Rhizosphere. Berlin, Heidelberg: Springer-Verlag, 1986; pp.288.
12 Hale MG, Foy CL, Shay FJ. Adv Agron 1971; 23: 89-109.
13 Hale MG, Moore LD. Adv Agron 1979; 31: 93-124.
14 Kaiser P, Reber H. Meded Fac Landouwwet Rijksuniv Gent 1970; 35: 689-705.
15 Ashton FM, Crafts AS. The Mode of Action of Herbicides. New York: John Wiley and Sons, 1981; pp.525.
16 Herring R, Bering CL. Bull Environ Contam Toxicol 1988; 40: 626-632.
17 Anderson TA, Guthrie EA, Walton BT. Environ Sci Technol 1993; 27: 2630-2636.
18 Shimp JF, Tracy JC, Davis LC, Lee E, Huang W, Erickson LE. Crit Rev Environ Sci Technol 1993; 23: 41-77.
19 Seibert K, Fuehr F, Cheng HH. Theory and Practical Use of Soil Applied Herbicides Symposium. European Weed Resource Society. Paris, France. 1981; 137-146.
20 Reddy BR, Sethunathan N. Appl Environ Microbiol 1983; 45: 826-829.
21 Lappin HM, Greaves MP, Slater JH. Appl Environ Microbiol 1985; 49: 429-433.
22 Sato K. In: Vancura V, Kunc F, eds. Interrelationships between Microorganisms and Plants in Soil. New York: Elsevier, 1989; 335-342.
23 Gunner HB, Zuckerman BM, Walker RW, Miller CW, Deubert KH, Longley RE. Plant Soil 1966; 25: 249-264.
24 Federle TW, Schwab BS. Appl Environ Microbiol 1989; 55: 2092-2094.
25 Federle TW, Ventullo RM. Appl Environ Microbiol 1990; 56: 333-339.
26 Wolverton BC, McDonald-McCaleb RC. J Miss Acad Sci 1986; 31: 79-89.
27 Aprill W, Sims RC. Chemosphere 1990; 20: 253-265.
28 Rasolomanana JL, Balandreau J. Rev Ecol Biol Sol 1987; 24: 443-457.
29 Anderson TA, Walton BT. Comparative plant uptake and microbial degradation of trichloroethylene in the rhizospheres of five plant species: Implications for bioremediation of contaminated surface soils. ORNL/TM-12017. Oak Ridge, TN. 1992; pp.186.
30 Walton BT, Anderson TA. Appl Environ Microbiol 1990; 56: 1012-1016.

31 Vestal JR, White DC. Bio Science. 1989; 39: 535-541.
32 Balkwill DL, Leach FR, Wilson JT, McNabb JF, White DC. Microbial Ecol 1988; 16: 73-84.
33 Tunlid A, Baird BH, Trexler MB, Olsson S, Findlay RH, Odham G, White DC. Can J Microbiol 1985; 31: 1113-1119.
34 Smith GS, Nickels JS, Kerger BD, Davis JD, Collins SP, Wilson JT, McNabb JF, White DC. Can J Microbiol 1986; 32: 104-111.
35 Nichols PD, Henson JM, Antworth CP, Parsons J, Wilson JT, White DC. Environ Toxicol Chem 1987; 6: 89-97.
36 Guckert JB, Hood MA, White DC. Appl Environ Microbiol 1986; 52: 794-801.
37 Moriarty DJW, White DC, Wassenberg TJ. J Microbiol Meth 1985; 3: 321-330.
38 McKinley VL, Vestal JR. Appl Environ Microbiol 1984; 47: 933-941.
39 White DC, Bobbie RJ, Morrison SJ, Oosterhof DK, Taylor CW, Meeter DA. Limnol Oceanogr 1977; 22: 1089-1099.
40 Vestal JR. Appl Environ Microbiol 1988; 54: 960-965.
41 White DC, Davis WM, Nickels JS, King JD, Bobbie RJ. Oecologia 1979; 40: 51-62.
42 Barnhart CL, Vestal JR. Appl Environ Microbiol 1983; 46: 970-977.
43 Bligh EG, Dyer WJ. Can J Biochem Phys 1959; 37: 911-917.
44 Findlay RH, White DC. J Microbiol Meth 1987; 6: 113-120.
45 Findlay RH, White DC. Appl Environ Microbiol 1983; 45: 71-78.
46 Guckert JB, Antworth CP, Nichols PD, White DC. FEMS Microbiol Ecol 1985; 31: 147-158.
47 Heipieper H-J, Diefenbach R, Keweloh H. Appl Environ Microbiol 1992; 58: 1847-1852.
48 Wilson JT, Wilson BH. Appl Environ Microbiol 1985; 49: 242-243.
49 Walton BT, Anderson TA. Curr Opin Biotechnol 1992; 3: 267-270.
50 Ampova G. Bulgarski Tyutyun 1971; 16: 39-43.
51 Gudin C. Proc International Symposium on Ground Water Pollution by Oil Hydrocarbons. Prague, Czechoslovakia, 1978; 411-417.
52 Knaebel DB, Vestal JR. Can J Microbiol 1992; 38: 643-653.
53 Katayama A, Matsumura F. Environ Toxicol Chem 1993; 12: 1059-1065.
54 Donnelly PK, Hegde RS, Fletcher JS. Chemosphere 1994; 28: 981-988.
55 Ferro AM, Sims RC, Bugbee B. J Environ Qual 1994; 23: 272-279.

Biotransformations: Microbial Degradation of Health Risk Compounds
Ved Pal Singh, editor
© 1995 Elsevier Science B.V. All rights reserved.

Microbial degradation of styrene

S. Hartmans

Division of Industrial Microbiology, Department of Food Science, Wageningen Agricultural University, P.O. Box 8129, 6700 EV Wageningen, The Netherlands

INTRODUCTION

Styrene is produced in large quantities by the chemical industry, mainly as a starting material for synthetic polymers, such as polystyrene and styrene-butadiene rubber. It is also used as a solvent in the polymer processing industry. Most of the styrene lost to the environment is released in a gaseous form, but large quantities also enter waters and soils [1]. The solubility of styrene in water is quite low at 320 mg/l (25°C). This is also reflected by the relatively high air/water partition coefficient of 0.13 at 25°C [2].

Styrene is presumably also formed in the environment from natural precursors, such as cinnamic acid. With pure cultures, the decarboxylation of cinnamic acid has been demonstrated with numerous microorganisms, and recently a *Penicillium* species was shown to synthesize styrene as a secondary metabolite [3]. An obvious pathway for styrene formation would be the transformation of phenylalanine to cinnamic acid, followed by decarboxylation to styrene.

Mammalian metabolism of styrene has been studied very extensively in view of the intensive industrial use of styrene and its possible toxic and carcinogenic properties [4]. The first step in the major pathway of mammalian metabolism is the oxidation of styrene to styrene-7,8-oxide (phenyl oxirane) by hepatic microsomal cytochrome P-450 monooxygenases. This epoxide is either hydrolyzed by microsomal epoxide hydrolase to phenyl-1,2-ethanediol, which is further oxidized to mandelic acid [5], or transformed into glutathione conjugates by glutathione-S-transferases [6]. Other (minor) products formed by the microsomal P-450 monooxygenases include styrene-3,4-oxide, which chemically rearranges to 4-vinylphenol and phenylacetaldehyde [4].

AEROBIC DEGRADATION OF STYRENE IN NATURE

Styrene mineralization was recently studied in samples from various natural environments, using [U-*ring*^{14}C]-styrene [1]. Samples were incubated at 22°C in the dark with an initial styrene concentration of 1 mg/l or 1 mg/kg. In sewage, 20% mineralization was observed within 3 days. In Lima loam, soil with a pH of 7.2 and 7.5% organic matter, ground water and lake water, it took 5, 8 and 10 days, respectively, for 20% mineralization to take place. This indicates that microorganisms, capable of degrading styrene, are widespread. At concentrations between 10 and 1000 mg/kg, the CO_2 evolution curve was sigmoidal, indicating that growth of styrene-degrading microorganisms took place. Experiments to determine the sorption of styrene to soil particles showed that between 75 and 95% of the styrene added at concentrations varying between 1 and 100 mg/l to suspensions of 25% (w/v) was sorbed on the solid material within 30 h. Therefore, the concentration of styrene, that is readily available in such systems, is much lower than calculated on the basis of the amount added and the volume of the system. Despite the high degree of sorption, extensive mineralization took place [1]. Sorption could actually be advantageous in view of the toxic nature of styrene.

TOXICITY OF STYRENE

I have studied the toxicity of styrene and a number of other compounds by determining their effect on the growth rate of *Rhodococcus* S5 and *Pseudomonas putida* S12 (unpublished results). For both strains, the styrene concentration, that gave a 50% decrease in the growth rate with acetate as growth substrate, was about 180 mg/l. The styrene-utilizing black yeast, *Exophiala jeanselmei* was much more sensitive towards styrene. In liquid culture, the maximum concentration that supported growth was 37 mg/l [7].

The styrene-utilizing *Pseudomonas putida* S12 initially could not grow in the presence of supersaturating amounts of styrene (1% v/v). However, after incubation for 20 h under these conditions, the culture started to evolve CO_2 [8]. When cells grown under these conditions were used to inoculate medium with 1% or 0.008% (v/v) styrene, no lag phase was observed and the growth rate in both cultures was the same (0.6 h^{-1}). This adaptation to high styrene concentrations is probably a result of an alteration in the membrane composition.

A similar or the same pathway is probably operative in *Xanthobacter* 124X and strain S3 (Table 1).

These strains can also grow on toluene, suggesting that the oxidation of styrene to 2,3-dihydroxystyrene (3-vinylcatechol), followed by *meta*-cleavage of this compound, would be a realistic option. Further metabolism of the cleavage product would then yield, by analogy with the toluene degradative pathway, acrylic acid. Incubation of these strains with styrene and an inhibitor of acrylic acid metabolism could be a simple approach to confirm this.

Table 1
Characteristics of styrene-utilizing bacteria

Strain	124X	S1	S3	S5	S6	S8	S12	S14
Gram-stain	+	+	+	+	−	−	−	nd
Isolation method[a]	enr	enr	dir	dir	enr	enr	enr	enr
Doubling time on styrene (h)	19	4.6	5.6	3.4	4.4	2.5	1.0	7.5
Growth on styrene	+	+	+	+	+	+	+	+
Growth on styrene oxide	+	+	+	+	+	+	+	±
Growth on 2-phenylethanol	+	+	+	+	+	+	+	+
Growth on 1-phenylethanol	+	+	+	−	−	−	−	−
Growth on acetophenone	−	+	+	+	−	−	−	−
Growth on styrene glycol	−	±	+	−	−	−	+	−
Growth on ethylbenzene	+	−	+	−	−	−	−	−
Growth on toluene	+	−	+	+	−	−	−	−
Growth on benzene	−	−	+	−	−	−	−	−
SMO in cell-free extract[b]	−	+	−	+	+	+	+	+
SOI in cell-free extract	+	nd	nd	+	nd	nd	+	nd

[a]All strains, except for *Xanthobacter* 124X, were isolated at very low concentrations of styrene, either from enrichment cultures (enr) or by directly (dir) plating dilutions of soil or water samples on agar plates, that were incubated in desiccators, containing a low concentration of styrene vapour.
[b]Styrene monooxygenase activity with NADH or NADPH, except for strain S14 which only exhibited NADPH-dependent SMO activity.
nd = Not determined.

ANAEROBIC DEGRADATION OF STYRENE

There is one report, describing the degradation of styrene by anaerobic consortia [8]. In various enrichment cultures incubated under anaerobic conditions, styrene mineralization was observed. No methane formation was observed, implicating that styrene transformation was brought about by fermentative microorganisms. Indeed, the facultative anaerobic *Enterobacter cloacae* DG-6 was also shown to degrade styrene. Although small amounts of a number of reduced compounds were detected, such as 2-ethylphenol and methylcyclohexane, most of the carbon appeared to be recovered as carbon dioxide after 4 to 8 months of incubation, suggesting that, perhaps, some oxygen had entered the bottles. Although no details were given, 2-phenylethanol was proposed to be one of the major initial transformation products [9].

Pure cultures capable of growth on styrene

The first published attempt to isolate styrene-degrading microorganisms from more than 100 soil samples was unsuccessful [10]. Omori et al. [10] incubated soil samples of 0.1 gram in 10 ml of mineral salts medium containing 0.5 g/l of yeast extract, with 0.2 ml of an aromatic hydrocarbon. No microorganisms were isolated with styrene, α-methylstyrene or β-methylstyrene as carbon source. With isopropylbenzene, a number of strains were isolated. Two of the isopropylbenzene-utilizing *Pseudomonas* strains, that were further studied, also utilized α-methylstyrene and β-methylstyrene, but could not grow with styrene.

Subsequently, Sielicki et al. [11] described a styrene-utilizing mixed culture enriched from landfill soil in mineral salts medium with 1% (w/v) styrene. Thin-layer chromatography of ether extracts of the styrene-degrading cultures showed phenylacetic acid and a second compound, which was tentatively identified as 2-phenylethanol to be present.

Also in 1978, Andreoni et al. [12] briefly reported the isolation of pure cultures capable of degrading styrene. A *Pseudomonas* sp. degraded styrene via phenylacetic acid, and a *Nocardia* sp. degraded styrene via initial oxygenation of the aromatic nucleus, but no experimental details were reported [12]. A year later, Shirai and Hisatsuka [13] described the isolation of a number of styrene-degrading strains. For the initial enrichment cultures, soil columns were used, through which mineral salt medium with yeast extract (0.1 g/l) and saturated with styrene was percolated. In 10 of the 29 soil samples tested, significant growth was observed. From these enrichment cultures, a number of strains, all identified as *Pseudomonas* sp., were isolated, that could utilize styrene. Some of the isolates developed a rose-like fragrance, when grown on the above medium containing 0.1 ml of styrene per 5 ml, suggesting the formation of 2-phenylethanol. One of these strains, *Pseudomonas* 305-STR-1-4, was studied in more detail for the production of 2-phenylethanol from styrene [14]. A number of media and different inhibitors were tested to optimize the production of 2-phenylethanol. Low concentrations of pyrazole, an inhibitor of alcohol dehydrogenase, led to slightly higher yields of 2-phenylethanol. However, at a concentration of 1 mg pyrazole per ml, the 2-phenylethanol yield was lower, and styrene oxide accumulated up to concentrations of 0.34 mg/ml. Subsequent experiments with styrene-grown washed cell suspensions demonstrated that the degradation of styrene was oxygen-dependent, whereas the transformation of styrene oxide to 2-phenylethanol was not significantly affected by the presence or absence of oxygen. Based on these experiments it was proposed that, in this *Pseudomonas* strain, styrene is degraded via styrene oxide, and that this compound could be transformed to 2-phenylethanol by "styrene oxide reductase" [14].

The first study focusing on styrene metabolism concerned a *Pseudomonas fluorescens* strain [15]. During growth of this strain, under a styrene saturated atmosphere, the absorbance at 260 nm increased, suggesting the accumulation of intermediates or dead-end products. After extraction and separation on TLC, two compounds were identified with mass-spectroscopy: phenylacetic acid and 2-hydroxyphenylacetic acid. The latter could also be isolated from cultures growing on phenylacetic acid. Further evidence for the involvement of 2-hydroxyphenylacetic acid in the catabolism of styrene was sought in experiments with cell-free extracts. No phenylacetic acid or 2-hydroxyphenylacetic acid oxidative activity could be detected, which is not surprising, considering the instability of this type of oxygenase. Homogentisate (2,5-dihydroxyphenylacetate)-dependent oxygen uptake was, however, detected in extracts of cells grown on styrene, phenylacetate, or 2-hydroxyphenylacetate, but not in extracts from cells grown on 4-hydroxyphenylacetate. The latter extract did contain homoprotocatechuate (3,4-dihydroxyphenylacetate)-dependent oxygen-uptake activity. None of the extracts oxidized 2,3-dihydroxyphenylacetate. Based on these observations, the pathway shown in Figure 1 was proposed by Baggi et al. [15].

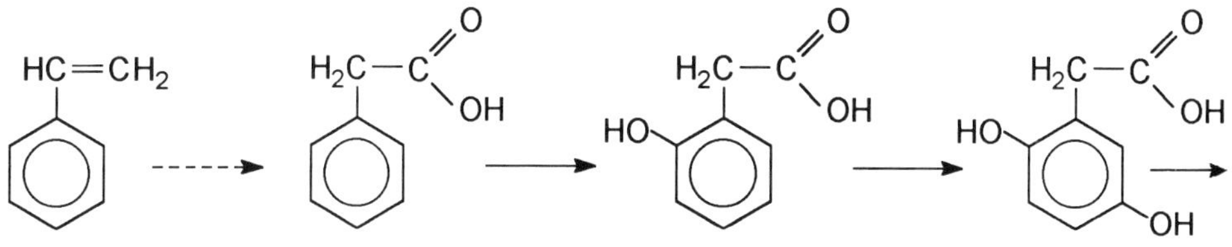

Figure 1. Styrene degradative pathway in *Pseudomonas fluorescens* [15].

Xanthobacter 124X, isolated from an enrichment culture with sewage as inoculum and styrene as carbon source [16], also grows on styrene oxide, 2-phenylethanol, 1-phenylethanol, and phenylacetate. Incubation of washed cells grown on styrene oxide or 2-phenylethanol with the growth substrate revealed a transient accumulation of phenylacetate [17]. No intermediates could be detected upon incubation of styrene-grown cells with styrene. During growth on styrene and 1-phenylethanol, a transient formation of a yellow colour was observed, but never during growth on styrene oxide or 2-phenylethanol. Based on these observations, it was concluded that in this strain the initial step in the metabolism of styrene was probably an oxidation of the aromatic ring, and not the formation of styrene oxide. Interestingly, strain 124X cells grown on styrene and styrene oxide contained a novel enzymatic activity, styrene oxide isomerase (SOI), that transformed styrene oxide into phenylacetaldehyde [17]. Styrene oxide isomerase (EC 5.3.99.7) was partially purified and appeared to have a very high substrate specificity [17]. Although SOI did not appear to play a role in the metabolism of styrene in *Xanthobacter* 124X, this enzymatic activity fits in very well in the pathway proposed by Baggi et al. [15] for *Pseudomonas fluorescens* (Figure 1).

Subsequently, we set out to isolate a number of styrene-utilizing microorganisms for further metabolic studies [18]. As we anticipated that styrene is toxic to many microorganisms at elevated concentrations, we set up enrichment cultures with a separate organic phase as a substrate reservoir, giving low concentrations of styrene in the aqueous phase. For example, with 50 ml of mineral salts medium and a tube with 25 μl of styrene in 5 ml of dibutyl phthalate in a closed flask of 300 ml, only 0.5% of the styrene is present in the aqueous phase, after equilibrium is reached. This corresponds with a concentration of 22 μM (2.3 mg/l) in the aqueous phase. The same method was also used to maintain a constant low concentration of styrene in desiccators for the isolation of styrene-utilizing strains directly from diluted soil suspensions spread on agar plates. Both methods resulted in the isolation of numerous strains. Fourteen bacterial isolates, that appeared to differ morphologically, were studied [18]. From enrichment cultures at pH 4.5, two fungi were also isolated, one of which was further studied by Cox et al. [7]. Seven of the newly isolated bacteria are briefly described in Table 1. Isolates, that appeared to be morphologically identical to strain S1, were present in almost all enrichment cultures. Isolates, that appeared to be similar to strain S8 or strain S12, were present in many of the enrichment cultures. The new isolates grew faster than the *Xanthobacter* 124X, that was isolated at a much higher initial concentration of styrene (1 g/l), possibly suggesting that this strain was capable of surviving relatively high concentrations of styrene. All the isolates could grow on styrene, styrene oxide, and 2-phenylethanol and, with the exception of strain S3, all strains oxidized phenylacetic acid, after having been grown on styrene. This could indicate that, in these strains, styrene is degraded via initial oxidation of the aliphatic side-chain, as was proposed by Baggi et al. [15] for *Pseudomonas fluorescens*.

Strain S5, tentatively identified as a *Rhodococcus* sp. (DSM 6697) by the Deutsche Sammlung von Mikroorganismen und Zellkulturen in Braunschweig, Germany, was studied in more detail. With washed cells of this strain, we observed a transient accumulation of phenylacetic acid from styrene. Styrene degradation by whole cells was oxygen-dependent, suggesting an oxygenase type of reaction. Indeed, in cell-free extracts, a styrene degrading activity was detected, that required NADH or NADPH and oxygen. Dialysis showed that FAD was also required. Incubation of dialyzed cell extracts with styrene in the presence of NADH, FAD and molecular oxygen, revealed styrene-dependent accumulation of three aromatic compounds. These were identified as phenylacetaldehyde, 2-phenylethanol, and phenylacetic acid. No styrene oxide (phenyloxirane) or any hydroxy styrenes, which would result from chemical rearrangement of arene oxides formed as a result of epoxidation of the aromatic ring, could be detected. Crude extracts of styrene-grown cells also contained styrene oxide isomerase, NAD^+-dependent 2-phenylethanol dehydrogenase, and NAD^+-dependent phenylacetaldehyde dehydrogenase. These three enzymatic activities would result in the transformation of any styrene oxide formed from styrene into the three aromatic compounds detected.

Styrene monooxygenase (SMO) from *Rhodococcus* S5 differs from many other monooxygenases in its inability to oxidize propene. Only alkenes with an aromatic substituent on the carbon-carbon double bond (e.g., α- and β–methylstyrene) appear to be substrates for the enzyme [18]. Other FAD-dependent monooxygenases are generally only known for their capacity to hydroxylate the aromatic nucleus, suggesting SMO to be a flavoprotein with a novel catalytic activity.

The utilization pattern of the hydroxyphenylacetic acid isomers by strain S5 was complementary to that of *Xanthobacter* 124X. *Rhodococcus* S5 utilized 2- and 3-hydroxyphenylacetic acid as sole source of carbon and energy, but could not grow on 4-hydroxyphenylacetic acid. The *P. fluorescens,* that was studied by Baggi et al. [15], grew with 2- and 4-hydroxyphenylacetic acid, but could not utilize 3-hydroxyphenylacetic acid.

The above results suggest a degradative pathway of styrene involving an initial epoxidation to styrene oxide, which is subsequently isomerized to phenylacetaldehyde and oxidized to phenylacetic acid. The proposed pathway for styrene degradation in strain S5 is shown in Figure 2, and is based on simultaneous adaptation experiments, the transient accumulation of phenylacetic acid and the presence of the required enzymatic activities [18]. After this pathway was published, the present author became aware of a publication by Chapman [19] in which the same pathway is proposed, although no experimental results were presented.

$$\text{C}_6\text{H}_5\text{-CH=CH}_2 \longrightarrow \text{C}_6\text{H}_5\text{-HC(O)CH}_2 \longrightarrow \text{C}_6\text{H}_5\text{-H}_2\text{C-C(=O)H} \longrightarrow \text{C}_6\text{H}_5\text{-H}_2\text{C-C(=O)OH} \longrightarrow$$

Figure 2. Styrene degradative pathway in *Rhodococcus* S5 [18].

Recently, two publications appeared describing the styrene-utilizing *Pseudomonas* sp. Y2 [20] and *Pseudomonas putida* R1 [21]. Incubation of strain Y2 with styrene led to the accumulation of a number of compounds, including 1-phenylethanol, 2-phenylethanol, phenylacetic acid, and salicylic acid. Based on adaptation experiments, strain Y2 was proposed to oxidize styrene via phenylacetic acid. In contrast to *Rhodococcus* S5 and *P. fluorescens* [15], this strain did not utilize 2-hydroxyphenylacetic acid.

For *Pseudomonas putida* R1 [21], an interesting modification of the previously suggested pathways is proposed. Based on the transient accumulation, in the beginning of the exponential phase of phenyl-1,2-ethanediol, and during the exponential phase of mandelic acid, the styrene degradative pathway, shown in Figure 3, is proposed. Interestingly, this strain could be grown with an initial styrene concentration of 2 g/l.

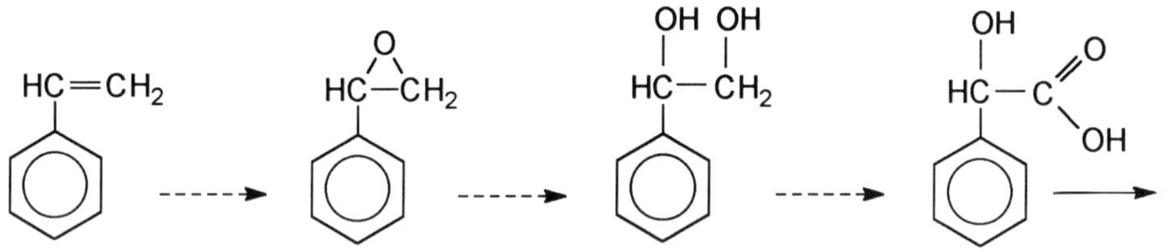

Figure 3. Styrene degradative pathway in *Pseudomonas putida* R1 [21].

Cox et al. [7] report that styrene-degrading fungi can also be isolated quite readily, provided that the styrene concentration is kept low. One strain, isolated by van der Werf at pH 4.5 [18], was identified as a black yeast, *Exophiala jeanselmei* and studied in more detail. Based on the growth-substrate utilization pattern and oxygen-uptake experiments, Cox et al. [7] suggest that styrene oxide, 2-phenylethanol, and phenylacetate could be intermediates of the styrene degradative pathway in this eukaryote.

2-Phenylethanol has been suggested to be involved in the degradation pathway of styrene by a number of authors, based on the accumulation of this compound during growth on styrene, or during the incubation of non-growing cells with styrene [11,14,20]. Accumulation of this compound can, however, probably be attributed to the presence of 2-phenylethanol dehydrogenase activity in these organisms, in combination with the pathway that we have proposed for *Rhodococcus* S5 (Figure 2). If phenylacetaldehyde accumulates during styrene degradation, it is probably reduced to 2-phenylethanol (Figure 4).

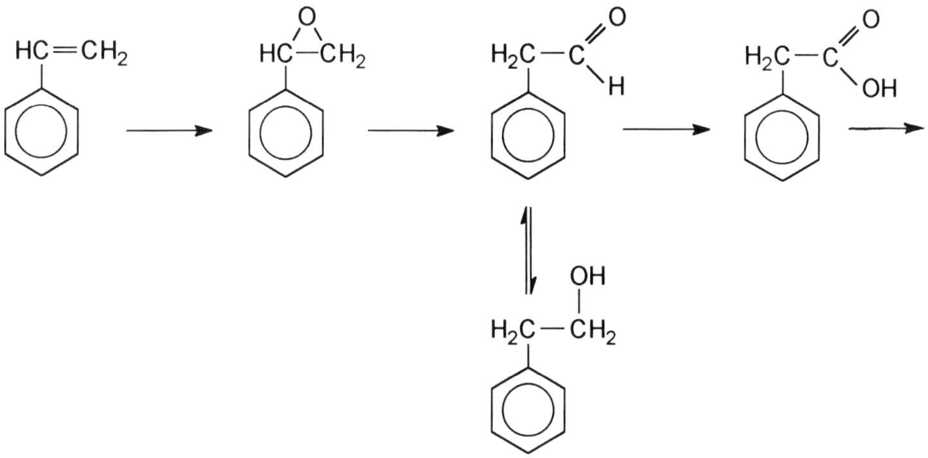

Figure 4. Formation of 2-phenylethanol as a side product in styrene metabolism.

Oxidation of the aromatic nucleus

Besides the two pathways, that appear to operate via an initial oxidation of styrene to styrene oxide (Figures 2 and 3), it is clear that at least one other pathway exists, involving initial oxidation of the aromatic ring. Such a pathway apparently operates in the *Nocardia* sp. described by Andreoni et al. [12], who reported formation of a dihydroxyderivative, followed by extra-diol fission of the aromatic ring.

GENETICS OF STYRENE DEGRADATION

Work on the genetics of styrene degradation has focused on the *Pseudomonas fluorescens* strain described by Baggi et al. [15]. This strain was shown to contain a plasmid of 37 kb [22]. The plasmid (pEG) was self-transmissible between *Pseudomonas* strains. Conjugation experiments performed with *P. fluorescens* ST as donor and *P. putida* PaW 340, a plasmid-free streptomycin resistant tryptophan auxotroph mutant of *P. putida,* mt-2, that can grow on phenylacetic acid as recipient, gave exconjugants (str^- trp^- $styrene^+$) at a frequency of $1x10^{-3}$. Subsequently, it was shown that pEG contains inverted repeat sequences, and that a 3 kb region located on the chromosome was homologous to sequences located at one end of the plasmid repeats [23].

Recently, it was shown that the catabolic genes for styrene degradation in *P. fluorescens* ST are located on the chromosome [24]. A genomic library of the ST strain was constructed and screened in *P. putida* PaW

340. One recombinant utilized styrene with a similar efficiency as strain ST. This cosmid contained a 27 kb insert, but by Tn5 mutagenesis, a 9 kb fragment was localized that contained the catabolic genes. This fragment is presently being sequenced [24]. The plasmid pEG appears to play a role in the regulation of the catabolic genes. By integrating into the chromosome, the expression of the catabolic genes is switched off.

In another styrene-utilizing strain, *Pseudomonas putida* CA3, in which styrene also appears to be degraded via the pathway shown in Figure 2, the styrene catabolic genes also appear to be located on the chromosome [25]. Transposon mutants of this strain, that no longer grow on styrene, could be complemented for growth on styrene, using a chromosomal DNA library, derived from the wild-type strain CA3.

Based on these developments, isolation of the structural genes for the enzymes involved in the first steps of the pathway, shown in Figure 2, can be expected in the near future.

STEREOCHEMISTRY OF STYRENE OXIDE METABOLISM

The stereospecificity of styrene oxide formation has been studied with a number of microorganisms, containing monooxygenase activity, in view of a potential application in the synthesis of optically pure styrene oxide. Many alkane-oxidizing microorganisms have been shown to oxidize styrene to styrene oxide. The alkane-utilizing *Nocardia corallina* B-276 predominantly forms R-styrene oxide with an enantiomeric excess of 69% [26]. Ethene-grown cells of *Mycobacterium* E3 also formed R-styrene oxide with an enantiomeric excess of about 98% [27]. In contrast, a mutant of *Pseudomonas putida* S12 devoid of styrene oxide isomerase activity only formed S-styrene oxide. The styrene oxide isomerase of this strain showed a slight preference for the S-isomer [27].

The mandelic acid, that accumulated during growth of *Pseudomonas putida* R1 [21] on styrene was D-(R)-mandelic acid (personal communication, S.A. Rustemov). This could indicate that this strain forms R-styrene oxide which, subsequently, is hydrolyzed to R-phenyl-1,2-ethanediol and oxidized to R-mandelic acid. However, besides the epoxide hydrolases, that hydrolyze epoxides with retention of the absolute configuration, such as mammalian microsomal and cytosolic hydrolases, recently the fungus *Beauveria sulfurescens* was shown to preferentially hydrolyze S-styrene oxide to R-phenyl-1,2-ethanediol [28]. Therefore, no conclusion can be made with respect to the stereospecificity of the monooxygenase, oxidizing styrene on the basis of the configuration of the mandelic acid, that accumulated transiently. Furthermore, mandelate racemase activity has also been detected in a number of *Pseudomonas putida* species [5]. The transient accumulation of D-mandelic acid by *P. putida* R1 could, therefore, also be caused by mandelate racemase activity, with L-mandelic acid being the true intermediate in the styrene degradative pathway.

CONCLUDING REMARKS

Although, initially, isolation of styrene-utilizing microorganisms was not always successful, probably due to the relatively high concentrations of styrene that were applied in the enrichment cultures, it is now clear that styrene-degrading microorganisms are ubiquitous. With respect to the studies concerning degradative pathways of styrene, one should remain critical when pathways are proposed, based on the accumulation of "intermediates". Quite often, these "intermediates" could be *dead-end* products or the result of side-reactions. Conclusive proof of anaerobic styrene degradation is still lacking, but in view of the recent developments in the field of the anaerobic degradation of aromatic hydrocarbons, new insights can be expected in the coming years.

A lot of research remains to be done, especially on strains that appear to degrade styrene via an initial oxidation of the aromatic nucleus, but also on the novel enzymes involved in the oxidation of the side-chain of styrene. An interesting aspect, from an ecological point of view, could be the comparison of the kinetic parameters of strains containing styrene monooxygenase with strains that attack styrene via a dioxygenase. However, at present, most of the research effort in the area of styrene degradation appears to be focusing on genetic aspects [24,25]. This work should result in a further understanding of styrene metabolism and its regulation in the coming years, and can possibly facilitate the purification and characterization of the enzymes involved in styrene metabolism.

REFERENCES

1 Fu MH, Alexander M. Environ Sci Technol 1992; 26: 1540-1544.
2 Amoore JE, Hautala J Appl Toxicol 1983; 3: 272-290.
3 Spinnler HE, Grosjean O, Bouvier I. J Dairy Res 1992; 59: 533-541.
4 Vainio H, Tulsi F, Belvedere G. In: Hietanen E, Laitinen M, Hänninen O, eds. Cytochrome P-450, Biochemistry, Biophysics and Environmental Implications. Amsterdam: Elsevier Biomedical Press, 1982; 679-687.
5 Fewson CA. FEMS Microbiol Rev 1988; 54: 85-110.
6 Delbressine LPC, van Bladeren PJ, Smeets FLM. Seutter-Berlage F. Xenobiotica 1981; 11: 589-594.

7 Cox HHJ, Houtman JHM, Doddema HJ, Harder W. Appl Microbiol Biotechnol 1993; 39: 372-376.
8 Weber FJ, Ooijkaas LP, Schemen RMW, Hartmans S, de Bont JAM. Appl Environ Microbiol 1993; 59: 3502-3504.
9 Grbic'-Galic' D, Churchman-Eisel N, Mrakovic' I. J Appl Bacteriol 1990; 69: 247-260.
10 Omori T, Jigami Y, Minoda Y. Agric Biol Chem 1975; 39: 1775-1779.
11 Sielicki M, Focht DD, Martin JP. Appl Environ Microbiol 1978; 35: 124-128.
12 Andreoni V, Baggi G, Galli E, Treccani V. Soc Gen Microbiol Quart 1978; 6: 18-19.
13 Shirai K, Hisatsuka K. Agric Biol Chem 1979; 43: 1595-1596.
14 Shirai K, Hisatsuka K. Agric Biol Chem 1979; 43: 1399-1406.
15 Baggi G, Boga MM, Catelani D, Galli E, Treccani V. System Appl Microbiol 1983; 4: 141-147.
16 van der Tweel WJJ, Janssens RJJ, de Bont JAM. Antonie van Leeuwenhoek 1986; 52: 309-318.
17 Hartmans S, Smits JP, van der Werf MJ, Volkering F, de Bont JAM. Appl Environ Microbiol 1989; 55: 2850-2855.
18 Hartmans S, van der Werf MJ, de Bont JAM Appl Environ Microbiol 1990; 56: 1347-1351.
19 Chapman PJ. In: Bourquin AW, Pritchard PH, eds. Proceedings of the Workshop: Microbial Degradation of Pollutants in Marine Environments. Gulf Breeze, Florida: EPA-600/9-79-012, 1979; 28-66.
20 Utikin IB, Yakimmov MM, Matveeva LN, Kozlyak EI, Rogozhin IS, Solomon ZG, Bezborodov AM. FEMS Microbiol Lett 1991; 77: 237-242.
21 Rustemov SA, Golovleva LA, Alieva RM, Baskunov BP. Microbiology 1992; 61: 1-5.
22 Bestetti G, Galli E, Ruzzi M, Baldacci G, Zennaro E, Frontali L. Plasmid 1984; 12: 181-188.
23 Ruzzi M, Zennaro E. FEMS Microbiol Lett 1989; 59: 337-344.
24 Marconi AM, Solinas F, Ruzzi M, Bestetti G, Zennaro E. In: Sixth European Congress on Biotechnology. Book of Abstracts, Vol IV, 1993; p. TH121. Florence, Italy.
25 O'Connor K, Buckley CM, Dobson ADW. In: Fourth International Symposium on Pseudomonads: Biotechnology and Molecular Biology. Book of Abstracts, 1993; p. 169. Vancouver, Canada.
26 Furuhashi K. In: Collins AN, Sheldrake GN, Crosby J, ed. Chirality in Industry. Chichester: John Wiley and Sons, 1992; 167-186.
27 Nöthe Ch, Hartmans S. Biocatalysis 1994; (submitted for publication).
28 Pedragosa-Moreau S, Archelas A, Furstoss R. J Organ Chem 1993; 58: 5533-5536.

Biotransformations: Microbial Degradation of Health Risk Compounds
Ved Pal Singh, editor
© 1995 Elsevier Science B.V. All rights reserved.

Microbial degradation of vinyl chloride

S. Hartmans

Division of Industrial Microbiology, Department of Food Science, Wageningen Agricultural University, P.O. Box 8129, 6700 EV Wageningen, The Netherlands.

INTRODUCTION

Vinyl chloride is produced on a very large scale by the chemical industry, mainly for use in the production of the polymer, polyvinyl chloride (PVC). In the major vinyl chloride production process, 1,2-dichloroethane is used as the precursor. Inevitably associated with these large scale processes are losses to the environment. Vinyl chloride is a gas at ambient conditions (boiling point, 14°C at 1 atm), and consequently a large percentage of the industrial losses are to the atmosphere. However, due to its relatively short half-life of 20 h in the troposphere, vinyl chloride does not accumulate in the atmosphere [1]. Vinyl chloride could only be measured incidently, with maximum concentrations up to 200 $\mu g/m^3$, downwind of plants producing vinyl chloride or using it as a raw material. The air/water partition coefficient of vinyl chloride is 1.25 at 30°C [2].

In contaminated groundwater, concentrations as high as 1000 mg/m^3 [3], may be encountered sometimes. Groundwater, contaminated with vinyl chloride, usually contains *cis*-1,2-dichloroethene also. Both compounds are formed as a result of the *microbial reduction* of tetra- and trichloroethene via *cis*-1,2-dichloroethene to vinyl chloride [4]. This is an undesirable process in view of the carcinogenic properties of vinyl chloride and the large number of locations, where the groundwater is contaminated with tetra- and trichloroethene.

Evidence that vinyl chloride is carcinogenic in experimental animals and humans [5,6], led to numerous studies concerning mammalian vinyl chloride metabolism and resulted in classification of vinyl chloride as a *priority pollutant* by the U.S. Environmental Protection Agency.

MAMMALIAN VINYL CHLORIDE METABOLISM

In the mid 1970's, a number of publications appeared that demonstrated that vinyl chloride induced angiosarcoma of the liver in workers in the polyvinyl chloride manufacturing industry [5,7]. Subsequently, a number of biochemical studies, focussing on vinyl chloride metabolism in the liver, were published. Barbin et al. [8] demonstrated that, in the presence of oxygen and NADPH, vinyl chloride was transformed to chloroethylene oxide (chlorooxirane) by a fraction from mouse-liver microsomes. Chlorooxirane is an extremely reactive molecule with an approximate half-life of one min in 100 mM potassium phosphate buffer of pH 7.7 at 23°C [9]. The major rearrangement product of chlorooxirane, is chloroacetaldehyde, but hydrolysis of the epoxide, followed by spontaneous dehydrochlorination, also occurs. The rate of epoxide hydrolysis can be enhanced by the addition of microsomal epoxide hydratase [9]. Due to their reactive nature, the vinyl chloride oxidation products, such as chlorooxirane and chloroacetaldehyde, have not been detected as metabolites in vivo. Chloroacetate and glycolate have, however, been detected in laboratory animals, administered with [^{14}C]-vinyl chloride [7]. Figure 1 summarizes the reactions, that have been proposed to be involved in the initial steps of vinyl chloride metabolism in mammalian liver. The ratios between enzymatic hydrolysis of chlorooxirane to glycolaldehyde and chemical rearrangement of chlorooxirane to chloroacetaldehyde in vivo are not known, but both processes are significant.

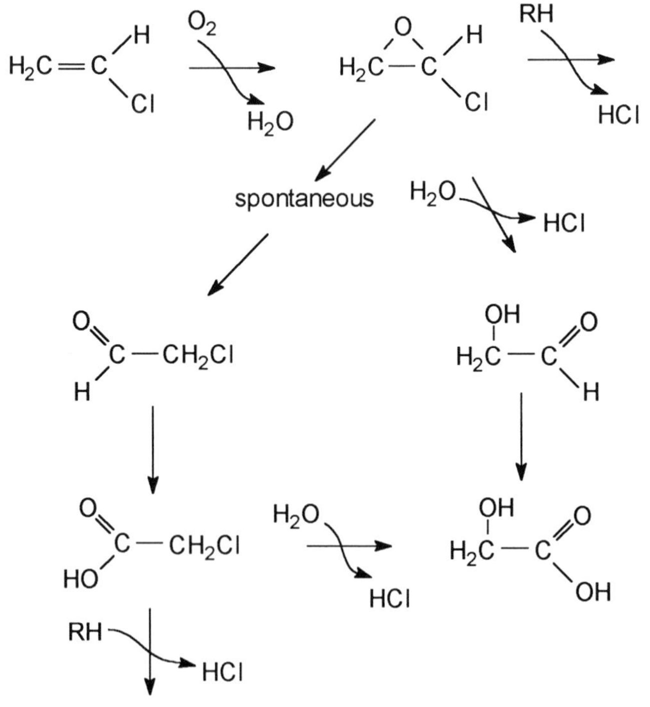

Figure 1. Mammalian vinyl chloride metabolism.

ANAEROBIC FORMATION AND DEGRADATION OF VINYL CHLORIDE

The presence of vinyl chloride in groundwater is attributed mainly to the biological reduction of polychlorinated ethenes [10]. Complete dechlorination of tetrachloroethene (PCE) and trichloroethene (TCE) to ethene has been observed under methanogenic conditions [11]. The rate-limiting step is the conversion of vinyl chloride to ethene. In anaerobic cultures with methanol as electron donor, 69% of the added [^{14}C]-PCE was recovered as [^{14}C]-ethene and 27% as [^{14}C]-vinyl chloride, whereas with glucose as electron donor, 8% was recovered as [^{14}C]-ethene and 88% as [^{14}C]-vinyl chloride. In both cases, no significant amounts of [^{14}C]-methane or [^{14}C]-CO_2 could be detected [11]. More recently, reductive dechlorination of PCE to ethene via vinyl chloride, in the absence of methanogenesis, has also been reported [12]. As a result of these reductive dechlorination reactions, vinyl chloride concentrations of more than 1 mg/l have been detected in groundwater, contaminated with PCE and TCE [3,4,13].

AEROBIC TRANSFORMATION OF VINYL CHLORIDE

Aerobic vinyl chloride transformation, or cometabolic degradation, has been observed with a wide range of microorganisms, exhibiting monooxygenase activity. Ethene- [14], methane- [15,16], propane- [17,18], isoprene- [19], ammonia- [20], and propene- [21] utilizing bacteria have been shown to oxidize vinyl chloride. No oxidation products were identified. By using purified soluble methane monooxygenase from *Methylosinus trichosporium* OB3b, the oxidation product of vinyl chloride was identified as chlorooxirane [22].

Aerobic *mineralization* of vinyl chloride by groundwater [23] and by a Gram-positive, propane-grown bacterium [24] has also been reported. Incubation of [^{14}C]-vinyl chloride (initial concentration of 1 mg per litre) with groundwater resulted in the recovery of 65% of the carbon as [^{14}C]-CO_2 after 108 days of incubation [23]. For a number *Rhodococcus* strains, Malachowsky et al. [18] showed that, after incubating 400-800 mg (dry weight) of propane-grown cells per litre for 7 days with [^{14}C]-vinyl chloride (initial concentration of 1 mg per litre), 5-10% of the ^{14}C form was recovered in the biomass, 20-25% as water soluble products and 70-80% as [^{14}C]-CO_2.

Presumably, the propane monooxygenase activity, present in *Rhodococcus* strains, oxidized vinyl chloride to chlorooxirane, which subsequently chemically rearranged to products that could be further oxidized to carbon dioxide. The data on the initial oxidation rates of various experiments, described by Malachowsky et al. [18], were contradictory. Probably, they are in same range (0.3 nmol vinyl chloride per mg dry weight per min), as reported earlier by the same group [24].

ISOLATION OF VINYL CHLORIDE-DEGRADING BACTERIA

To our knowledge, *Mycobacterium* strain L1 [14] was the first bacterial strain described, that utilizes vinyl chloride aerobically as a sole source of carbon and energy for growth. *Mycobacterium* strain L1 was isolated from soil that had been contaminated with vinyl chloride-containing water for a number of years, by using the enrichment technique with vinyl chloride added at 5% (v/v). This corresponds to a vinyl chloride concentration in the liquid phase of about 100 mg/l. Strain L1 has been deposited at the German Collection of Microorganisms and Cell Cultures, Braunschweig, Germany, under accession number DSM 6695.

Subsequently, a series of inocula from different sources, without any known history of vinyl chloride contamination, were used in enrichment cultures with vinyl chloride as the carbon source, using the following procedure [25]. Soil (20 g) or water (20 ml) was mixed with 10 ml of mineral salts medium and put it in serum bottles (about 130 ml) which were sealed with rubber septa. After the addition of 5 ml of vinyl chloride, the serum bottles were incubated statically in the dark at 30°C. After about 1 month, the content was diluted five-fold with mineral salts medium and 30 ml of this diluted suspension was once again incubated under the same conditions. After a total of 2 months, 1 ml of these enrichment cultures was added to 30 ml of mineral salts medium in serum bottles of 130 ml together with 2 ml of vinyl chloride, and cultures were incubated with gentle shaking at 30°C. The vinyl chloride concentration was determined weekly by analyzing headspace samples. Dilutions from cultures, showing vinyl chloride degradation, were plated on agar plates of mineral salts medium which were incubated in a desiccator to which vinyl chloride (1% v/v) was added. After 2 and 4 weeks, the plates were inspected, and colonies which appeared to grow on vinyl chloride were streaked to purity. In this manner, three vinyl chloride-degrading strains designated VC2, VC3, and VC4 were isolated from 20 different inocula used. They were isolated from soil which had been treated with 1,3-dichloropropene for 20 years (VC2), from the sludge of an aerobic wastewater treatment plant mainly treating domestic wastewater (VC3) and from the River Rhine sampled at Wageningen (VC4), respectively. All the new isolates formed yellow colonies on agar plates, as did the previously isolated *Mycobacterium* strain L1. All the four strains were tentatively identified as *M. aurum*. The colour and colony morphologies of the new isolates differed somewhat, indicating that they were different strains.

In contrast to the previously isolated strain L1, the three new strains were isolated from environments not known to be contaminated with vinyl chloride. Prolonged contamination with vinyl chloride is, therefore, apparently not a prerequisite for the evolution of the vinyl chloride degradative pathway in mycobacteria.

Very recently, Meier [26] described the application of a vinyl chloride-degrading *Mycobacterium aurum* strain. Although Meier gave no further details, describing his isolate, it is intriguing that all isolates, both from the Wageningen area in the Netherlands and from Berlin in Germany, have been identified as *Mycobacterium aurum* strains.

A similar situation has, however, been observed when ethene is used as the carbon source in enrichment cultures. Until now, all strains isolated from enrichment cultures with ethene as sole carbon and energy source were identified as mycobacteria [27].

However, none of the ethene- or propene-utilizing strains of the genera *Mycobacterium* and *Xanthobacter*, previously isolated in our laboratory [28, 29], could grow with vinyl chloride as the sole carbon source. Alkene-grown cells did, however, oxidize vinyl chloride at initial rates similar to those of vinyl chloride-grown cells of strain L1.

VINYL CHLORIDE DEGRADATION KINETICS

The growth rate of *Mycobacterium aurum* L1 with vinyl chloride was 0.04 h^{-1} [2], which corresponds with a culture doubling time of about 17 hours. The growth yield determined in batch culture was 0.22 gram dry weight of biomass formed per gram of vinyl chloride consumed [25]. These data correspond nicely with the vinyl chloride oxidation rate of 55 nmol min^{-1} mg of dry $weight^{-1}$ by vinyl chloride-grown and washed cells of strain L1 [25]. These data also illustrate that an oxidation rate of 0.3 nmol min^{-1} mg of dry $weight^{-1}$, as reported by Phelps et al. [24] for a propane-utilizing strain, is too low to support growth on vinyl chloride as a sole source of carbon and energy.

VINYL CHLORIDE METABOLISM IN *MYCOBACTERIUM AURUM*

Freshly harvested, vinyl chloride-grown, washed cells of *Mycobacterium aurum* strain L1 oxidized vinyl chloride at an initial rate of 55 nmol min^{-1} mg of dry $weight^{-1}$, but very rapid inactivation of vinyl chloride degradation was observed (Figure 2) [25]. Incubation of cells at the same densities for 1 h at 30°C, before vinyl chloride was added, gave an almost identical vinyl chloride degradation curve, suggesting that the observed inactivation was not due to instability of vinyl chloride transforming enzyme. Inactivation could, however, be delayed by the

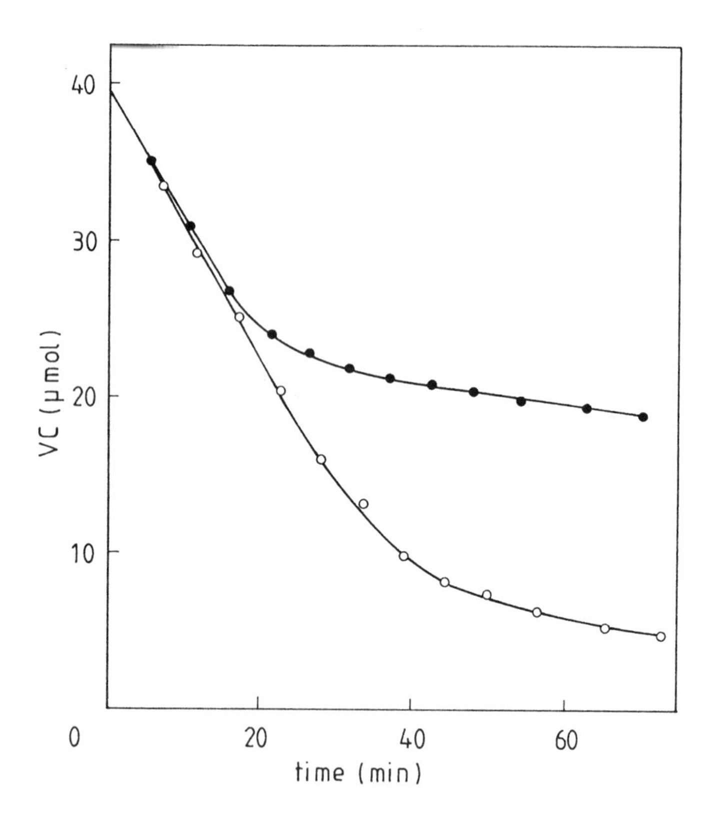

Figure 2. Vinyl chloride degradation by washed cells of *M. aurum* L1. (●) Total volume of liquid phase (3 ml) containing 16 mg of freshly harvested vinyl chloride-grown cells in phosphate buffer; (o) same composition plus 16 mg of heat-inactivated cells [25].

addition of boiled cells (Figure 2). No vinyl chloride degradation was observed when only boiled cells were present (results not shown). From this experiment, it was concluded that the inactivation was caused by a toxic metabolite of vinyl chloride metabolism which could accumulate extracellularly [25].

The most obvious toxic product, that could be formed from vinyl chloride, is chlorooxirane, due to oxidation of vinyl chloride by a monooxygenase.

Indeed, vinyl chloride- and ethene-grown cells of strain L1 have been shown to oxidize propene to 1,2-epoxypropane [30], indicating that alkene monooxygenase was present in cells grown on both the substrates.

NADH- and oxygen-dependent alkene monooxygenase was also detected in dialyzed crude extracts of ethene- and vinyl chloride-grown cells [25]. No monooxygenase activity was detected in cells grown on acetate or succinate.

The presence of alkene monooxygenase activity in extracts of vinyl chloride-grown *M. aurum* L1, combined with the observation that vinyl chloride degradation can be competitively inhibited by the addition of ethene or propene, which are both oxidized to the corresponding epoxides, is strong evidence that the initial step in vinyl chloride metabolism is indeed catalyzed by alkene monooxygenase (Figure 3) [25].

$$H_2C{=}CHCl \xrightarrow[NADH \rightarrow NAD]{O_2 \rightarrow H_2O} H_2C{-}CHCl\ (\text{epoxide, } O) \longrightarrow$$

Figure 3. The initial step in vinyl chloride metabolism of *M. aurum* L1, catalyzed by alkene monooxygenase.

The presumed product of vinyl chloride oxidation, chlorooxirane, is a very reactive and unstable compound (rearranging to chloroacetaldehyde with a half-life of 1.6 min in Tris-HCl buffer, pH 7.4 at 37°C [8]) indicating that, during growth with vinyl chloride, the epoxide must be metabolized very effectively to prevent the accumulation of toxic levels within the cell. However, on the basis of the inactivation of washed cells, degrading vinyl chloride (Figure 1), this very effective enzyme would also appear to be rather unstable.

This hypothesis was confirmed by an experiment in which the vinyl chloride supply to a chemostat culture was interrupted for only 43 min. After the vinyl chloride supply was restored, chlorooxirane was detected in the air from the chemostat [25]. No chlorooxirane could be detected prior to the interruption. Apparently, this time period was already long enough to allow (some) loss of activity of the chlorooxirane transforming enzyme, resulting in the accumulation of the inhibitory epoxide [25]. Apparently this enzyme is very unstable or perhaps its activity is very sensitive to changes in the intracellular environment (e.g., cofactor levels) due to the absence of an exogenous carbon and energy source.

In studying the inhibitory effects of the less reactive 1,2-epoxypropane, it was previously demonstrated that the inhibitory effect on the monooxygenases examined was much stronger than on other physiological functions of the cell [31]. Although not examined for strain L1, this is probably also the case with alkene monooxygenase inactivation by chlorooxirane.

The apparent instability of the chlorooxirane-degrading enzyme, in combination with the reactivity and instability of the epoxide itself, have hampered further elucidation of the vinyl chloride degradative pathway until now.

The presence of epoxyethane dehydrogenase [32] activity in the extracts of vinyl chloride-grown cells of strain L1 [25] does not necessarily indicate the involvement of this enzyme in vinyl chloride metabolism. For example, in the ethene-utilizing *Mycobacterium* strain E3, the monooxygenase and the ethene epoxide dehydrogenase are both induced simultaneously by epoxyalkanes (unpublished results). A number of other possible enzymatic transformations of epoxides can, however, be ruled out. Hydrolysis or isomerization [33] of the epoxide would result in glycolaldehyde and chloroacetaldehyde, respectively. As strain L1 does not grow on ethanediol, glycolate, or chloroethanol, this would seem unlikely to occur. Only very low levels of glutathione could be detected in strain L1 grown on various substrates, indicating that glutathione-dependent transformation of chlorooxirane is also unlikely [25].

CONCLUSIONS

Although vinyl chloride is carcinogenic in man, specific mycobacteria have been isolated, that are capable of aerobic growth on vinyl chloride as a sole source of carbon and energy. The reason, why these mycobacteria have (developed) the capacity to degrade this truly *xenobiotic* compound, remains unknown.

Interestingly, the initial step in vinyl chloride metabolism in *Mycobacterium aurum* L1 is the same as in mammalian liver. Unfortunately, as is the case in mammals, the further metabolism of the extremely toxic and reactive chlorooxirane is still obscure. It will probably, also prove to be very difficult to elucidate this step in strain L1, in view of the very low stability of the enzymatic activity responsible for the further transformation of chlorooxirane.

REFERENCES

1 Guicherit R, Schulting FL. Science of the Total Environment 1985; 43: 193-219.
2 Hartmans S, Kaptein A, Tramper J, de Bont JAM. Appl Microbiol Biotechnol 1992; 37: 796-801.
3 Brauch H-J, Kühn W, Werne, P. Vom Wasser 1987; 68: 23-32.
4 Milde G, Nerger M, Mergler R. Water Sci Technol 1988; 20: 67-73.

5 Creech JL, Johnson MN. J Occup Med 1974; 16: 150-151.
6 Maltoni C, Lefemine G. Environ Res 1974; 7: 387-405.
7 Plugge H, Safe S. Chemosphere 1977; 6: 309-325.
8 Barbin A, Brésil H, Croisy A, Jacquignon P, Malaveille C, Montesano R, Bartsch H. Biochem Biophys Res Commun 1975; 67: 596-603.
9 Guengerich FP, Crawford WM, Watanabe PG. Biochemistry 1979; 18: 5177-5182.
10 Vogel TM, McCarty PL. Appl Environ Microbiol 1985; 49: 1080-1083.
11 Freedman DL, Gossett JM. Appl Environ Microbiol 1989; 55: 2144-2151.
12 DiStefano TD, Gossett JM, Zinder SH. Appl Environ Microbiol 1991; 57: 2287-2292.
13 Kästner M. Appl Environ Microbiol 1991; 57: 2039-2046.
14 Hartmans S, de Bont JAM, Tramper J, Luyben KCAM. Biotechnol Lett 1985; 7: 383-388.
15 Fogel MM, Taddeo AR, Fogel S. Appl Environ Microbiol 1986; 51: 720-724.
16 Tsien H-C, Brusseau GA, Hanson RS, Wackett LP. Appl Environ Microbiol 1989; 55: 3155-3161.
17 Wackett LP, Brusseau GA, Householder SR, Hanson RS. Appl Environ Microbiol 1989; 55: 2960-2964.
18 Malachowsky KJ, Phelps TJ, Teboli AB, Minnikin DE, White DC. Appl Environ Microbiol 1994; 60: 542-548.
19 Ewers J, Freier-Schröder D, Knackmuss H-J. Arch Microbiol 1990; 154: 410-413.
20 Vannelli T, Logan M, Arciero DM, Hooper AB. Appl Environ Microbiol 1990; 56: 1169-1171.
21 Ensign SA, Hyman MR, Arp DJ. Appl Environ Microbiol 1992; 58: 3038-3046.
22 Fox BG, Borneman JG, Wackett LP, Lipscomb JD. Biochemistry 1990; 29: 6419-6427.
23 Davis JW, Carpenter CL. Appl Environ Microbiol 1990; 56: 3878-3880.
24 Phelps TJ, Malachowsky KJ, Schram RM, White DC. Appl Environ Microbiol 1991; 57, 1252-1254.
25 Hartmans S, de Bont JAM. Appl Environ Microbiol 1992; 58: 1220-1226.
26 Meier T. In: Biologische Abgasreinigung, Düsseldorf: VDI Berichte VDI-Verlag, 1994; 1104: 325-331.
27 Hartmans S, de Bont JAM, Harder W. FEMS Microbiol Rev 1989; 63: 235-264.
28 Habets-Crützen AQH, Brink LES, van Ginkel CG, de Bont JAM, Tramper J. Appl Microbiol Biotechnol 1984; 20: 245-250.

29 van Ginkel CG, de Bont JAM. Arch Microbiol 1986; 145: 403-407.
30 Weijers CAGM, van Ginkel CG, de Bont JAM. Enz Microb Technol 1988; 10: 214-218.
31 Habets-Crützen AQH, de Bont JAM. Appl Microbiol Biotechnol 1985; 22: 428-433.
32 de Bont JAM, Harder W. FEMS Microbiol Lett 1978; 3: 89-93.
33 Hartmans S, Smits JP, van der Werf MJ, Volkering F, de Bont JAM. Appl Environ Microbiol 1989; 55: 2850-2855.

Biotransformations: Microbial Degradation of Health Risk Compounds
Ved Pal Singh, editor
© 1995 Elsevier Science B.V. All rights reserved.

Isolation and characterization of neurotoxin-degrading gene

I. M. Santha and S. L. Mehta

Division of Biochemistry, Indian Agricultural Research Institute, New Delhi-110 012, India

INTRODUCTION

In nature, a variety of microorganisms are known to exist, which are endowed with the property of degrading a large spectrum of simple and complex molecules [1-4]. *Detoxification* of natural and man-made pollutants are carried out with the help of various microorganisms. Their involvement in the degradation of sewage, breakdown of oil waste from industry and degradation and recycling of the products and inputs of agriculture, namely cellulose and pesticides, are well known.

A simple way to obtain bacterial strains, with increased degradative capabilities, is by selective enrichment, in which a known or unknown population of bacteria is grown in the presence of the chemical of interest. Subsequently, either physiological or genetic changes, or both may occur, permitting the chemical to be degraded, and the new strain is isolated. Cook obtained bacteria, capable of degrading S-triazines by enrichment of cultures from sewage [5]. A continuous chemostate enrichment with a *Pseudomonas putida* strain (toluene degrader) and *Pseudomonas* strain B13 (a 3-chlorobenzoate degrader) was used [6] to select for a strain capable of degrading 4-chlorobenzoate and 3,5-dichlorobenzoate. The first degradative phenotypes, attributed to *catabolic plasmids* found in *Pseudomonas* [7], were for camphor and octane. Catabolic plasmids are widespread in nature, and an increase in their frequency has been observed in pollutant - stressed environment [8-11]. Having come across these reports as well as many other related published work, we got the idea to isolate a microbe from sewage sludge to degrade the neurotoxin from *Lathyrus sativus*.

Lathyrus sativus or chickling vetch or khesari dal, as is commonly known in various parts of India, is grown despite ban on its cultivation by Government of India in 1961 [12]. The ban is based on the reports that prolonged consumption of this pulse causes lower limb paralysis, a disease known as *neurolathyrism* in human beings, specially in children [12,13]. The active principle causing this disease has been isolated from *Lathyrus sativus,* and is identified as an amino acid derivative called β-N-oxalyl L-α, β-diamino propionic acid (BOAA, also designated as ODAP by some authors) [14-16]. This toxin is present in all parts of the plant, but its concentration is maximum in leaves during vegetative phase and in the embryo during reproductive phase [17]. This grain legume is very rich in proteins, and does not require any agronomical management practices for growing. Other than India, it is also widely grown in countries like Bangladesh, Ethiopia, Nepal, Pakistan, Spain etc. Cases of neurolathyrism continue to be reported from Ethiopia. Various plant breeding efforts over the last two decades or more by various workers have not succeeded in developing any cultivar of *L. sativus* with low BOAA content.

Recently, genetic engineering techniques, coupled with proper and efficient regeneration protocols, have made it possible for introducing and expressing foreign genes in crop plants without any species barriers. With the successful development of efficient protocols for in vitro plant regeneration from leaf [18], internode [19], and root explants [20] in *Lathyrus sativus*, the decks are cleared for application of plant genetic engineering techniques for toxin removal. For this purpose, one of the strategies planned was to isolate a microbe with the capability of BOAA degradation and to isolate the gene, degrading BOAA, for its eventual transfer to *Lathyrus sativus* plants. In the present chapter we discuss the isolation and characterization of a microbial gene that can degrade *Lathyrus sativus* toxin.

MATERIALS AND METHODS

Culture conditions

Sewage sludge, collected from an open drainage canal near the Indian Agricultural Research Institute in New Delhi, was filtered through two layers of Mira cloth to remove undissolved solid particles, and 10 μl of the filtrate was used to inoculate 1 ml of a medium containing 0.5 g/l K_2HPO_4, 0.1 g/l $MgSO_4$ and 0.01 g/l $FeSO_4$ and 200 μg/ml of *Lathyrus* toxin, BOAA (β-N-oxalyl L-α,β-diamino propionic acid) and incubated at 37°C with constant shaking (300 rpm). At 24 h intervals, the BOAA - supplemented media was diluted 1:100 with fresh media and continued till 8 serial transfers. At the end of 8th serial dilution, the growth of the cells was determined by measuring optical density at 600 nm, and the remaining BOAA was determined by the method described earlier [21]. Aliquots of the culture were then plated on various LB agar plates containing different antibiotics.

Plasmid isolation, restriction, cloning DNA sequencing, etc. were carried out, using standard protocols described by Sambrook et al. [22].

Plasmid isolation

Plasmid isolation was done by the alkaline lysis method of Brinboim and Doly, as described by Sambrook et al. [22]. For the preparation of a library of the plasmid, the purified plasmid was restricted with restriction enzyme *Sau*3AI and cloned into the vector pUC18 linearized with *Bam*HI. The ligated DNA was used for transforming *E. coli* strain DH5a. The recombinant clones were selected on LB agar plates containing 100 μg/ml ampicillin, 80 μg/ml X-gal, and 0.5 mM IPTG. The recombinants were further plated on minimal agar plates containing 100 μg/ml ampicillin and 800 μg/ml BOAA as carbon and nitrogen sources.

Sequencing was done by the dideoxy sequencing method of Sangers, using T_7 Deaza DNA sequencing kit from Pharmacia, according to the manufacturers protocols. Sequence analysis was performed by using software of Microgenei and DNAsis.

RESULTS AND DISCUSSION

Filterate of sludge samples, when grown in minimal media with BOAA, as sole source of carbon and nitrogen, was serial transferred to fresh media eight times at 24 h regular intervals. At the end of eighth serial transfer, the utilization of BOAA by the surviving bacteria was very rapid; complete utilization of BOAA in the media was observed by 6 to 12 h after transfer. Increasing concentrations of BOAA in the media were found to stimulate the growth further. The bacteria grown in the BOAA media on further plating on LB agar plates, having ampicillin (100 μg/ml), tetracycline (10 μg/ml), and kanamycin (50 μg/ml), showed growth on all the three plates and were designated as BYA1, BYT1, and BYK1, respectively. BYA1 was sensitive to tetracycline and kanamycin; BYT1 showed resistance to ampicillin and tetracycline, but sensitivity to kanamycin, and BYK1 showed ampicillin and kanamycin resistance and tetracycline sensitivity (Table 1). Yadav et al. [23] have further characterized BYA1 strain as an *Enterobacter cloacae* strain. Since this strain showed better BOAA utilization than the other two strains, it was further studied in detail.

Table 1
Antibiotic resistance/sensitivity characteristics of various BOAA-utilizing bacterial strains

Strain	Antibiotics
BYA1	Amp^R, Tet^S, Kan^S, Ery^R, Pen^R, Neo^{S100}, Str^S, Spc^S
BYT1	Amp^R, Tet^S, Kan^{S100}, Ery^R, Pen^R, Neo^S, Str^R, Spc^R
BYK1	Amp^R, Tet^S, Kan^R, Ery^R, Pen^R, Neo^S, Str^S, Spc^R

The degradative property of the strain BYA1 was found to reside on a mega plasmid carried by it, which was confirmed by transforming an *E. coli* strain DH5a with the plasmid isolated from BYA1. The transformed *E. coli* could utilize BOAA as sole source of carbon and nitrogen, whereas the non-transformed cells could not. An overlapping restriction map of this plasmid, with respect to various enzymes, like *Hind*III, *Kpn*I, *Pst*I, and *Not*I [24], has been prepared (Figure 1).

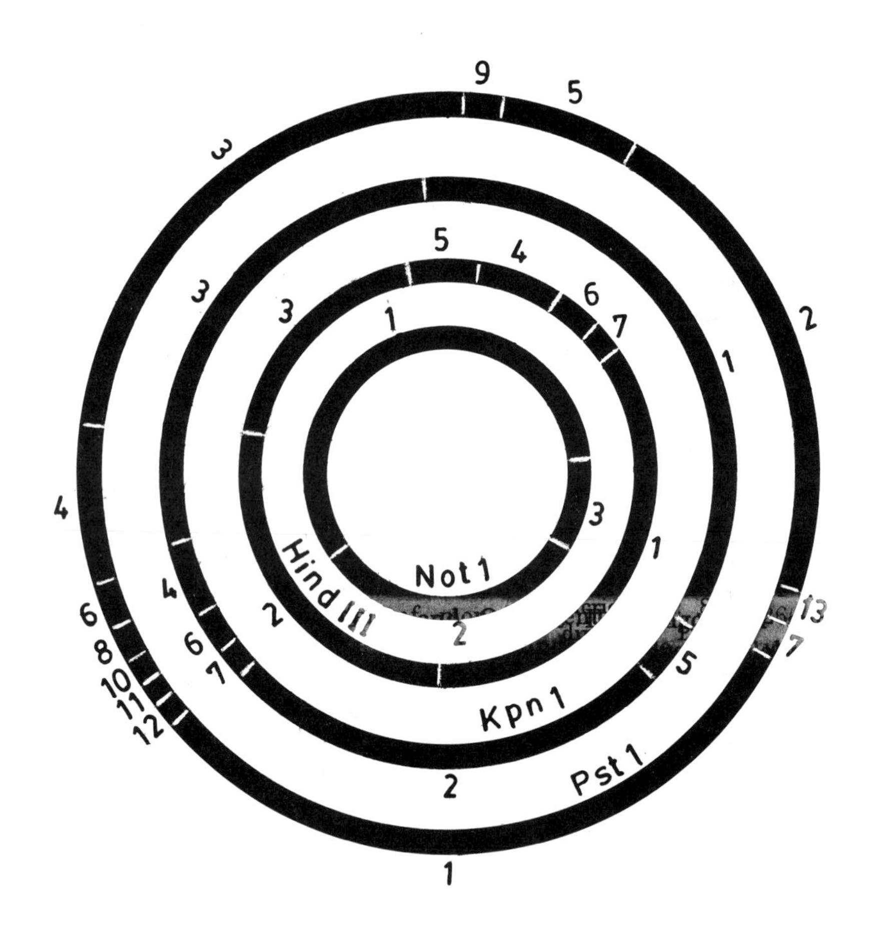

Figure 1. An overlapping restriction map of the BOAA-degrading plasmid, pBYA1 with respect to restriction enzymes *NotI*, *Kpn*I *Hind*III, and *Pst*I.

For the purpose of isolating the DNA fragment of the pBYA1 plasmid, which is involved in the degradation of BOAA, the plasmid was partially restricted with *Sau*3AI and cloned into pUC18, linearized with *Bam*HI and used for transforming *E. coli* DH5a. The recombinant clones selected on X-gal IPTG plates were further transferred to minimal agar plates containing 800 μg/ml BOAA and 100 μg/ml ampicillin. One such clone, growing on BOAA, was selected and further analysed. The clone BM1 showed a plasmid with an insert size of about 1.8 kb (Figure 2). The study of the growth pattern of BM1 clone in minimal media, with 800 μg/ml BOAA as sole source of carbon and nitrogen, indicated an initial lag phase of 12 h, followed by a log phase of 36 h, by which time the entire BOAA in the medium was utilized (Table 2).

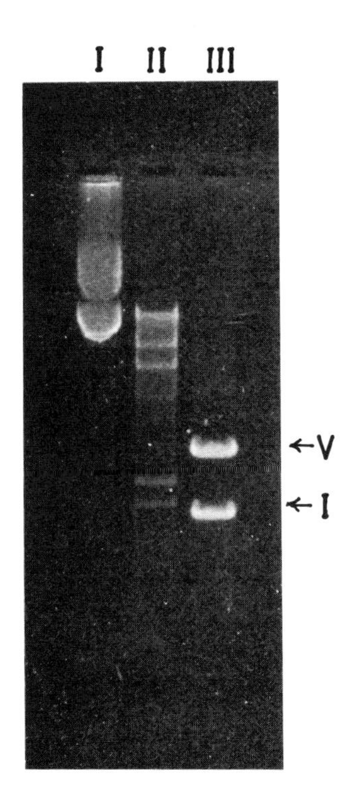

Figure 2. Restriction analysis of pBM1, the plasmid carrying the BOAA-degrading gene present in the clone BM1 to determine the size of the insert. Approximately 1 μg of purified pBM1 was restricted with *Eco*RI (5U) and *Pst*I (5U) and analysed on 0.7% agarose gel. Lane I, uncut pBM1; Lane II, *Hind*III marker; Lane III, *Eco*RI + *Pst*I restricted pBM1. I - Insert; V - Vector.

Table 2
BOAA utilization and growth of BYA1 and BM1 strains as well as cell protein content of BM1

Time(h)	BOAA utilization (mg/ml)		Growth (OD_{600})		Cell protein content of BM1 (g/ml)
	BYA1	BM1	BYA1	BM1	
0	0.000	0.000	0.007	0.004	10.00
12	0.070	0.350	0.100	0.010	16.60
24	0.780	0.375	0.120	0.015	33.20
36	0.800	0.375	0.159	0.035	66.00
48	0.800	0.800	0.220	0.168	119.20

The original strain, on the other hand, had a very short lag phase of 6 h, followed by log phase of 42 h, by which time the entire BOAA in the medium was utilized. The growth was also confirmed by measuring the optical density and cell protein content at regular intervals. An increase in cell protein content, with an increase in absorbance, and a concomitant decrease in BOAA concentration, confirmed that the clone utilizes BOAA as carbon and nitrogen source for its growth (Table 2).

The 1.8 kb insert from the pUC18 recombinant clone was taken out by restriction with *Eco*RI and *Pst*I flanking enzymes of *Bam*HI, on the vector and cloned into a phagemid pBluescript SK-vector. This 1.8 kb fragment, having the property of BOAA degradation, was sequenced after generating deletions by *Bal*31 and subcloning. The sequences of deletions on an overlapping restriction map gave two stretches of nucleotide sequence, one with 1140 nucleotides and the other with 703 nucleotides, with a small stretch missing in between. These sequences were further analyzed by using DNAsis and Microgenei software packages. The 1140 nucleotide stretch, on analysis, showed a largest reading frame of 631 nucleotides in only one of the reading frames, and the other two reading frames had frequent termination codons. The coding frame of 631 nucleotide stretch (Figure 3), on analysis, showed it to have sequences similar to that of prokaryotic regulatory sequences, namely the "-43", "-35", and "-10" sequences at its 5' end, encoding 192 amino acid. The initiation codon is proposed to be `GUG', in this case, with the corresponding deoxy nucleotide being `GTG'.

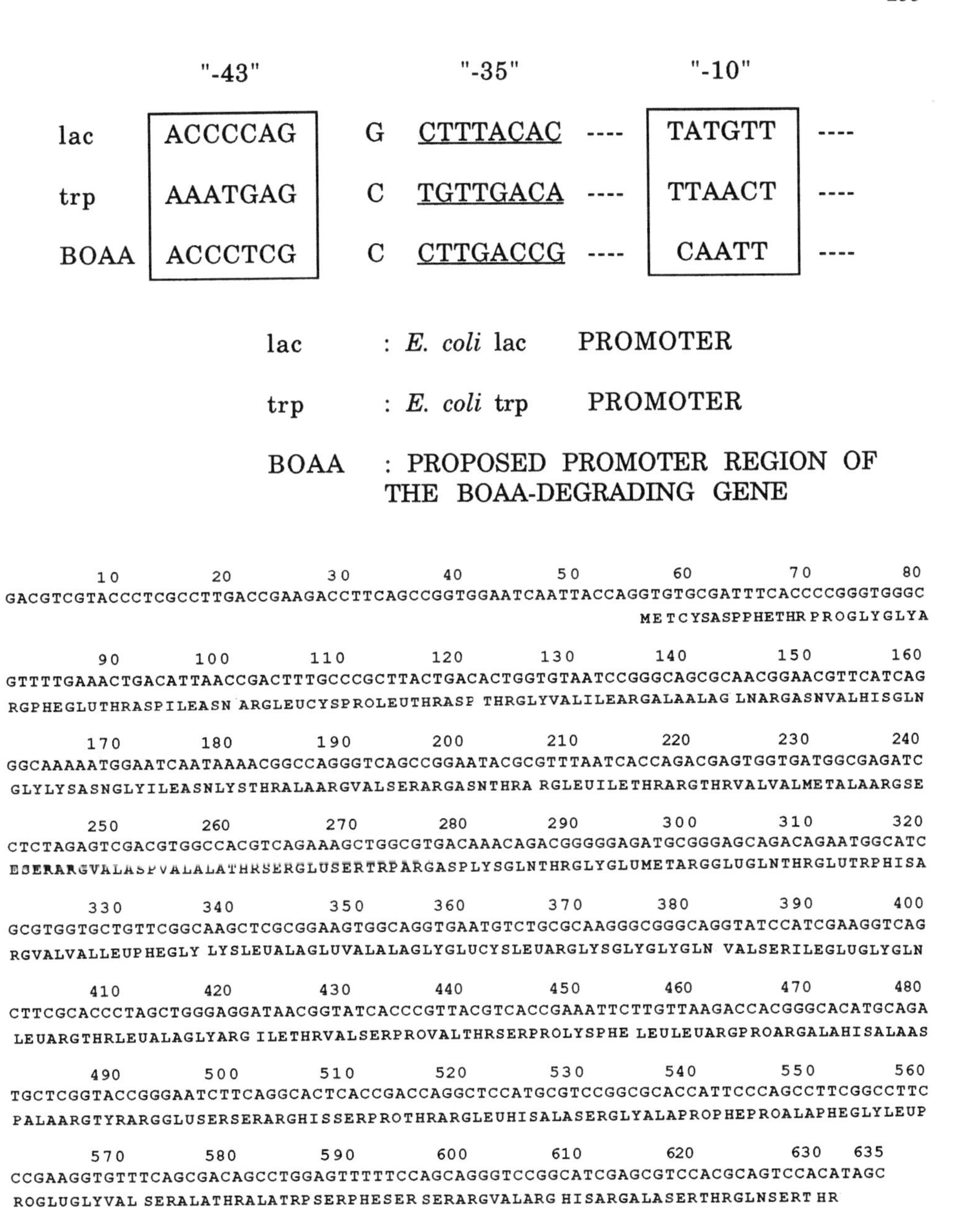

Figure 3. Coding frame of 631 nucleotide stretch.

The 631 nucleotide region, on analysis, showed restriction sites for a large number of restriction enzymes like *Kpn*I, *Sau*3A, *Sma*I, *Xba*I, *Xma*I, *Sal*I (one site each), etc. The sequence had higher 'GC' content, with 28.0% 'C' and 29.3% 'G' residues, than 'AT' content; the percentages of 'A' and 'T' were 23.3 and 19.4, respectively. It codes for a polypeptide of about 21 kD. However, the nature of the protein and the mechanism of degradation of BOAA is not understood. Homology search did not reveal any homology with any of the known bacterial decarboxylase or deaminase which are present in our data base.

In order to study the nature of the polypeptide involved in BOAA degradation, the fragment was excised with *Eco*RI and *Pst*I. The fragment was further restricted with *Xmn*I and cloned into an expression vector $pMLC_2$ (NEB, USA), restricted with *Eco*RI and *Xmn*I, without affecting the reading frame. The protein coded by this fragment is expressed as a fusion protein with a maltose-binding protein. The fusion protein, on isolation and purification using an affinity column, followed by cleavage with an endoprotease factor Xa, gives the protein coded by the toxin-degrading fragment as a single polypeptide of ~21 kD and the 42 kD maltose-binding protein [25]. The nucleotide sequence analysis had also shown the protein to be of ~21 kD. Further work is in progress, for studying the nature of protein involved in the degradation of the neurotoxin. It will also lead us to understand the mechanism of BOAA degradation by the bacteria.

The next step is to introduce BOAA-degrading gene into *Lathyrus sativus* and to get it expressed. Work is already going on in this direction. The BOAA-degrading fragment has been cloned into an *Agrobacterium*-based binary vector under CaMV 35S constitutive promoter and GUS as *reporter gene* and introduced into disarmed *Agrobacterium tumefaciens* strain LBA 4404 by triparental mating. The transformed *Agrobacterium* strain is now being used for transforming *Lathyrus sativus* to develop *transgenic* plants.

CONCLUSIONS

As reported earlier, microorganisms have, time and again, shown to possess the property of degrading simple and complex materials, often found associated with plasmids [1-4, 26-28]. There are earlier reports also of isolation of *Not*I [23] and *Pst*I [29] fragments from the same plasmid pBYA1 of *Enterobacter cloacae* strain with the property of BOAA degradation. The 0.5 kb *Pst*I fragment has been sequenced [29], and did not find any homology with the 1.8 kb fragment reported in this paper. The proteins encoded by these two fragments may be of two different enzymes, involved in alternative degradation pathways. The 0.5 kb *Pst*I fragment also had no complete sequence, as it did not have any sequences similar to prokaryotic promoter region. The 1.8 kb fragment, on sequencing, also possessed "-43", "-35", and "-10" sequences in the largest coding frame of 631 nucleotide stretch, and showed similarity to that of *E. coli lac* promoter region. The study of the properties of the protein, coded by this fragment, will help in understanding the degradation pathway of the toxin.

If our present attempt to develop transgenic *Lathyrus sativus* with neurotoxin-degrading gene expression is successful, it will be of great help to the farmers of not only India but also to those from other arid regions of the world.

REFERENCES

1 Chakrabarty AM. J Bacteriol 1972; 112: 815-823.
2 Wong CL, Dunn NW. Genet Res 1974; 23: 227-232.
3 Williams PA, Murray K. J Bacteriol 1974; 120: 416-423.
4 Pemberton JM, Fisher RR. Nature 1977; 268: 732-733.
5 Cook AM. FEMS Microbiol Rev 1987; 46: 93-116.
6 Hartman J, Reineke W, Knackmuss HJ. Appl Environ Microbiol 1979; 37: 421-428.
7 Chakrabarty AM, Gunsalus IC. Bacterial Process 1971; 1971: 4-6.
8 Hada HS, Sizemore RK. Appl Environ Microbiol 1981; 44: 199-202.
9 Burton NF, Day NJ, Bull AT. Appl Environ Microbiol 1982; 44: 1026-1029.
10 Spain JC, Nishino SF. Appl Environ Microbiol 1987; 53: 1010-1019.
11 Haigler BE, Nishino SF, Spain JC. Environ Microbiol 1988; 54: 294-301.
12 Ganapathy KT, Dwivedi MP. Lathyrism Enquiry Field Unit, ICMR Gandhi Memorial Hosp, Rewa, Madhya Pradesh 1961.
13 Kessler A. Monatsschr Psychiat Neurol 1947; 113: 76-92.
14 Adiga PR, Rao SLN, Sarma PS. Curr Sci 1963; 32: 153-155.
15 Murti VVS, Seshadri TR, Venkitasubramanian TA. Phytochem 1964; 3: 73-78.
16 Rao SLN, Adiga PR, Sarma PS. Biochemistry 1964; 3: 432-436.
17 Prakash S, Misra BK, Adsule RN, Barat GK. Biochem Physiol Pflazen 1977; 171: 369-374.
18 Roy PK, Singh B, Mehta SL, Barat GK, Gupta N, Kirti PB, Chopra VL. Ind J Exptl Biol 1991; 29: 327-330.
19 Roy PK, Ali K, Gupta A, Bharati GK, Mehta SL. J Plant Biochem Biotechnol 1993; 2: 9-13.
20 Roy PK, Barat GK, Mehta SL. Plant Cell Tissue Organ Cult 1992; 29: 135-138.

21 Rao SLN. Analytical Biochem 1978; 86: 386-395.
22 Sambrook J, Fritsch EF, Maniatis T. Molecular Cloning : Laboratory Manual, 2nd Edition, Cold Spring Harbour Laboratory Press, 1989.
23 Yadav VK, Santha IM, Timko MP, Mehta SL. J Plant Biochem Biotechnol 1992; 1: 87-92.
24 Sukanya R, Santha IM, Mehta SL. J Plant Biochem Biotechnol 1993; 2: 77-82.
25 Nair AJ, Khatri GS, Santha IM, Mehta SL. J Plant Biochem Biotechnol 1994; 3: (in Press)
26 Shields MS, Hooper SW, Sayler GS. J Bacteriol 1985; 163: 882-889.
27 Brandsch R, Faller W, Schneider K. Mol Gen Genet 1986; 202: 96-101.
28 Stalker DM, McBride KE. J Bacteriol 1987; 169: 955-960.
29 Arti P. Molecular Biology of Neurotoxin β-N-Oxalyl L-α, β-Diaminopropionic Acid (ODAP) Utilizing Gene. Ph.D. Thesis, PG School, IARI, New Delhi, India.

Biotransformations: Microbial Degradation of Health Risk Compounds
Ved Pal Singh, editor
© 1995 Elsevier Science B.V. All rights reserved.

Microbial degradation of tannins

R.K. Saxena[a], P. Sharmila[a] and Ved Pal Singh[b]

[a]Department of Microbiology, University of Delhi, South Campus, New Delhi-110021, India

[b]Department of Botany, University of Delhi, Delhi-110007, India

INTRODUCTION

Tanning industry is one of the major industries in India. Tannery effluents are the most complex industrial wastes which are toxic to plants, animals, and soil microorganisms [1]. Tannery wastes contain vegetable tannins in addition to soluble organic matter, suspended solids, chromium, high chloride and sulphide concentration, and high pH (Table 1) [2].

Table 1
Composition of tannery waste

Colour	Yellow to brown
Total solid	479,000 ppm
Suspended solid	12,166 ppm
Dissolved solid	35,733.33 ppm
Oil and grease	5.15 ppm
pH	8.00
COD	6113.3 ppm
BOD	2502 ppm
Total nitrogen as N	590.98 ppm
Cl^-	11,258.7 ppm
Phenol	7.5 mg/litre
SO_4^{2-}	1499.2 ppm
NO_3^-	5.34 ppm
F^-	1.04 ppm
PO_4^{3-}	0.77 ppm
Cr(VI)	Nil
Cr(III)	0.6

Tannins are defined as water soluble polyphenols which differ from most other natural polyphenolic compounds in their ability to precipitate proteins, such as gelatin [3]. Tannins are the most abundant plant constituents, following cellulose, hemicellulose, and lignin [4]. In tribal pulse (*Bauhinia malabarica* Roxb.), tannins have been found to be one of the antinutritional factors, in addition to phenols, L-DOPA, and haemagglutinating activity [5]. However, cotton condensed tannin has been shown to interact with *Heliothis virescens* larvae and the cry IA(C) δ-endotoxin of *Bacillus thuringiensis* [6]. They are added to the soil in the form of leaves, twigs, and fruits. According to Rice and Pancholy [7], 85 kg/ha of tannins are added every year to the forest soils.

Haslam [8] classified tannins into two groups, according to their structures : (i) Condensed tannins and (ii) Hydrolyzable tannins.

Condensed tannins

They are polymers of catechin or similar flavans, that are connected by carbon to carbon linkages, and are more resistant to microbial attack as compared to hydrolyzable tannins [9]. The non-hydrolyzable tannins are formed by flavan 3,4-diols (leucoanthocyanin and proanthocyanin). Catechin is a flavan-3-ol with two hydroxyl groups in the side ring. The catechins include gallic acid esters with the acid moieties attached to the hydroxyl groups. Flavan-3,4-diols are also termed as leucoanthocyanidins (Figure 1) [1].

Hydrolyzable tannins

Esters of sugars and phenolic acids or their derivatives are referred to as hydrolyzable tannins. They are composed of a molecule of carbohydrate, generally glucose, to which gallic acid or similar acids are attached by ester linkages (Figure 1) [10,11]. Hydrolyzable tannins are subdivided into gallotannins and ellagitannins. Gallotannins are esters of glucose or a polysaccharide and gallic acid or *m*-digallic acid. On hydrolysis, they yield gallic acid. The chief commercial tannin is tannic acid. Gallotannin contains 8 to 10 moles of gallic acid and glucose. On hydrolysis, ellagitannin yields gallic acid and its derivatives, especially ellagic acid which is formed by lactonization of hexahydroxy-diphenic acid [1].

MICROBIAL DEGRADATION OF TANNINS

Tannins are *recalcitrant* molecules and resist microbial attack [12]. Condensed tannins are more resistant than hydrolyzable tannins [13], and are toxic to a variety of microorganisms. However, there are several reports by various workers regarding the role of fungi in biodegradation of tannins.

TANNIN DEGRADATION BY FUNGI

As early as in 1900, Fernback and Pottevin [14,15] independently discovered the enzymatic nature of tannin hydrolysis using cell-free

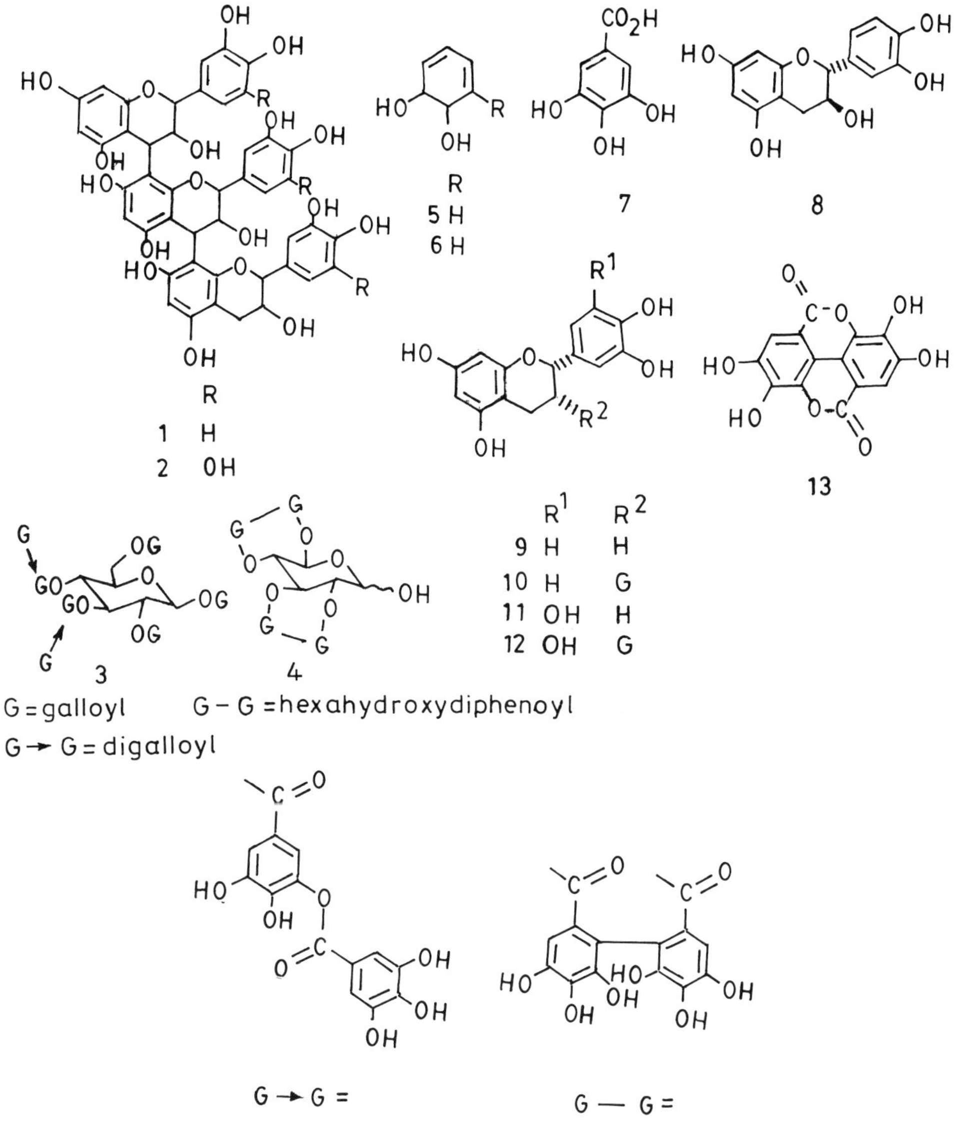

Figure 1. Chemical structures of proanthocyanicidins, hydrolyzable tannins and other structurally related phenols. 1, procyanidin trimer; 2, prodelphinidin trimer; 3, heptagalloylglucose; 4, pedunculagin; 5, catechol; 6, pyrogallol; 7, gallic acid; 8, (+)-catechin; 9, (-)-epicatechin; 10, (-)-epicatechin gallate; 11, (-)-epigallocatechin; 12, (-)-epigallocatechin gallate; 13, ellagic acid.

preparation of *Aspergillus niger*. Filamentous fungi, especially species of the genera *Penicillium* and *Aspergillus*, have been implicated in tannin decomposition [16-18] Lewis and Starkey [19] reported that some microoganisms grown as pure culture were shown to develop on media, containing tannins as sole source of carbon (Table 2). Both condensed and hydrolyzable tannins were used as substrates, and several kinds of tannins were compared. *Aspergillus, Penicillium*, *Fomes, Polyporus, Poria*, and *Trametes* species were shown to grow better on gallotannin than on chestnut tannin (ellagitannin) or wattle tannin (condensed tannin).

Table 2
Decomposition of tannins by microorganisms

Microorganisms	% Decomposition of			
	Gallotannin	Chestnut tannin	Catechin	Wattle tannin
Achromobacter sp.	74	13	N	N
Streptomyces sp.	18	20	N	N
Aspergillus flavus	84	N	69	N
A. fumigatus	80	N	50	N
A. niger	100	10	100	N
Penicillium citrinum	90	43	93	N
P. frequentans	81	10	N	N
P. janthinellum	74	N	75	N
P. purpurogenum	83	23	N	N
P. thomii	74	30	50	N
Fusarium sp. 1	96	N	100	N
Fusarium sp. 2	85	N	77	N

Most of the fungal species, that have been used for biodegradation of tannery effluent, belong to the genera *Aspergillus* and *Penicillum*. Other fungi, including *Chaetomium, Fusarium, Rhizoctonia, Cylindrocarpon, Trichoderma,* and *Candida* spp., are capable of degrading tannery waste constituents [20]. *Psalliata campestris* can oxidize catechin and *A. niger* can degrade gallic acid [1]. The intermediates of this degradation were *cis*-aconitic acid, α-ketoglutaric acid and citric acid. *A. fumigatus* decomposed gallotannins to gallic acid. Catechin, at 0.3% level, was degraded by *A. fumigatus* without intermediate formation in 6-8 days. Chandra et al. [21] reported that catechin was degraded by *A. fumigatus*, *A. niger, A. flavus*, and *Penicillium* sp. Dalvesco et al. [22] reported that species of *Aspergillus* and *Penicillium* were found to utilize gallotannin and gallic acid as carbon sources.

Tannins have been found to be degraded rapidly in the presence of other metabolizable substances. Chandra et al. [23] found that *A. niger* and *Penicillium* sp. grew profusely in wood apple tannin medium containing glucose, with wattle tannin at 0.3% and glucose at 166.7 mM concentration, growth of *A. niger* improved. Additional carbon [24] and nitrogen sources favoured rapid production of tannase which, in turn, cleaved tannins providing a continuous supply of carbon source for growth. In another study with *A. niger* isolated from rice fields, it was found that maximum growth occurred in 0.3% wattle tannin. However, a marked increase in growth was observed in the presence of glucose (Table 3).

Table 3
Effect of wattle tannin on the mycelial growth of *A. niger*

Wattle tannin concentration (%)	Mycelial dry weight (mg)	
	Wattle tannin samples	Wattle tannin + glucose (166.7mM) samples
Control	29	454
0.2	52	505
0.3	61	505
0.4	55	500
0.5	47	463

Jacob and Pignal [25] reported the growth and hydrolytic action of tannins by 6 strains of yeast which were isolated from tanning liquors and xylophagous insects in culture media, containing various concentrations of tannic acid. Hydrolyzable tannins, except gallotannins, were not hydrolyzed.

Otuk and Deschamps [26], for the first time, reported the capacity of yeasts to degrade a condensed tannin. The biodegradation of wattle tannin with six strains of yeasts, isolated from decaying bark samples, was tested.

A strain of *Candida guilliermondii* degraded flavanol groups rapidly without release of phenolic units. *Candida tropicalis* degraded leucoanthocyanin groups with the liberation of catechin.

While studying the effects of certain factors, such as temperature, pH, and carbon sources on the decomposition of tannic acid and gallic acid, a decomposition product of tannic acid by *Penicillium chrysogenum* has been studied by Suseela and Nandy [27]. According to them, the decomposition of both tannic and gallic acids was found to be maximum in shake cultures at 28°C. Tannic acid and gallic acid were found to be completely decomposed in 3 days.

Sugars present as additional carbon source at 3% level retarded the decomposition of tannic and gallic acids. Glucose at 10% concentration inhibited the decomposition of tannic and gallic acids (Table 4).

Table 4
Effect of various carbon sources on the decomposition of tannic acid and gallic acid by *P. chrysogenum*

Carbon source (3%)	Per cent decomposition of	
	Tannic acid	Gallic acid
Tannic acid+	85.3	-
Gallic acid+	-	72.1
Glucose	68.1	29.1
Rhamnose	61.8	25.3
Sucrose	72.5	27.0
Melibiose	63.0	24.3
Lactose	65.0	26.2
Raffinose	62.1	24.8
Cellobiose	58.9	21.0

Inhibition studies throw more light· on the functioning of tannic acid and gallic acid degrading enzymes. EDTA was found to have no inhibitory effect on either of the enzymes, however, the metal binding KCN affected both the enzymes (Table 5).

Table 5
Effect of specific inhibitors on the decomposition of tannic acid and gallic acid by *P. chrysogenum*

Inhibitors	Per cent inhibition Tannic acid	Decomposition of Gallic acid
EDTA	0	0
KCN	17	20
Indolacetate	28	85
$CuSO_4$	22	7
Phenyl methyl sulphonyl fluoride	61	2
Mercaptoethanol	18	3
2,4-Dinitrophenol	78	43
Fluoroacetate	12	15

The observations by Suseela and Nandy [28] led to the conclusion that two separate enzymes are involved in tannic acid decomposition. Further studies on *P. chrysogenum* tannase by Suseela and Nandy[26] showed the enzyme to have a molecular weight of 300,000 on Sephadex G-200, which is higher as compared to the tannases from other microbial sources. Upon denaturation, the molecular weight of the enzyme was found to be 158,000, suggesting that the enzyme consists of two subunits of same molecular weight.

TANNIN DEGRADATION BY BACTERIA

The degradation of tannic acid and tannins by fungi and yeasts has been reported several times [27-30]. However, reports relating to the ability of bacteria to degrade tannins are few. It is well known that tannic acid and tannins are bacteriostatic and toxic compounds making non-reversible reactions with proteins. Nevertheless, bacteria may degrade many phenolic compounds, including natural ones like catechol, proto-catechuic acid and many others. Deschamps et al. [31], for the first time, reported the biodegradation of tannins by bacteria. They isolated fifteen bacterial strains belonging to the genera *Bacillus, Staphylcoccus,* and most probably *Klebsiella* by enrichment culture technique, using tannic acid as the sole source of carbon. Nine of the isolated strains grew both on tannic acid as well as gallic acid which are obtained upon hydrolysis. However, only four strains degraded catechol or catechin. Deschamps and Lebeault [32] isolated several bacterial strains, capable of degrading hydrolyzable and condensed tannins, including chestnut, wattle and Quebracho commercial tannin extracts by enrichment. Wattle tannin degrading strains were identified as *Enterobacter aerogenes, Enterobacter agglomerans, Cellulomonas* sp., *Staphylococcus* sp., while Quebracho tannin-degrading strains were *Cellulomonas* sp., *Arthrobacter* sp., *Bacillus* sp., *Micrococcus* sp., *Corynebacterium* sp., and *Pseudomonas* sp. Phenolic acid was not detected in uninoculated and inoculated samples of wattle tannin; however, gallic acid was detected in samples inoculated with Quebracho tannin. The production of extracellular tannase by bacterial cultures, with simultaneous release of gallic acid and glucose, was reported for the first time by Deschamps et al. [33]. Strains of *Bacillus pumilus*, *B. polymyxa, Corynebacterium* sp., and *Klebsiella pneumoniae* produced tannase with chestnut bark as the sole source of carbon. Three of the strains degraded tannin to gallic acid, but one of the strains, *B. pumilus* yielded two intermediates considered to be di- and tri- gallic structures, probably bonded to glucose.

EFFECT OF TANNINS

Polyphenolic compounds are often regarded as inhibitors of microbial growth [34-36]. In addition, it has been shown that numerous enzymes are inactivated by tannins [36-39]. However, polyphenolic compounds can induce stimulatory effects on the growth, respiration, fermentation, and excretion of amino acids. The effect of condensed tannins on the growth and cellulolytic activity of *Trichoderma viride* has also been studied. Condensed tannins, catechin, and epicatechin do not inhibit cellulose degradation and have a stimulating effect on the growth and cellulase production. The effect of wattle (condensed) and myrobalan (hydrolyzable) tannins on soil respiration and glucose oxidation by *Pseudomonas solanacearum* and *Penicillium* sp. was studied by Muthukumar and Mahadevan [40]. *P. solanacearum* oxidised myrobalan tannin, as revealed by the oxygen uptake. However, glucose oxidation was inhibited by 58% (Figure 2a) by myrobalan tannin. Wattle tannin completely inhibited glucose oxidation (Figure 2b). Oxidation of glucose was inhibited slightly by myrobalan tannin (Figure 3a). However, *Penicillium* sp. was able to oxidize wattle tannin only after an initial lag of 20 min. Wattle tannin inhibited glucose oxidation by *Penicillium* (Figure 3b).

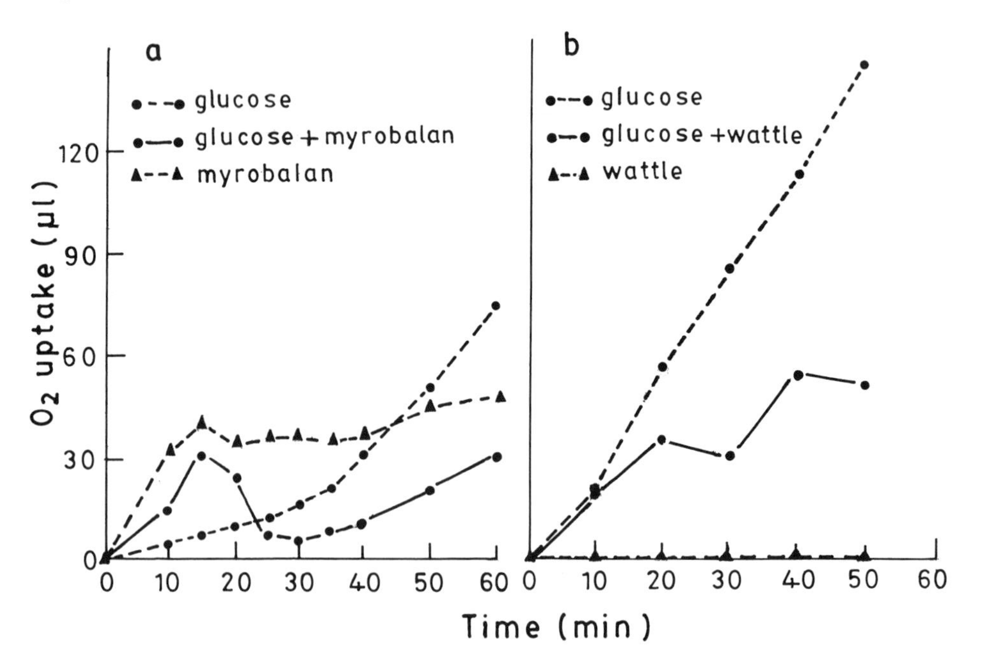

Figure 2. Oxidation of (a) myrobalan tannin and glucose and (b) wattle tannin and glucose by *Pseudomonas solanacearum*.

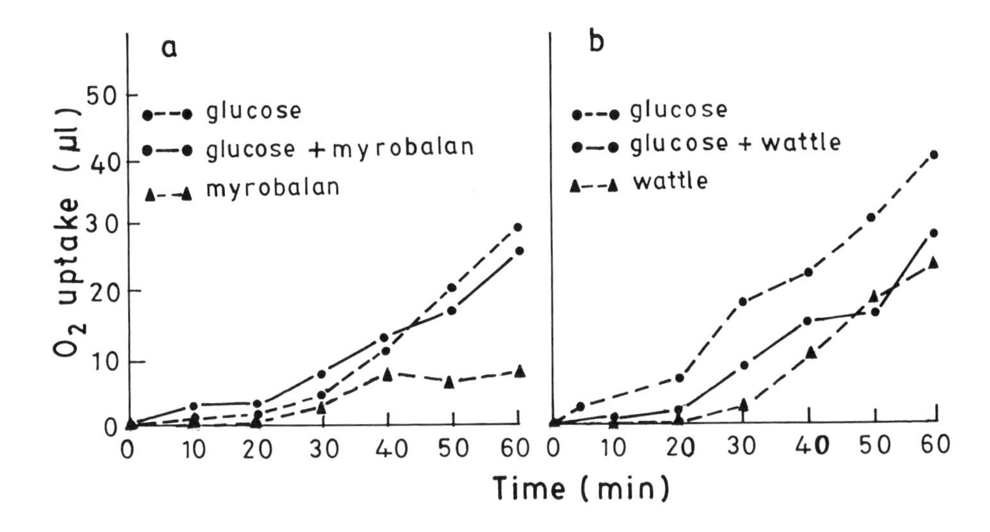

Figure 3. Oxidation of (a) myrobalan tannin and glucose and (b) wattle tannin and glucose by *Penicillium* sp.

Chaetomium cupreum was isolated from rice field soils by Sivaswamy and Mahadevan [12] and the effect of tannins on growth was studied. The growth of *C. cupreum* was found to be maximum at pH 3, but in the presence of tannins maximum growth took place at pH 4 and above.

Gallic acid at 15 mM; tannic acid, 1 mM; wattle tannin, 0.3%, and catechin, 12 mM were most effective in favouring growth (Table 6) [12]. Enhanced mycelial growth was observed with glucose (Table 7).

Table 6
Effect of tannins on growth of *Chaetomium cupreum*

Gallic acid		Tannic acid		Wattle tannin		Catechin	
Conc. (mM)	Dry wt. (mg)	Conc. (mM)	Dry wt. (mg)	Conc. (%)	Dry wt. (mg)	Conc. (mM)	Dry wt. (mg)
Control	205	Control	223	Control	187	Control	116
5	280	1	144	0.2	294	4	166
10	326	2	102	0.3	378	8	154
15	445	4	124	0.4	375	12	288
20	214	6	106	0.5	332	16	253

Table 7
Effect of glucose and tannins on growth of *Chaetomium cupreum*

Glucose conc. (mM)	Dry wt. of mycelium (mg)				
	Control	Gallic acid (15 mM)	Tannic acid (1 mM)	Wattle tannin (0.3%)	Catechin (12 mM)
83.3	109	249	136	226	277
166.7	149	303	142	457	372
250.0	199	231	200	651	413
333.3	217	154	166	648	445
416.7	203	220	197	-	554

The presence of tannins, urea, and sodium nitrate at 35mM induced profuse growth, and toxicity was found to decline in the presence of tannins.

ENZYMATIC ASPECTS OF TANNIN DEGRADATION

Tannase (tannin acyl hydrolase EC 3.1 1:20), the enzyme capable of cleaving tannin, has been isolated and characterized from various sources. Knudson [41] reported the production of tannase by *Aspergillus niger* with increasing concentration of tannin. The enzyme has also been isolated from *Penicillium* sp. [42,4], *A. niger* [44], and *Candida* sp. [29]. Tannase from *A. niger* was inducible and hydrolyzed the ester linkages of gallic acid [45]. *Aspergillus* sp. and *Penicillium* sp., isolated from soil, were able to produce tannase even in its absence, thereby suggesting that an inducer is not required for tannase production [46].

BIOTECHNOLOGY

In recent years, the application of biotechnology and genetic engineering in the fields of agriculture, medicine, and environmental programmes for human welfare has gained considerable importance. Since microorganisms can readily adapt to adverse and changing environmental conditions, researchers have explored the possibilities of using them as tools to help control environmental pollution. Chakrabarty [47] demonstrated that enzymes of hydrocarbon degradation pathways are plasmid specific. Since then, plasmids have been implicated in the degradation of camphor [50] and several other hydrocarbons [51]. Plasmids can, therefore, facilitate the production of organisms specific for a particular substrate.

REFERENCES

1 Mahadevan A, Sivaswamy SN. In: Mukerji KG, Pathak NC, Singh VP, eds. Frontiers in Applied Microbiology. Lucknow: Print House, 1985; 327-347.
2 Sastry CA. In: Sundarsen BB, ed. Proc National Workshop on Microbial Degradation of Industrial Wastes. Nagpur: NEERI, 1982; 122-144.
3 Spencer CM, Cai Y, Martin R, Gaffney SM, Goulding PN, Magnolato D, Lilley TH, Haslam E. Phytochemistry 1988; 27:2397.
4 White T. J Sci Food Agric 1957; 8:377.
5 Vijaykumari K, Siddhuraju P, Janardhanan K. Plant for Human Nutrition (Dordrecht) 1993; 44(3): 291-298.
6 Navon A, Hare JD, Federici BA. J Chem Ecol 1993, 19(11): 2485-2499.
7 Rice EL, Pancholy SK. Am J Bot 1973; 60(7): 691-702.
8 Haslam E. Plant Polyphenols. Cambridge: Cambridge University Press, 1989.
9 Lewis JA, Starkey RL. Soil Sci 1968; 106: 241-247.
10 Hathway DE. In: Hills WE, ed. Wood Extractives. New York: Acadomic Press, 19C2, 191-228.
11 Jurd L. In: Hills WE, ed. Wood Extractives. New York: Academic Press, 1962; 229-260.
12 Sivaswamy SN, Mahadevan A. J Indian Bot Soc 1986; 65: 95-100.
13 Haslam E. In: Priss J, ed. The Biochemistry of Plants. New York: Academic Press, 1981; 527-556.
14 Fernbach MA. Compt Rend 1900; 131: 1214-1215.
15 Pottevin M. Compt Rend 1900; 131: 1215-1217.
16 Friedrich M. Arch Mikrobiol 1956; 25: 297-306.
17 Cowley GT, Wittingham WF. Mycologia 1961; 53: 539-542.
18 Nishira H. Chem Abstr 1959-62; 54: 6864; 58: 11702; 60: 3298.
19 Lewis JA, Starkey RL. Soil Sci 1969; 107: 235-241.
20 Mahadevan A, Muthukumar G. Microbiologia 1980; 72: 73-79.
21 Chandra T, Madhavakrishna W, Neyudamma Y. Can J Microbiol 1969; 15: 303-306.
22 Dalvesco G, Fiusello N, Vettipamus M. Allionia 1972; 17: 25-40.
23 Chandra T, Krishnamurthy V, Mahadevakrishna W, Nayudamma Y. Leather Sci 1973; 20: 269-273.

24 Ganga PS, Suseela G, Nandy SC, Santappa M. Leather Sci 1977; 25: 203-209.
25 Jacob FH, Pignal MC. Mycopathologia 1975; 57: 139-148.
26 Otuk G, Deschamps AM. Mycopathologia 1983; 83: 107-111.
27 Suseela RG, Nandy SC. Leather Sci 1985; 32: 249-254.
28 Suseela RG, Nandy SC. Leather Sci 1985; 32:278-280.
29 Haslam E, Haworth RD, Jones K, Rogers MJ. J Chem Soc 1961; 8: 1829-1835.
30 Yamada K, Iibuchi S, Minoda Y. J Ferment Technol 1967; 45: 233-240.
31 Aoki K, Shinke R, Nishira H. Agri Biol Chem 1976; 40: 79-85.
32 Grant WD. Science 1976; 193: 1137-1139.
33 Deschamps AM, Mohudeau G, Conti M, Lebeault JM. J Ferment Technol 1980; 58:93-97.
34 Deschamps AM, Lebeault JM. In: Moo-Young M, Robinson CW, eds. Advances in Biotechnology. New York: Pergamon Press, 1981; 639-643.
35 Deschamps AM, Otuk G, Lebeault JM. J Ferment Technol 1983; 61: 55-59.
36 Henis Y, Tagari A, Volcani R. Appl Microbiol 1964; 12: 204.
37 Basarba J. Can J Microbiol 1966; 12: 787.
38 Benoit RF, Starkey RL. Soil Sci 1968; 105: 291.
39 Siu RGH. In: Microbial Decomposition of Cellulose. New York: Reinhold, 1951.
40 Hathway DE, Seakins JWT. Biochem J 1958; 70: 158.
41 Goldstein JL, Swain T. Phytochem 1965; 4: 185.
42 Muthukumar G, Mahadevan A. Ind J Exp Bot 1981; 19: 1083-1085.
43 Knudson L. J Biol Chem 1913; 14: 159-206.
44 Nishira H. J Ferment Technol 1961; 39: 137-146.
45 Nishira H. Sci Rep Hyogo Univ Agric 1962; 5: 117-123.
46 Dhar SC, Bose SM. Leather Sci 1964; 11: 27-38.
47 Yamada H, Adachi O, Watanabe M, Gato O. Agric Biol Chem 1968; 32: 257-258.
48 Basarba J. Plant Soil 1964; 21: 8-16.
49 Chakrabarty AM. J Bacteriol 1972; 112: 815-823.
50 Rheinwald JG, Chakrabarty AM, Gunsalus IC. Proc Natl Acad Sci USA 1973; 70: 885-889.
51 Friello DA, Mylorie JR, Chakrabarty AM. In: Sharpley JM, Kapalan AM, eds. Proc 3rd Int Biodegradation Symposium. London: Applied Science, 1976; 205-214.

INDEX

B

C

D

E

F

G

H

I

K

L

M

N

O

P

R

S

T

U

V

W

X